AN APPETITE FOR WONDER

The Making of a Scientist

A Memoir

Richard Dawkins

BLACK SWAN

TRANSWORLD PUBLISHERS
61–63 Uxbridge Road, London W5 5SA
A Random House Group Company
www.transworldbooks.co.uk

AN APPETITE FOR WONDER
A BLACK SWAN BOOK: 9780552779050

First published in Great Britain
in 2013 by Bantam Press
an imprint of Transworld Publishers
Black Swan edition published 2014

Diagrams by Patrick Mulrey

A CIP catalogue record for this book
is available from the British Library.

Addresses for Random House Group Ltd companies outside the UK
can be found at: www.randomhouse.co.uk
The Random House Group Ltd Reg. No. 954009

The Random House Group Limited supports the Forest Stewardship Council® (FSC®),
the leading international forest-certification organisation. Our books carrying the FSC
label are printed on FSC®-certified paper. FSC is the only forest-certification scheme
supported by the leading environmental organisations, including Greenpeace. Our paper
procurement policy can be found at www.randomhouse.co.uk/environment

Typeset in Minion by Falcon Oast Graphic Art Ltd.
Printed and bound by CPI Group (UK) Ltd, Croydon, CR0 4YY.

2 4 6 8 10 9 7 5 3 1

To my mother and my sister, who shared the years with me,
and in memory of my father, missed by all.

CONTENTS

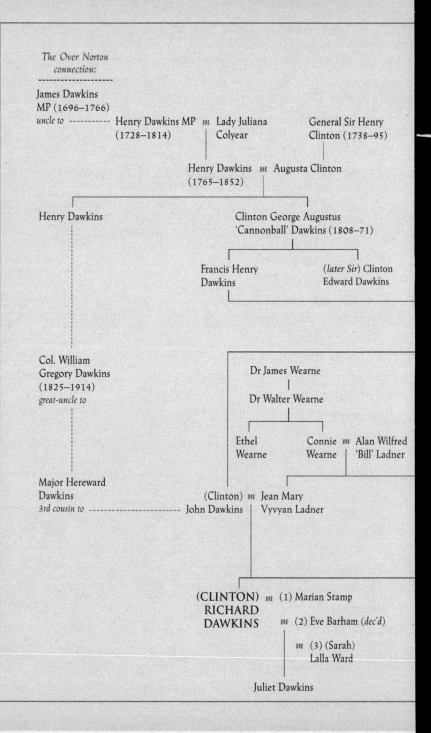

The Over Norton
connection:

James Dawkins
MP (1696–1766)
uncle to ---------- Henry Dawkins MP *m* Lady Juliana
                      (1728–1814)   Colyear

                                                                General Sir Henry
                                                                Clinton (1738–95)

                              Henry Dawkins *m* Augusta Clinton
                              (1765–1852)

Henry Dawkins   Clinton George Augustus
                                      'Cannonball' Dawkins (1808–71)

                              Francis Henry   (*later Sir*) Clinton
                              Dawkins   Edward Dawkins

Col. William   Dr James Wearne
Gregory Dawkins
(1825–1914)   Dr Walter Wearne
great-uncle to

                              Ethel   Connie *m* Alan Wilfred
                              Wearne   Wearne   'Bill' Ladner

Major Hereward
Dawkins   (Clinton) *m* Jean Mary
3rd cousin to ----------------------- John Dawkins   Vyvyan Ladner

                      (CLINTON) *m* (1) Marian Stamp
                      RICHARD
                      DAWKINS   *m* (2) Eve Barham (*dec'd*)

                                 m (3) (Sarah)
                                      Lalla Ward

                              Juliet Dawkins

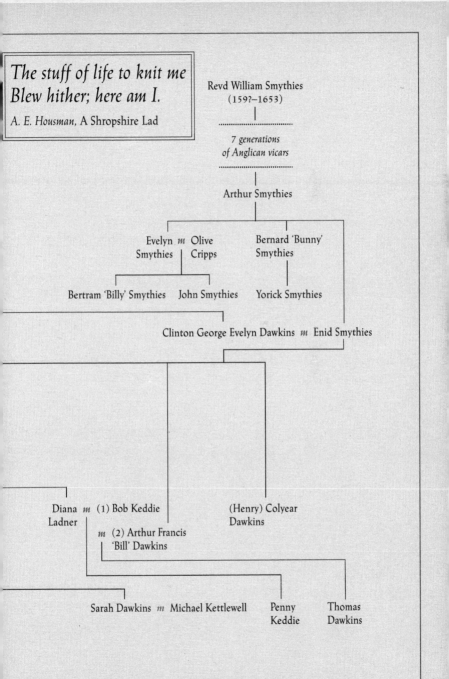

> *The stuff of life to knit me*
> *Blew hither; here am I.*
>
> A. E. Housman, A Shropshire Lad

Revd William Smythies
(159?–1653)

7 generations
of Anglican vicars

Arthur Smythies

Evelyn *m* Olive
Smythies   Cripps

Bernard 'Bunny'
Smythies

Bertram 'Billy' Smythies   John Smythies

Yorick Smythies

Clinton George Evelyn Dawkins *m* Enid Smythies

Diana *m* (1) Bob Keddie
Ladner

m (2) Arthur Francis
'Bill' Dawkins

(Henry) Colyear
Dawkins

Sarah Dawkins *m* Michael Kettlewell

Penny
Keddie

Thomas
Dawkins

GENES AND
PITH HELMETS

'GLAD to know you, Clint.' The friendly passport controller was not to know that British people are sometimes given a family name first, followed by the name their parents wanted them to use. I was always to be Richard, just as my father was always John. Our first name of Clinton was something we forgot about, as our parents had intended. To me, it has been no more than a niggling irritation which I would have been happier without (notwithstanding the serendipitous realization that it gives me the same initials as Charles Robert Darwin). But alas, nobody anticipated the United States Department of Homeland Security. Not content with scanning our shoes and rationing our toothpaste, they decreed that anyone entering America must travel under his first name, exactly as written in his passport. So I had to forgo my lifelong identity as Richard and rebrand myself Clinton R. Dawkins when booking tickets to the States – and, of course, when filling in those important forms: the ones that require you explicitly to deny that you are entering the USA in order to overthrow the constitution by force of arms. ('Sole purpose of visit' was the British broadcaster Gilbert Harding's response to that; nowadays such levity will see you banged up.)

Clinton Richard Dawkins, then, is the name on my birth

3

certificate and passport, and my father was Clinton John. As it happened, he was not the only C. Dawkins whose name appeared in *The Times* as the father of a boy born in the Eskotene Nursing Home, Nairobi, in 1941. The other was the Reverend Cuthbert Dawkins, Anglican missionary and no relation. My bemused mother received a shower of congratulations from bishops and clerics in England, unknown to her but kindly calling down God's blessings upon her newborn son. We cannot know whether the misdirected benedictions intended for Cuthbert's son had any improving effect on me, but he became a missionary like his father and I became a biologist like mine. To this day my mother jokes that I might be the wrong one. I am happy to say that more than just my physical resemblance to my father reassures me that I am not a changeling, and was never destined for the church.

Clinton first became a Dawkins family name when my great-great-great-grandfather Henry Dawkins (1765–1852) married Augusta, daughter of General Sir Henry Clinton (1738–95), who, as Commander-in-Chief of British forces from 1778 to 1782, was partly responsible for losing the American War of Independence. The circumstances of the marriage make the commandeering of his name by the Dawkins family seem a bit cheeky. The following extract is from a history of Great Portland Street, where General Clinton lived.

> In 1788 his daughter eloped from this street in a hackney-coach with Mr Dawkins, who eluded pursuit by posting half a dozen other hackney-coaches at the corners of the street leading into Portland Place, with directions to drive off as rapidly as possible, each in a different direction . . .[1]

---

[1] H. B Wheatley and P. Cunningham, *London Past and Present* (London, Murray, 1891), vol. 1, p. 109.

I wish I could claim this ornament of the family escutcheon as the inspiration for Stephen Leacock's Lord Ronald, who '... flung himself upon his horse and rode madly off in all directions'. I'd also like to think that I inherited some of Henry Dawkins's resourcefulness, not to mention his ardour. This is unlikely, however, as only one 32nd part of my genome is derived from him. One 64th part is from General Clinton himself, and I have never shown any military leanings. *Tess of the D'Urbervilles* and *The Hound of the Baskervilles* are not the only works of fiction that invoke hereditary 'throwbacks' to distant ancestors, forgetting that the proportion of genes shared is halved with every generation and therefore dies away exponentially – or it would if it were not for cousin-marriage, which becomes ever more frequent the more distant the cousinship, so that we are all more or less distant cousins of each other.

It is a remarkable fact, which you can prove to yourself without leaving your armchair, that if you go back far enough in a time machine, any individual you meet who has any living human descendants at all must be an ancestor of everybody living. When your time machine has travelled sufficiently far into the past, everybody you meet is an ancestor either of everybody alive in 2013 or of nobody. By the method of *reductio ad absurdum* beloved of mathematicians, you can see that this has to be true of our fishy ancestors of the Devonian era (my fish has to be the same as your fish, because the absurd alternative is that your fish's descendants and my fish's descendants stayed chastely separate from each other for 300 million years yet are still capable of interbreeding today). The only question is how far back you have to go to apply that argument. Clearly not as far as our fishy forebears, but how far? Well, hurdling swiftly over the detailed calculation, I can tell you that if the Queen is descended from William the Conqueror, you quite probably are too (and – give or take the odd illegitimacy – I know I am, as does almost everybody with a recorded pedigree).

Henry and Augusta's son, Clinton George Augustus Dawkins (1808–71) was one of the few Dawkinses actually to use the name Clinton. If he inherited any of his father's ardour he nearly lost it in 1849 during an Austrian bombardment of Venice, where he was the British consul. I have a cannonball in my possession, sitting on a plinth bearing an inscription on a brass plate. I don't know whose is the authorial voice and I don't know how reliable it is, but, for what it is worth, here is my translation (from French, then the language of diplomacy):

> One night when he was in bed, a cannonball penetrated the bed covers and passed between his legs, but happily did him no more than superficial damage. I first took this to be a tall story, until I learned for certain that it was based on the exact truth. His Swiss colleague met him later in the funeral procession of the American consul and, when asked about it, he laughingly confirmed the facts and told him it was precisely for this reason he was limping.

This narrow escape of my ancestor's vital parts took place before he was to put them to use, and it is tempting to attribute my own existence to a stroke of ballistic luck. A few inches closer to the fork of Shakespeare's radish and . . . But actually my existence, and yours, and the postman's, hangs from a far narrower thread of luck than that. We owe it to the precise timing and placing of everything that ever happened since the universe began. The incident of the cannonball is only a dramatic example of a much more general phenomenon. As I have put it before, if the second dinosaur to the left of the tall cycad tree had not happened to sneeze and thereby fail to catch the tiny, shrew-like ancestor of all the mammals, we would none of us be here. We all can regard ourselves as exquisitely improbable. But here, in a triumph of hindsight, we are.

C. G. A. ('Cannonball') Dawkins's son Clinton (later Sir Clinton)

Edward Dawkins (1859–1905) was one of many Dawkinses to attend Balliol College, Oxford. He was there at the right time to be immortalized in the Balliol Rhymes, originally published as a broadsheet called *The Masque of Balliol* in 1881. In the spring term of that year, seven undergraduates composed and printed scurrilous rhymes about personalities of the college. Most famous is the verse that celebrates Balliol's great Master, Benjamin Jowett, composed by H. C. Beeching, later Dean of Norwich Cathedral:

> First come I, my name is Jowett.
> There's no knowledge but I know it.
> I am Master of this College,
> What I don't know isn't knowledge.

Less witty, but intriguing to me, is the rhyme on Clinton Edward Dawkins:

> Positivists ever talk in s-
> Uch an epic style as Dawkins;
> God is naught and Man is all,
> Spell him with a capital.

Freethinkers were much less common in Victorian times, and I wish I had met great-great-uncle Clinton (as a child I did meet two of his younger sisters in advanced old age, one of whom had two maids called – I found the surname convention weird – Johnson and Harris). And what should we make of that 'epic style'?

I believe Sir Clinton later paid for my grandfather, his nephew Clinton George Evelyn Dawkins, to go to Balliol, where he seems to have done little but row. There is a photograph (reproduced in the picture section) of my grandfather preparing for action on the river that is wonderfully evocative of Edwardian high summer in Oxford.

It could be a scene from Max Beerbohm's *Zuleika Dobson*. The behatted guests are standing on the college barge, the floating boat-house which all the college rowing clubs maintained until living memory. Today, alas, they have been replaced by serviceable brick boathouses on the shore. (One or two of the barges are still afloat – or at least aground – as houseboats, having been towed to watery resting places amid moorhens and grebes in the backwaters and rivers around Oxford.) The resemblance between Grandfather and two of his sons, my father and my Uncle Colyear, is unmistakable. Family resemblances fascinate me, although they die away rapidly as the generations march on.

Grandfather was devoted to Balliol and contrived to stay there far beyond the normally allotted span of an undergraduate – solely, I suspect, in order to carry on rowing. When I used to visit him in old age, the college was his main topic of conversation, and he repeatedly wanted to know whether we still used (I repeatedly had to tell him we didn't) the same Edwardian slang: 'Mugger' for Master; 'wagger pagger' for wastepaper basket; Maggers' Memogger for the Martyrs' Memorial, the landmark cross outside Balliol that commemorates the three Anglican bishops who were burned alive in Oxford in 1555 for their attachment to the wrong flavour of Christianity.

One of my last memories of Grandfather Dawkins was of delivering him to his final Balliol gaudy (reunion dinner for former members, where each year a different age cohort is entertained). Surrounded by old comrades pushing Zimmer frames ('walkers') and festooned with ear trumpets and pince-nez, he was recognized by one of them who indulged the obvious sarcasm: 'Hello, Dawkins, you still rowing for Leander?' I left him looking a trifle forlorn among the boys of the old brigade, some of whom must surely have fought in the Boer War and were, therefore, rightful dedicatees of Hilaire Belloc's famous poem 'To the Balliol Men Still in Africa':

Years ago, when I was at Balliol,
Balliol men – and I was one –
Swam together in winter rivers,
Wrestled together under the sun.
And still in the heart of us, Balliol, Balliol,
Loved already, but hardly known,
Welded us each of us into the others:
Called a levy and chose her own.
Here is a House that armours a man
With the eyes of a boy and the heart of a ranger
And a laughing way in the teeth of the world
And a holy hunger and thirst for danger:

Balliol made me, Balliol fed me,
Whatever I had she gave me again:
And the best of Balliol loved and led me.
God be with you, Balliol men.

With difficulty I read this at my father's funeral in 2011, and then again in 2012 when I gave a eulogy for Christopher Hitchens, another Balliol man, at the Global Atheist Convention in Melbourne. With difficulty because, even on happier occasions, I become tearful with embarrassing ease when reciting loved poetry, and this particular poem by Belloc is one of the worst offenders.

After leaving Balliol, Grandfather made his career, like so many of my family, in the Colonial Service. He became Conservator of Forests in his district of Burma, where he spent much time in remote corners of the hardwood forests, supervising the heavy work of the highly trained elephant lumberjacks. He was up-country among the teak trees when the news reached him – I like to fancy by hand of runner with cleft stick – of the birth, in 1921, of his youngest son Colyear (named after Lady Juliana Colyear, mother of the enterprising Henry who eloped with Augusta Clinton). He was so excited that, without waiting for other transport to be available, he bicycled 50 miles to be at his wife Enid's bedside, where he proudly opined that the new boy had the 'Dawkins nose'. Evolutionary psychologists have noted the particular eagerness with which new babies are scanned for resemblances to their paternal, as opposed to maternal relatives – for the obvious reason that it is harder to be confident of paternity than maternity.

Colyear was the youngest and John, my father, the eldest of three brothers, all of whom were born in Burma to be carried around the jungle in Moses baskets slung from poles by trusty bearers, and all of whom eventually followed their father into the Colonial Service, but in three different parts of Africa: John in Nyasaland (now Malawi), the middle brother, Bill, in Sierra Leone, and Colyear in Uganda. Bill was christened Arthur Francis after his two grandfathers, but was always called Bill for a childhood resemblance to Lewis Carroll's Bill the Lizard. John and Colyear looked alike as young men, to the extent that John was once stopped in the street and asked: 'Are you you or your brother?' (That story is true, which is perhaps more than can be said of the famous legend that W. A. Spooner, the only Warden (head) of my present Oxford college to qualify for an 'ism', once greeted a young man in the quad with the question: 'Let me see, I never can remember, was it you or your brother was killed in the war?') As they aged, Bill and Colyear grew more alike (and like their father) and John less so, to my eyes. It often happens that family resemblances appear and disappear at different stages during a life history, which is one reason I find them fascinating. It is easy to forget that genes continue to exert their effects throughout life, not just during embryonic development.

There was no sister, to the regret of my grandparents, who had intended that their youngest would be Juliana but had to settle for her noble surname instead. All three brothers were talented. Colyear was the cleverest academically, and Bill the most athletic: I was proud to see his name on the roll of honour at the school I attended later, as holder of the school record for the hundred yards sprint – an ability which doubtless served him well at rugby when he scored a dashing touchdown for the Army against Great Britain early in the Second World War. I share none of Bill's athleticism, but I like to think that I learned how to think about science from my father, and how to explain it from my Uncle Colyear. Colyear became an Oxford don after leaving Uganda and was widely revered as a brilliant teacher of statistics, a notoriously difficult subject to convey to biologists. He died too young, and I dedicated one of my books, *River Out of Eden*, to him in the following terms:

> To the memory of Henry Colyear Dawkins (1921–1992), Fellow of St John's College, Oxford: a master of the art of making things clear.

The brothers died in reverse order of age and I sadly miss them all. I spoke the eulogy at the funeral of Bill, my godfather and uncle, when he died at the age of 93 in 2009.[1] I tried to convey the idea that, although there was much that was bad in the British Colonial Service, the best was very good indeed; and Bill, like his two brothers, and like Dick Kettlewell whom I'll mention later,[2] was of the best.

If the three brothers could be said to have followed their father into the Colonial Service, they had a similar heritage on their mother's side too. Their maternal grandfather, Arthur Smythies, was Chief Conservator of Forests in his district of India; his son Evelyn became Chief Conservator of Forests in Nepal. It was my Dawkins grandfather's friendship with Evelyn, forged while both were

---

[1] See web appendix: www.richarddawkins.net/afw.
[2] And whose obituary I wrote: see web appendix.

reading forestry at Oxford, that led to his meeting and marrying Evelyn's sister Enid, my grandmother. Evelyn was the author of a noted book on *India's Forest Wealth* (1925) as well as various standard works on philately. His wife Olive, I am sorry to say, was fond of shooting tigers and published a book called *Tiger Lady*. There is a picture of her standing on a tiger and under a solar topee, with her husband proudly patting her on the shoulder, captioned: 'Well done, little woman.' I don't think she would have been my type.

Olive and Evelyn's eldest son, my father's taciturn first cousin Bertram ('Billy') Smythies, was also in the forest service, in Burma and later Sarawak: he wrote the standard works *Birds of Burma* and *Birds of Borneo*. The latter became a kind of bible to the (not at all taciturn) travel writer Redmond O'Hanlon, on his hilarious journey *Into the Heart of Borneo* with the poet James Fenton.

Bertram's younger brother John Smythies departed from family tradition and became a distinguished neuroscientist and authority on schizophrenia and psychedelic drugs, living in California, where he is credited with inspiring Aldous Huxley to take mescaline and cleanse his 'doors of perception'. I recently asked his advice on whether to accept the kind offer of a friend to mentor me through an LSD trip. He advised against. Yorick Smythies, another first cousin of my father, was a devoted amanuensis of the philosopher Wittgenstein.[1] Peter Conradi, in his biography of the novelist Iris Murdoch, identifies Yorick as the 'holy fool' upon whom she based one of the characters in *Under the Net*, Hugo Belfounder. I must say it is hard to see the resemblance.

> Yorick wished to become a bus conductor but, [Iris Murdoch] noted, was the only person in the history of the bus company to fail the theory test ... During his single driving-lesson

---

[1] http://wab.uib.no/ojs/agora-alws/article/view/1263/977

the instructor left the car as Yorick drove on and off the pavement.

Having failed to make the grade as a bus conductor, and dissuaded by Wittgenstein (along with most of his other pupils) from a career in philosophy, Yorick worked as a librarian in the Oxford forestry department, which may have been his only connection with the family tradition. He had eccentric habits, took to snuff and Roman Catholicism, and died tragically.

Arthur Smythies, grandfather to the Dawkins and Smythies cousins, seems to have been the first in my family to enter Imperial service. His paternal ancestors for seven unbroken generations back to his great-great-great-great-great-grandfather (the Reverend William Smythies, born in the 1590s) were Anglican clergy to a man. I suppose it is not unlikely that, had I lived in any of their centuries, I might have been a clergyman too. I have always been interested in the deep questions of existence, the questions that religion aspires (and fails) to answer, but I have been fortunate to live in a time when such questions are given scientific rather than supernatural answers. Indeed, my interest in biology has been largely driven by questions about origins and the nature of life, rather than – as is the case for most young biologists I have taught – by a love of natural history. I might even be said to have let down the family tradition of devotion to outdoor pursuits and field natural history. In a brief previous memoir published in an anthology of autobiographical chapters by ethologists, I wrote:

> I should have been a child naturalist. I had every advantage: not only the perfect early environment of tropical Africa but what should have been the perfect genes to slot into it. For generations, sun-browned Dawkins legs have been striding in khaki shorts through the jungles of Empire. Like my father and his two

younger brothers, I was all but born with a pith helmet on my head.[1]

Indeed, my Uncle Colyear was later to say, on seeing me in shorts for the first time (he habitually wore them himself, held up by two belts): 'Good God, you've got authentic Dawkins knees.' I went on to write of my Uncle Colyear that the worst thing he could say of a young man was:

'Never been in a youth hostel in his life'; a stricture, which, I am sorry to say, describes me to this day. My young self seemed to let down the traditions of the family.

I received every encouragement from my parents, both of whom knew all the wildflowers you might encounter on a Cornish cliff or an Alpine meadow, and my father amused my sister and me by throwing in the Latin names for good measure (children love the sound of words even if they don't know their meanings). Soon after arriving in England, I was mortified when my tall, handsome grandfather, by now retired from the Burma forests, pointed to a blue tit outside the window and asked me if I knew what it was. I didn't and miserably stammered, 'Is it a chaffinch?' Grandfather was scandalized. In the Dawkins family, such ignorance was tantamount to not having heard of Shakespeare: 'Good God, John' – I have never forgotten his words, nor my father's loyal exculpation – 'is that *possible*?'

To be fair to my young self, I had only just set foot in England, and neither blue tits nor chaffinches occur in east Africa. But in any case I learned late to love watching wild creatures, and I have never

---

[1] 'Growing up in ethology', ch. 8 in L. Drickamer and D. Dewsbury, eds, *Leaders in Animal Behavior* (Cambridge, Cambridge University Press, 2010).

been such an outdoor person as either my father or my grandfather. Instead:

> I became a secret reader. In the holidays from boarding school, I would sneak up to my bedroom with a book: a guilty truant from the fresh air and the virtuous outdoors. And when I started learning biology properly at school, it was still bookish pursuits that held me. I was drawn to questions that grown-ups would have called philosophical. What is the meaning of life? Why are we here? How did it all start?

My mother's family came from Cornwall. Her mother, Connie Wearne, was the daughter and grand-daughter of Helston doctors (as a child I imagined them both as Dr Livesey in *Treasure Island*). She was herself fiercely Cornish, referring to the English as 'foreigners'. She regretted having been born too late to speak the now extinct Cornish language, but she told me that when she was a girl the old Mullion fishermen could understand the Breton fishermen 'who came to pinch our crabs'. Of the Brythonic languages, Welsh (alive), Breton (dying) and Cornish (dead), Breton and Cornish are sister species on the language family tree. A number of Cornish words survive in the Cornish dialect of English, for example *quilkin* for frog, and my grandmother could do the dialect well. We, her grandchildren, repeatedly persuaded her to recite a lovely rhyme about a boy who 'clunked a bully' (swallowed a plumstone). I even recorded one of these recitations, and sadly regret that I have lost the tape. Much later, Google helped me to track down the words,[1] and I can still hear her squeaky voice saying them in my head.

---

[1] From *Randigal Rhymes*, ed. Joseph Thomas (Penzance, F. Rodda, 1895).

There was an awful pop and towse[1] just now down by the hully,[2]
For that there boy of Ben Trembaa's, aw went and clunked[3] a bully,[4]
Aw ded'n clunk en fitty,[5] for aw sticked right in his uzzle,[6]
And how to get en out again, I tell ee 'twas a puzzle,
For aw got chucked,[7] and gasped, and urged,[8] and rolled his eyes,
    and glazed;
Aw guggled, and aw stank'd[9] about as ef aw had gone mazed.[10]

Ould Mally Gendall was the fust that came to his relief,–
Like Jimmy Eellis 'mong the cats,[11] she's always head and chief;
She scruffed 'n by the cob,[12] and then, before aw could say 'No,'
She fooched her finger down his throat as fur as it would go,
But aw soon catched en 'tween his teeth, and chawed en all the
    while,
Till she screeched like a whitneck[13]–you could hear her 'most a mile;

And nobody could help the boy, all were in such a fright,
And one said: 'Turn a crickmole,[14] son; 'tes sure to put ee right;'
And some ran for stillwaters,[15] and uncle Tommy Wilkin
Began a randigal[16] about a boy that clunked a quilkin;[17]
Some shaked their heads, and gravely said: "Twas always clear to them

---

[1] Fuss.
[2] Store for live bait.
[3] Swallowed.
[4] Pebble, though my grandmother translated it as plumstone, which makes more sense.
[5] Properly.
[6] Throat.
[7] Choked.
[8] Retched.
[9] Stamped.
[10] Mad.
[11] Local proverb.
[12] Forelock.
[13] Stoat, weasel.
[14] Somersault.
[15] Medicine distilled from peppermint.
[16] Nonsensical story.
[17] Swallowed a frog.

That boy'd end badly, for aw was a most anointed lem,[1]
For aw would minchey,[2] play at feaps,[3] or prall[4] a dog or cat,
Or strub[5] a nest, unhang a gate, or anything like that.'

Just then Great Jem stroathed[6] down the lane, and shouted out so
    bold:
'You're like the Ruan Vean men, soase, don't knaw and waant be
    told;'
Aw staved right in amongst them, and aw fetched that boy a clout,
Just down below the nuddick,[7] and aw scat the bully out;
That there's the boy that's standing where the keggas are in
    blowth:[8]
Blest! If aw haven't got another bully in his mouth!'

I am fascinated by the evolution of language, and how local versions diverge to become dialects like Cornish English and Geordie and then imperceptibly diverge further to become mutually unintelligible but obviously related languages like German and Dutch. The analogy to genetic evolution is close enough to be illuminating and misleading at the same time. When populations diverge to become species, the time of separation is defined as the moment when they can no longer interbreed. I suggest that two dialects should be deemed to reach the status of separate languages when they have diverged to an analogously critical point: the point where, if a native speaker of one attempts to speak the other it is taken as a compliment rather than as an insult. If I went into a

---

[1] Mischievous imp.
[2] Truant.
[3] Pitch and toss.
[4] Tie a tin can or something to an animal's tail.
[5] Rob.
[6] Briskly strode.
[7] Back of the head.
[8] Cow parsleys are in bloom.

Penzance pub and attempted to speak the Cornish dialect of English I'd be asking for trouble, because I'd be heard as mockingly imitating. But if I go to Germany and attempt to speak German, people are delighted. German and English have had enough time to diverge. If I am right, there should be examples – maybe in Scandinavia? – where dialects are on the cusp of becoming separate languages. On a recent lecturing trip to Stockholm I was a guest on a television talk show which was aired in both Sweden and Norway. The host was Norwegian, as were some of the guests, and I was told that it didn't matter which of the two languages was spoken: audiences on both sides of the border effortlessly understand both. Danish, on the other hand, is difficult for most Swedes to understand. My theory would predict that a Swede visiting Norway would probably be advised not to attempt to speak Norwegian for fear of being thought insulting. But a Swede visiting Denmark would probably be popular if she attempted to speak Danish.[1]

When my great-grandfather Dr Walter Wearne died, his widow moved out of Helston and built a house overlooking Mullion Cove on the west side of the Lizard peninsula, which has remained in the family ever since. A lovely cliff walk among the sea pinks from Mullion Cove takes you to Poldhu, site of Guglielmo Marconi's radio station from which the first ever transatlantic radio transmission was sent in 1901. It consisted of the letter 's' in Morse code, repeated over and over. How could they be so dull, on such a momentous occasion, as to say nothing more imaginative than s s s s s s?

My maternal grandfather, Alan Wilfred 'Bill' Ladner, was Cornish too, a radio engineer employed by the Marconi company. He joined too late to be involved in the 1901 transmission but he was sent to work at the same radio station at Poldhu around 1913,

---

[1] I've consulted an expert on Scandinavian languages, Professor Björn Melander, and he agreed with my theory of 'insult or flattery' but added that there are, inevitably, complications of context.

shortly before the First World War. When the Poldhu Wireless Station was finally dismantled in 1933, my grandmother's elder sister Ethel (known simply as 'Aunt' to my mother, although she wasn't her only aunt) was able to acquire some large slate slabs that had been used as instrument panels, with holes drilled in them in patterns that traced out their use – fossils of a bygone technology. These slates now pave the garden of the family house at Mullion (see the picture section), where they inspired me, as a boy, to admiration of my grandfather's honourable profession of engineer – honoured less in Britain than in many other countries, which may go some way towards explaining my country's sad decline from a once great manufacturing power to the indignity of being a provider of (often, as we now sadly know, rather dodgy) 'financial services'.

Before Marconi's historic transmission, the distance across which radio signals could be received was believed to be limited by the curvature of the Earth. How could waves that travelled in a straight line be picked up beyond the horizon? The solution proved to be that waves could bounce off the Heaviside Layer in the upper atmosphere (and modern radio signals, of course, bounce off artificial satellites instead). I am proud that my grandfather's book, *Short Wave Wireless Communication*, went through many editions from the 1930s to the early 1950s as the standard textbook on the subject, until it was eventually superseded around the time when valves[1] were replaced by transistors.

That book was always legendary in the family for its in-comprehensibility, but I have just read the first two pages and find myself delighted by its lucidity.

The ideal transmitter would produce an electrical signal which was a faithful copy of the impressed signal and would transmit

---

[1] 'Vacuum tubes' in American English.

this to the connecting link with perfect constancy and in such a manner that no interference was caused to other channels. The ideal connecting link would transmit the electric impulses through or over it without distorting them, without attenuation, and would collect no 'noise' on the way from extraneous electrical disturbances of whatever kind. The ideal receiver would pick up the required electrical impulses despatched through the connecting link by the transmitter of the channel and transform them with perfect faithfulness into the required form for visual or audible observation . . . As it is very unlikely that the ideal channel will ever be developed, we must consider in what directions we would prefer to compromise.

Sorry, Grandfather; sorry I was put off reading your book while you were still around to talk about it – and when I was old enough to understand it but was put off even trying. And you were put off by family pressure, put off ever divulging the rich store of knowledge that must have been there still in your clever old brain. 'No, I don't know anything about wireless,' you would mutter to any overture, and then resume your near ceaseless whistling of light opera under your breath. I would love to talk to you now about Claude Shannon and information theory. I would love to show you how just the same principles govern communication between bees, between birds, and indeed between neurones in the brain. I would love you to teach me about Fourier transforms and reminisce about Professor Silvanus Thompson, author of *Calculus Made Easy* ('What one fool can do, another can'). So many missed opportunities, gone for ever. How could I have been so short-sighted, so dull? Sorry, shade of Alan Wilfred Ladner, Marconiman and beloved grandfather.

It was my Uncle Colyear rather than my Grandfather Ladner who prompted me to try to build radios in my teens. He gave me a book by F. J. Camm, from which I took the plans to build first a crystal set (which just faintly worked) and then a one-valve set – with a large, bright red valve – which worked slightly better but still needed headphones rather than a loudspeaker. It was unbelievably badly made. Far from arranging the wires tidily, I took delight in the fact that it didn't matter how untidy were the pathways they took, stapled down on a wooden chassis, so long as each wire ended up in the right place. I won't say I went out of my way to make the course of each wire untidy, but I certainly was fascinated by the mismatch between the topology of the wires, which really mattered, and their physical layout, which didn't. The contrast with a modern integrated circuit is staggering. Many years later, when I gave the Royal Institution Christmas Lectures to children of about the same age as I was when I made my one-valve set, I borrowed the hugely magnified layout diagram of an integrated circuit from a modern computer company to show them. I hope my young auditors were awestruck and a bit bewildered by it. Experimental embryologists have shown that growing nerve cells often sniff out their correct end organs in something like the way I built my one-valve set, rather than by following an orderly plan like an integrated circuit.

Back to Cornwall before the First World War. It was my great-grandmother's habit to invite the lonely young engineers from the clifftop radio station to tea at Mullion, and that was how my grandparents met. They became engaged, but then the war broke out. Bill Ladner's skills as a radio engineer were in demand, and he was sent by the Royal Navy as a smart young officer to the southern tip of what was then Ceylon to build a radio station at that strategically vital staging post in the Empire's shipping lanes.

Connie followed him out in 1915, where she stayed in a local vicarage, from which they were married. My mother, Jean Mary Vyvyan Ladner, was born in Colombo in 1916.

In 1919, the war over, Bill Ladner brought his family back to

England: not to Cornwall in the far west of the country but to Essex in the far east, where the Marconi company had its headquarters in Chelmsford. Grandfather was employed teaching young trainee engineers at the Marconi College, an institution of which he later became head and where he was regarded as a very good teacher. At first the family lived in Chelmsford itself, but later they moved into the neighbouring countryside, to a lovely sixteenth-century Essex longhouse called Water Hall near the straggling village of Little Baddow.

Little Baddow was the site of an anecdote about my grandfather which I think tells us something revealing about human nature. It was much later, during the Second World War, and Grandfather was out on his bicycle. A German bomber flew over and dropped a bomb (bomber crews on both sides occasionally did this in rural areas when, for some reason, they had failed to find their urban target and shrank from returning home with a bomb on board). Grandfather mistook where the bomb had fallen, and his first desperate thought was that it had hit Water Hall and killed his wife and daughter. Panic seems to have sparked an atavistic reversion to ancestral behaviour: he leapt off his bike, hurled it into the ditch, and *ran* all the way home. I think I can imagine doing that in extremis.

It was to Little Baddow that my Dawkins grandparents retired from Burma in 1934, to a large house called The Hoppet. My mother and her younger sister Diana first heard of the Dawkins boys from a girlfriend, breathless with Jane-Austen-style gossip about eligible young newcomers to the neighbourhood. 'Three brothers have come to live at The Hoppet! The third one is too young, the middle one is pretty good news, but the eldest one is completely mad. He spends all his time throwing hoops around in a marsh and then lying on his stomach and looking at them.'

This apparently eccentric behaviour of my father was in fact thoroughly rational – not the first or the last time a scientist's motives were uncomprehendingly called into to question. He was doing postgraduate research based in the Department of Botany at Oxford, on the statistical distribution of tussocks in marshes. His work required him to identify and count plants in sample quadrats of marshland, and throwing 'hoops' (quadrats) at random was the standard method of sampling. His botanical interest turned out to be among the things that drew my mother to him after they met.

John's love of botany had begun early, during one of the holidays from boarding school which he and Bill spent with their Smythies grandparents. In those days it was quite common for colonial parents to send their children, especially sons, to boarding school in Britain, and at the ages of seven and six respectively John and Bill were despatched to Chafyn Grove, a boarding school in Salisbury which I too was later to attend. Their parents would remain in Burma for another decade and more, and with no air travel would not see their sons even during most school holidays. So between terms the two little boys stayed elsewhere, sometimes at professional boarding homes for boys of colonial parents, sometimes with their Smythies grandparents in Dolton, Devon, where they often had their Smythies cousins for company.

Nowadays, such long-term separation of children from their parents is regarded with something approaching horror, but it was quite common at the time, accepted as an inevitable concomitant of empire, and indeed diplomatic service, when international travel was long, slow and expensive. Child psychologists might suspect that it did lasting damage. Both John and Bill, as it happened, ended up well-adjusted, very personable characters, but there may have been others less robustly equipped to come through such childhood deprivation. Their cousin Yorick, as I have already mentioned, was eccentric and possibly unhappy; but then, he went to Harrow, which

– to say nothing of the pressures of association with Wittgenstein – might explain everything.

During one of these school holidays with the grandparents, old Arthur Smythies offered a prize to whichever of his grandchildren could make the best collection of wildflowers. John won, and that boyhood collection became the nucleus of his own herbarium, setting him on the road to becoming a professional botanist. As I have said, a love of wildflowers was one of the things he later found in common with Jean, my mother. They also shared a love of remote and wild places, and a dislike of noisy company: they were not fond of parties, unlike John's brother Bill and Jean's sister Diana (who later married each other).

At the age of thirteen, John and then Bill left Chafyn Grove and were sent to Marlborough College in Wiltshire, one of England's better-known 'public' (i.e. private) schools, originally founded for the sons of clergymen. The regime was spartan; cruel, according to John Betjeman in his verse autobiography. John and Bill don't seem to have suffered in the way the poet did – indeed, they enjoyed it – but it may be revealing that, when Colyear's turn came some six years later, their parents decided to send him to a gentler school, Gresham's, in Norfolk. For all I know, Gresham's might have suited John better too, except that Marlborough had a legendary teacher of biology, A. G. ('Tubby') Lowndes, who probably inspired him. Lowndes has a number of famous pupils to his credit, including the great zoologists J. Z. Young and P. B. Medawar and at least seven Fellows of the Royal Society. Medawar was an exact contemporary of my father, and they went on to Oxford together, Medawar to read zoology at Magdalen and my father to read botany at Balliol. I have reproduced, in the web appendix, a historical vignette which is a transcript of a monologue by Lowndes, recorded verbatim by my father and almost certainly heard by Medawar in the same Marlborough classroom. I think it is of interest as a kind of anticipation of the central idea of the 'selfish gene', although it didn't influence me as I didn't discover it in my father's notebook until long after *The Selfish Gene* was published.

After his degree at Oxford, my father stayed on to do a post-graduate research degree – the one on tussocks that I mentioned earlier. He then decided on a career in the agriculture department of the Colonial Service. This necessitated further training in tropical agriculture at Cambridge (where his landlady had the memorable name of Mrs Sparrowhawk) and then – after becoming engaged to Jean – at the Imperial College of Tropical Agriculture (ICTA) in Trinidad. In 1939 he was posted to Nyasaland (now Malawi) as a junior Agricultural Officer.

CAMP FOLLOWERS
IN KENYA

JOHN's posting to Africa hastened my parents' plans, and they were married on 27 September 1939 in Little Baddow church. John then left by ship for Cape Town, whence he travelled on to Nyasaland by train, and Jean followed in May 1940 in the flying boat *Cassiopeia*. Her rather dramatic journey took a week, with numerous landings for refuelling; one of them was in Rome, which caused some anxiety as Mussolini was teetering on the brink of entering the war on the German side, and had he done so at that point the *Cassiopeia*'s passengers would all have been interned for the duration of the war.

As soon as Jean arrived, John had to break it to her that he had been called up to join the King's African Rifles (KAR) in Kenya. The young couple had only a month of married life in Nyasaland (during which time, calculating backwards, I must have been conceived) before they had to leave. The Nyasaland Battalion was sending a convoy by road to Kenya, where they were to train. John somehow wangled permission to bypass the convoy and drive himself. What he did not have permission to do was take his bride with him. The colonial wives of Nyasaland were under strict orders to stay behind, or go to England or South Africa, when their husbands trekked north to the war. As far as she knows, my mother was the only one who disobeyed. My wonderful parents smuggled her into Kenya illegally – which caused problems later, as I shall tell.

On 6 July 1940, John and Jean, together with their servant Ali, who loyally accompanied them and was later to play a big part in my young life, drove off in 'Lucy Lockett', their old rattletrap Ford station wagon. They kept a joint diary of their journey, which I shall quote in what follows. They deliberately set off ahead of the convoy, in case they might break down and need rescuing: a prudent decision, given that the very first page of the diary mentions that the car had to be pushed by a gang of boys to get started at all. Day 4 of the journal records, after a successful bout of haggling for some gourds:

> This episode made us feel very cheerful, especially having won the battle and secured our gourds, and John was so hearty that he started the car before Ali was in the car & ripped off the door on a tree. This was very sad.

But even the mishap of losing a door didn't depress their young spirits, and the trio cheerfully made their way north, past ostriches and under giraffes, with Kilimanjaro on the horizon, sleeping by night in the back of the car, making a fire at each camp to scare the lions and cook delicious stews and pies in an improvised oven – the kind of ingenious invention my father delighted in throughout his life. From time to time they met up with the convoy. On one of these occasions the Commanding Officer, a

> big military gent . . . red hat and gold braid and minions, dived into an Indian shop, having commanded us to wait, and came out with a large bar of chocolate which he presented to me saying, 'A present for a little girl going on a big journey!' John ate the chocolate.

I wonder whether the chocolate was the genial commander's way of winking at the illegality of Jean's presence?

As they neared the Kenyan border,

> We were prepared to bury me under the bedding rolls, and have Ali sit on top when the Kenya frontier appeared. But the frontier never materialized, and after the most amazing and wonderful trip we found ourselves driving into Nairobi, and no one any the wiser. John deposited me in the Norfolk Hotel and drove off to join up – with Ali, who soon pinched an askari uniform and appointed himself a soldier.[1] Later he came out 'top' in an askari driver's course, thereby drawing attention to himself and causing John much embarrassment.

Despite this embarrassing triumph Ali never was officially a soldier, but he travelled around as my father's unofficial batman, accompanying him wherever he went, from training camp to training camp. At one of these, Nyeri, they coincided with the military funeral of Lord Baden-Powell, founder of the Boy Scouts. As a former scout himself, John was drafted in to be a pall-bearer and march beside the gun carriage. I have a photograph of him on this occasion (reproduced in the plate section) and I must say I think he looks very dashing in his KAR uniform, complete with khaki shorts, long socks, and the hat whose increasingly battered remains he was to wear for the rest of his life. Incidentally, the tall officer marching (out of step) next to him is Lord Errol of 'Happy Valley', soon to be murdered in the notorious and still officially unsolved 'White Mischief' case.

For Jean, the next three years were a time of more-or-less continuous migration as she camp-followed John's many postings,

---

[1] 'Askaris' was the name given to the African rank and file in the KAR.

in Uganda as well as Kenya. As she remarked in the private memoir that she wrote for the family much later,

> John was very clever at finding temporary homes for me near his different postings while he was training in the KAR. I did little jobs looking after people's children, and working in a couple of Prep schools, as well as being just a paying guest. Once John's commanding officer said when they had orders to go and take Addis Ababa that they'd better be quick or Jean Dawkins would be there first!

Among Jean's many kind hosts during this period were a Dr and Mrs McClean in Uganda, who took her in as a nursemaid for their toddler daughter 'Snippet'.

> The McCleans in Jinja were kind to me, and I trailed after Snippet doing this and that. The houses in Jinja were all around a golf course on the lake shore, and hippos used to play on the greens at night, belching and grunting, and marauding gardens too. There were droves of crocodiles, lazing in the water and basking at the shallow edges just below the falls, where I stupidly used to paddle. The crocs were funny keeping their jaws wide open so that their little pet bird friends could safely pick their teeth for them!

The symbiotic cleaner habit is now well described in coral reef fish. I wrote about it and the interesting evolutionary theory underpinning it in *The Selfish Gene*, but it hadn't occurred to me, until I read my mother's memoir much more recently, that there is a similar relationship between crocodiles and birds. I would expect the underlying evolutionary theory to be the same, best expressed in the mathematical language of game theory.

It was while staying with the McCleans that my mother got the first of her many bouts of malaria, which were to recur during her nine years in Africa and were one of the reasons for my parents' eventual decision to return to England. On one later occasion, when they were living in Nyasaland after the war, she has a vivid memory of hearing, through her fevered delirium, the urgent voice of Dr Glynn, senior physician of the Lilongwe hospital, saying: 'If they don't call John Dawkins quickly it may be too late.' Probably wrongly, she attributes her recovery to overhearing the doctor's fear that she was dying and her defiant resolve to prove him wrong.

However, one of her first suspected bouts of malaria at the McCleans' house turned out to have a different diagnosis:

> The doctor was a cheerful breezy chap and one day he said: 'You know what your trouble is, don't you?', and I said: 'malaria?', and he said, 'You're pregnant, my dear!' That was a shock, but we were delighted. Of course looking back it was very wrong of us in such an unpredictable and homeless situation. But then, had we been prudent and sensible and safe we would not have got Richard! So there! We took it in our stride, and I started making baby clothes and of course we were lucky. Luck stayed with us all the way. But now I realize it must have been hard for Richard later on being dragged all over the world, and may have been alarming to him. We made a list of how many times his little suit-case was packed in his first few years. Many nights were spent in the Kenya and Uganda Railway trains. Everywhere there were new faces and his early years must have been pitifully insecure.

I have found the list she made, covering my peregrinations during 1941 and 1942. She wrote it in a notebook, the 'blue book', now very tattered, in which she also recorded some of my childish sayings, and later those of my sister Sarah. The only place in the list that I remember is Grazebrook's Cottage, Mbagathi, near Nairobi, probably because we were there on two separate occasions. Here we were the guests of Mrs Walter, her war-widowed daughter-in-law Ruby, and her little grandsons.

My mother's memoir continues:

Kenya, Uganda and Tanganyika are full of memories, many very happy and wonderful. But a lot of sorrow and fear and anxiety and loneliness after John went away for long periods and there was no news. Letters were very far between and tended to come in bunches with very old dates. I was often frightened and lonely and always anxious but we did have a lot of good kind friends and I was so lucky in that. Most notable were the Walters at Mbagathi who totally adopted Richard and me.

I was there when the telegram came to say that [Mrs Walter's son] John, who had just been home on leave, had been killed. Mrs Walter had been through it all before with her husband in the first world war when John was a baby. It was very very bad.

So we all concentrated on young William Walter and then later on posthumous Johnny. Richard had them as brothers and Mrs Walter as a granny for quite a while. She was a remarkable and splendid lady and she kept busy and positive. She concentrated on giving happy holidays to servicemen on leave and I used to be sent into Nairobi, ferrying in and out batches of soldiers and sailors and airmen, in Juliana who was not a very predictable form of transport. Juliana had two fuel tanks, she started on petrol, and then with luck switched over to paraffin. Once I only just survived the 20 odd miles home. An enormously fat huge naval cook, badly drunk I soon realized, who I'd fetched from the New Stanley Hotel fell asleep across the seat and leaned against me so heavily I could barely steer the car and I couldn't move him. It was very difficult.

I think those men really enjoyed the Walter ménage. They played with the children, did lots of man-about-the house little jobs for Mrs Walter who treated them as boys and fed wonderful meals. It was a real home for us all.

Richard and I built another mud hut at Mbagathi, a splendid

double one of two rondavels [the traditional circular form] elongated with a straight length between them. It was lovely.

These two huts with a shared roof only took about a week to construct. They constitute what I believe is my earliest memory.

Mrs Walter had by then purchased a bit of land. One day when she was clearing bushes with an African there was a huge explosion and the poor man had the back of one lower leg blown clean off by (we presumed) a first world-war left-over-mine. She was a very tall strong person and she lifted him into her box-body old banger and brought him home. We propped him and covered him, and she took him to Nairobi. He remained totally cheerful and chattered throughout. We could not believe such amazing bravery!

It is easy to forget that the First World War reached far down into sub-Saharan Africa. Tanganyika (plus Rwanda and Burundi) was in those days German East Africa and there was fighting in the area, including even naval battles in Lake Tanganyika between German boats on one side and those of Britain and Belgium on the other (the west coast of the lake was in the Belgian Congo). Elspeth Huxley, in her truly great novel *Red Strangers*, an epic saga of Kikuyu life, portrays the war through Kikuyu eyes as a mysterious and unspeakable aberration of the white men, in which Africans became horrifically caught up. Not only was it horrible, it was completely pointless, because the winning side didn't end up driving home any of the losers' cattle or goats.

Not all the shocks of this time were to do with wars, current or past.

Sometimes I was sent on Ruby's horse called Bonnie to take a message to the Lennox Browns' neighbouring farm. The first time

I went the house boy showed me into their big drawing-room while he called the Memsahib. The room was dark with chintzy curtains drawn against the bright sun and as I waited I suddenly realized that I was not alone. There was an enormous lioness stretched full-length on a sofa, who yawned at me! I was fairly paralysed. When Mrs Lennox Brown came in she smacked it and pushed it off the sofa. I gave my message and left.

My mother's picture of the incident, painted recently from her memory, appears in the picture section.

Later, Richard and William Walter used to play with two pet lion-cubs at another farm. They were about the size and heaviness of full-grown big Labradors (with short legs) and very rough and powerful. But he and William seemed to find it fun. We used to go for picnics up into the Ngong hills driving over the short mountain grass – no roads. Cool and high and splendid. But we were certainly stupid because there were buffaloes there in huge droves over those hills.

My next two memories are both of injections: the first by Dr Trim in Kenya and the second (more painful) by a scorpion, later in Nyasaland. Dr Trim was fortuitously well named, for he was presumably the one responsible for having me circumcised. Obviously I wasn't asked for my consent, but it seems my parents weren't either! My father, away at the war, knew nothing of it. My mother was simply informed as a matter of routine by a nurse that it was time for me to go for my circumcision, and that was that. Apparently it was the default presumption in Dr Trim's nursing home – as it may have been in many British hospitals of the time: at my various boarding schools, the numbers circumcised and uncircumcised were about even, and there was no obvious correlation with religion, or social position, or indeed anything else that I could detect. The situation is different in Britain today, and I understand that America is beginning to move in the same direction. A recent landmark case in a German court ruled that even religious circumcision of infants is a violation of the rights of those too young to give their consent. This German verdict will probably be overruled because of the shrieks of protest that to prevent parents circumcising their children is a violation of the parents' rights to practise their religion. Significantly, no mention of the child's rights. Religion enjoys astonishing privileges in our societies, privileges denied to almost any other special interest group one can think of – and certainly denied to individuals.

As for the scorpion, it gave me a painful rebuke for my deficiencies as a budding naturalist. I saw it crawling across the floor and I misidentified it as a lizard. How *could* I? Lizards and scorpions don't resemble each other in any respect that I can now see. I thought it would be fun to feel the 'lizard' run over my bare foot, so I stuck it in the animal's path. The next thing I knew was a burning pain. I screamed the house down and then I think I passed out. My mother tells me that three Africans heard my screams and came rushing in. When they saw what had happened, they took turns at trying to suck the poison out of my foot. That is a recognized emergency procedure for snakebite. I have no idea whether it is effective with scorpion stings but I am touched that they tried. I now actually have a horror of scorpions, such that I would not pick one up even if it had had its sting removed. As for the eurypterids, the giant marine scorpions of the paleozoic era, some of which reached six feet long . . .

I am often asked whether my African childhood prepared me to become a biologist, and the episode of the scorpion is not the only indication that the answer is no. Another story suggests the same, and I blush to tell it. Close to Mrs Walter's house when we were living there, a pride of lions had made a kill and some neighbours offered to take the whole household to watch them. We drove in a safari car to within 10 yards of the kill where the lions were gnaw-ing, or in some cases lying around as if they had already eaten too much. The adults sitting in the vehicle were transfixed with excite-ment and wonder. But, my mother now tells me, William Walter and I stayed on the floor, totally absorbed with our toy cars, which we were driving around saying *vroom vroom*. We showed complete indifference to the lions, despite the adults' repeated attempts to arouse our interest.

What I lacked in zoological curiosity I seem to have made up for in human sociability. My mother says that I was exceptionally friendly, with no fear of strangers: an early talker with a love of words. And despite my shortcomings as a naturalist I do seem to have been an early sceptic. At Christmas 1942 a man called Sam dressed up as Father Christmas and entertained a children's party in Mrs Walter's house. He apparently fooled all the children, and finally took his departure amid much jovial waving and ho-ho-ho-ing. As soon as he had left, I looked up and breezily remarked, to general consternation, 'Sam's gone!'

My father came through the war unscathed. I guess he was lucky to be fighting not the Germans or Japanese but the Italians, who perhaps had by then seen through their preposterously vainglorious *Duce* and were sensible enough to have lost interest in winning. John played his subaltern's part in armoured cars in the Abyssinian and Somaliland campaigns and then, after the Italians were defeated, was sent for training to Madagascar with the East African Armoured Car Regiment, expecting to be posted to Burma. There he might have met his younger brother Bill, who was by then a major in the Sierra Leone Regiment, fighting the much more formidable Japanese and later to be mentioned in despatches. However, in 1943 the govern-ment gave higher priority to John's agricultural than to his military work, and he was recalled to civilian life, along with others of the Nyasaland Agriculture Department.

The welcome news of his demobilization so excited Jean when she read it that she was nearly run over in the street, carrying me. She was fetching her mail, as usual, from the *poste restante* box in Nairobi. John's letter purported to be a description of a cricket match. But she had no interest in cricket, as John knew well, and he would never have bored her with it. It had to have a secret meaning. The couple had previously worked out a private code, and had used it several times before, because mail from army personnel in wartime was routinely opened and read by censors. Their code was a simple one: read only the first word of each line and ignore the rest. And the first words of the next three lines about the cricket match were '*bowler . . . hat . . . soon*'. Unfortunately the letter doesn't survive, but it is easy to imagine. 'Bowler' ostensibly referred to the cricket bowler, and John must have worked 'hat' in somehow (perhaps the umpire's Panama; my mother doesn't recall) and then 'soon' in some plausible comment about the match. What did it mean? Well, a bowler hat was the epitome of civilian dress – demob kit, civvy street. '*Bowler hat soon*' could only mean one thing, and Jean didn't need to be a crossword expert to discern it. John was about to be demobbed, and Jean nearly got herself and me run over in her excitement at the realization.

Actually getting back to Nyasaland, however, was not so easy. The illegality of Jean's original entry to Kenya now came back to haunt her. The dundridges[1] of the colonial government couldn't give her a visa to leave Kenya because, as far as their records showed, she

---

[1] My wife's and my private word for heartlessly rule-loving bureaucrats, a word that I am trying to introduce into the English language. It comes from a comic novel by Tom Sharpe, in which J. Dundridge epitomized the type. It's such a suitable-sounding word. For a new word to qualify for the *Oxford English Dictionary* it must be used sufficiently often in the written language, without definition or attribution. I speak from experience and am delighted to say that an earlier coining, 'meme', has met the criterion and is safely perched among the Ms. Please use dundridge and give it currency.

had never arrived. And Jean and John couldn't drive down together in the way that they had driven up, because this time John was under strict orders to travel with the army: he was not officially demobbed until he reached the Nyasaland Battalion's headquarters in its home country. So the couple had to leave Kenya separately, and Jean couldn't leave because she wasn't there. Mrs Walter was wheeled out to vouch for her existence and Dr Trim to vouch for mine – as, having brought me into the world, he was in a position to do. Finally, it was my legal birth certificate that did the trick, and the reluctant dundridges grudgingly stamped Jean's leaving papers. She and I, aged two, set off in a small plane of the kind that would today be called a puddle-jumper – pretty exciting puddles, no doubt, filled with crocodiles and hippos, flamingos and bathing elephants. We lost all our luggage when changing planes in Northern Rhodesia (now Zambia) but it soon didn't matter. My parents were delighted to find that their trunks, shipped by sea from England at the beginning of the war, had finally arrived in Nyasaland, having survived, presumably, a navy-escorted convoy, and containing, as my mother happily recalled in her memoir–

All our half-remembered wedding-presents, and my new clothes. It was a tremendous home-coming, and Richard there to help explore the boxes.

THE LAND OF
THE LAKE

OUR life remained as peripatetic as it had been in Kenya. John and the other returnees from the army were used as stand-ins so that resident agricultural officers who had had no leave from their tropical duties since the beginning of the war could take a break in the balmy haven of South Africa. So John was posted to a different job, in a different part of Nyasaland, every few months. But, as my mother acknowledged, 'it was good fun and no doubt good experience for John, and we saw a lot of Nyasaland and lived in lots of interesting houses'.

Of this period, the house I remember best is the one at Makwapala, under Mount Mpupu near Lake Chilwa, where my father was in charge of an agricultural college and prison farm. The prisoners, who provided labour on the farm, seemed to have a good deal of freedom, and I remember watching them playing football with their toughened bare feet. My sister Sarah was born in Zomba hospital during this time, and my mother recalled that the Makwapala prisoners, some of them convicted murderers, 'used to queue up to be allowed to push her in her go-cart after tea'.

When we first arrived at Makwapala, we had to share the official Agricultural Officer's house with the outgoing family, whose return passage to England had been delayed a few weeks. They had two sons, the elder of whom, David, had the unpleasant habit of biting other children. My arms became covered with bite marks. On one occasion, at tea on the lawn, my father caught David at it and gently interposed his shoe to stop him. David's mother was outraged. She clasped the child to her bosom and roundly scolded my poor father. 'Do you have *no idea* of child psychology? Surely everybody knows that the very *worst* thing you can do to a biter is to stop him in mid-bite.'

Makwapala was a hot, humid, mosquito- and snake-infested place. It was too remote to enjoy a regular postal service, and the settlement had its own 'messenger', Saidi, whose daily job was to cycle the 15 miles to Zomba and back with the mail. One day Saidi didn't return; we learned that

> the unprecedented rain on Zomba mountain had roared down all the steep ravines washing great lumps of mountain and enormous rocks ahead of it. In Zomba Town, roads and bridges disappeared, and people in their cars, houses were marooned, and of course the road to Makwapala had washed away.

Saidi was safe, but I was sad that a nice man called Mr Ingram, who used to let me drive his car sitting on his lap, had been killed when a bridge that he was driving over washed away. 'Later', my mother wrote, 'we learned from local people that this sort of thing had happened before, though not in living memory. It was caused by some enormous snake-like creatures called Nyapolos, who got into the valleys and disrupted everything.'

I loved the rain. I think I perhaps picked up the sense of relief that people in a periodically dry country feel 'the day that the rains came down'. At the time of the great Nyapolos rain, having 'missed out on rain mostly', I was apparently, 'enchanted – he stripped off and rushed about in the downpour shouting with joy and going quite mad'. I still get a warm feeling of contentment in heavy rain, but I no longer like being out in it, perhaps because English rain is colder.

Makwapala is the site of my earliest coherent memories, and also of many of my parents' recordings of my sayings and activities. Here are just two of many:

Come and look Mummy. I've found where the night goes to sleep when it's sun-times [darkness under the sofa].

I measured Sally's bath with my ruler, and it said seven and ninepence, so she's very late for her bath.

Like all small children I was obsessed with pretending.

No, I think I'll be an accelerator.

Now you stop being the sea Mummy.

I am an angel, and you're Mr Nye, Mummy. You say Good morning Angel. But angels don't talk, they just grunt. Now this angel's going to sleep. They always go to sleep with their heads under their toes.

I also enjoyed second-order meta-pretends:

Mummy, let me be a little boy pretending to be Richard.

Mummy, I'm an owl being a water wheel.

There was a water wheel near where we lived, which fascinated me. My three-year-old self tried to put together some instructions for how to make a water wheel:

Tie a bit of string on the sticks all round, and have a ditch near and very fast water in it. Now get a bit of wood and put a bit of tin on it for a handle and use it for the water to come. Then get some bricks for the water to go rushy down, and get a bit of wood and make it round and make a lot of things sticking out of it, then put it onto a long stick and that's a water wheel and it goes round in the water and makes a big BANG BANG BANG noise.

I suppose the following is zero-order pretending, for my mother and I both had to pretend to be ourselves:

Now you be Mummy and I'll be Richard and we're going to London in this garrimotor [most likely this Anglo-Indianism entered my family through my colonial grandparents and great-grandparents, but it may have spread from India throughout the Empire].

In February 1945, when I was nearly four, my parents recorded that I had 'never been known to draw anything recognizable'. This may have been a disappointment to my artistically gifted mother, who had been hired to illustrate a book when she was sixteen, and later attended art school. To this day I remain quite extraordinarily inept where visual art is concerned, and I have a blind spot even for appreciating it. Music is another matter entirely, as is poetry. I can easily be moved to tears by poetry and (slightly less easily) by music, for example the slow movement of the Schubert String Quintet, or some songs of Judy Collins and Joan Baez. My parents' notes show an early fascination with the rhythms of speech. They would listen in when I was having my afternoon rest at Makwapala.

The wind blows in
The wind blows in

> The rain comes in
> The cold comes in
> The rain comes
> Every day the rain comes
> Because of the trees
> The rain of the trees

Apparently I talked or sang to myself all the time, often in nonsensical but rhythmic cadences.

> The little black ship was blowing in the sea
> A little black ship was blowing in the wind
> Down down down to the sea
> Down in the meadows, a little black ship
> The little black ship was down in the meadows
> The meadows were down to the sea
> Down to the meadows, and down to the sea
> The little black ship down in the meadows
> Down in the meadows, down to the sea

I think that this kind of soliloquizing, experimenting with rhythms and permuting words perhaps only half understood, is common among small children. There is a very similar example in Bertrand Russell's autobiography, when he tells of eavesdropping on his two-year-old daughter Kate talking to herself, and hearing her say:

> The North wind blows over the North Pole.
> The daisies hit the grass.
> The wind blows the bluebells down.
> The North wind blows to the wind in the South.

My best guess is that my garbled allusion to Ezra Pound

in the following must have come from my parents' reading aloud.

> The Askari fell off the ostrich
> In the rain
> Huge sing Goddamn
> And what became of the ostrich?
> Huge sing Goddamn

My parents also record that I had a large repertoire of songs, which I would render, always correctly in tune, pretending to be a gramophone, sometimes with 'jokes' such as getting stuck in a groove and singing the same word over and over until the 'needle' (my finger) was pushed out of the groove. We had a portable, wind-up clockwork gramophone, of exactly the kind immortalized in Flanders and Swann's 'Song of Reproduction'.

> I had a little gramophone
> I'd wind it round and round.
> And with a sharpish needle,
> It made a cheerful sound.
>
> And then they amplified it
> It was much louder then.
> And used sharpened fibre needles,
> To make it soft again.

My father didn't buy fibre needles. Characteristically, he improvised with the thorns at the end of sisal leaves.

Some of my songs I think I got from records, some were gibberish made up by me on the spur of the moment like those quoted above, and some were from my parents. My father, especially, delighted in teaching me nonsense songs, often derived from his own father, and many an evening rang to the strains of such gems as 'Mary had a William goat', 'Hi Ho Cathusalem, the harlot of Jerusalem' or 'Hoky Poky Winky Fum', which I learned was sung daily by my Smythies great-grandfather while lacing up his boots and at no other time. I was once temporarily lost on a Lake Nyasa beach, and was eventually discovered sitting between a pair of old ladies in deckchairs regaling them with the Gordouli song, bawled since 1896 by Balliol undergraduates as a mocking serenade over the wall to the neighbouring college, Trinity, and a favourite of my grandfather and father.

> Gordoooooooooli.
> He's got face like a ham.
> Bobby Johnson says so.
> And he ought to know.
> Bloody Trinity. Bloody Trinity.
> If I were a bloody Trinity man
> I would. I would.
> I'd go into the public rear,
> I would. I would.
> I'd pull the plug and disappear.
> I would. I would.
> Bloody Trinity. Bloody Trinity.

Well, it's scarcely great poetry and never normally sung sober, but I suppose it is slightly intriguing to wonder what the old ladies made of it. My mother reports that, despite being missionaries, they seemed to be enjoying it. When I eventually got to Balliol myself in 1959, by the way, I discovered that the tune had changed for the worse – having suffered a destructive memetic mutation and lost a subtlety – at some point during the twenty-two years since my father had left.

My gramophone metaphor was regularly pressed into service in a guileful attempt to postpone bedtime: the gramophone would run down, the song becoming slower and grinding down in pitch, and would need to be 'wound up'. This was indeed a part of everyday life, for we had no electricity and our clockwork gramophone had to be wound up at frequent intervals to play my father's collection of 78 rpm records: mostly Paul Robeson, whom I adore to this day, plus another great bass, Feodor Chaliapin, singing *Tom der Reimer* in German (I wish I could track down that recording, but iTunes has so far let me down) and some miscellaneous orchestral music including César Franck's *Symphonic Variations*, which I called the 'Dripping Water', presumably in reference to the piano part.

With no electricity, our houses were lit by paraffin pressure lamps. They had to be primed with methylated spirit to heat the mantle, then pumped up with paraffin vapour, whereupon they hissed comfortably through the evening. For most of our time in Nyasaland we didn't have a water closet either, and had to use an earth closet, sometimes in an outhouse. In other respects, however, we lived in great luxury. We always had a cook, a gardener and several other servants (known, I regret to say, as 'boys'), headed by Ali, who became my constant companion and friend. Tea was served on the lawn, with beautiful silver teapot and hot-water jug, and a milk jug under a dainty muslin cover weighted down with periwinkle shells sewn around the edges. And we had drop scones (Scotch pancakes) which, to this day, are my equivalent of Proust's *madeleine*.

We had bucket-and-spade holidays on the sandy beaches of Lake Nyasa, which is big enough to seem like the sea with no land visible on the horizon, staying in a nice hotel whose rooms were thatched beach huts. We also had a holiday in a borrowed cottage high up Zomba Mountain. One anecdote from this trip demonstrates my lack of critical faculty (and perhaps belies the story of my seeing through Sam's Father Christmas act when I was one). Playing hide and seek with a friendly African man, I searched one particular hut and he definitely wasn't there. Later I went back to the same hut and he was there, in a place where I had positively looked. He swore that he had been there all the time, but had made himself invisible. I accepted this explanation as more plausible than the now obvious alternative hypothesis that he was lying. I can't help wondering whether a diet of fairy stories filled with magic spells and miracles, including invisible men, is educationally harmful. But whenever I suggest such a thing today I get kicked around the room for seeking to interfere with the magic of childhood. I don't think I told my parents my Zomba Mountain hide-and-seek story, but I can't help feeling that I'd have been rather pleased if they had talked me through a version of Hume on miracles. Which do you think would be the greater miracle? The miracle that a man might tell a lie to amuse a gullible child? Or the miracle that he really did turn himself invisible? So, little Richard, now what do you think really happened in that hut, high on Zomba Mountain rearing up out of the plain?

Another illustration of childhood gullibility: someone had attempted to relieve my distress at the death of pets by telling me that animals, when they die, go to their own heaven called the Happy Hunting Ground. I believed this totally, and didn't even wonder whether it was also 'heaven' for the prey animals they hunted there. Once, in Mullion Cove, I met a dog and asked whose dog it was. I misheard the answer as 'Mrs Ladner's dog come back'. I knew that, before I was born, my grandmother had had a dog called Saffron, now long dead. I immediately presumed, with a credulous curiosity too mild to be even worth following up, that this dog was indeed Saffron, returned from the Happy Hunting Ground for a visit.

Why do adults foster the credulity of children? Is it really so obviously wrong, when a child believes in Father Christmas, to lead her in a gentle little game of questioning? How many chimneys would he have to reach, if he is to deliver presents to all the children in the world? How fast would his reindeer have to fly in order that he should finish the task by Christmas morning? Don't tell her point blank that there is no Father Christmas. Just encourage her in the unfaultable habit of sceptical questioning.

Christmas and birthday presents in wartime, thousands of miles from relatives and high streets, were inevitably limited, but my parents made up for it in ingenuity. My mother made me a magnificent teddy bear, as big as I was. And my father made me various ingenious contraptions including a lorry, which had under the bonnet (hood) a single real (and incongruously but delightfully not-to-scale) sparking plug. The lorry was my pride and joy when I was about four. My parents' notes show that I would pretend it 'broke down', whereupon I would:

Mend the puncture
Wipe the water off the stridibutor (distributor)
Fix the battery
Put water in the radiator
Tickle the carburettor
Pull the choke
Try the switch the other way
Fix the plugs
Put the spare batteries in properly
Put some oil in the engine
See if the steering is all right
Fill up with petrol
Let the engine get cool
Turn it over and have a look underneath
Test the pops by shorting the terminals [I now don't know what that meant]
Change a spring
Fix the brakes

Etc

Each item is followed through with appropriate motions and noises, and is followed by Ger er er er er Ger er er er er on the starter, which may, or usually may not, start the engine.

In 1946, the war having ended the previous year, we were able to go 'home' to England on leave (England was always 'home' even though I had never been there; I have met second-generation New Zealanders who follow the same nostalgic convention). We went by train to Cape Town where we were to board the *Empress* (I thought it was 'Emprist') *of Scotland*, bound for Liverpool. South African trains had an open walkway between carriages, with railings like a ship's that you could lean over to watch the world go by and catch the cinders from the horribly polluting steam engine. Unlike a ship's, however, these railings had to be telescopic so they could lengthen or shorten when the train went round a bend. Here was an accident waiting to happen, and indeed it did. I had hooked my left arm over the rail and didn't notice when the train started going round a bend. My arm was caught as the railings telescoped in, and there was nothing my stricken parents could do to free me until the long bend ended and the line straightened out again. At the next station, Mafeking, the train was halted while I was taken to hospital to have my arm stitched. I hope the other passengers were not annoyed by the delay. I have the scar still.

When we finally reached Cape Town, the *Empress of Scotland* turned out to be a dismal ship. It had been converted as a wartime troop-carrier: no cabins, but dungeon-like dormitories with three-high rows of bunks. There were dormitories for the men and separate dormitories for the women and children. There was so little space that they had to take turns doing things like getting dressed. In the women's dormitory, as my mother's diary records . . .

it was bedlam with so many small children. We dressed them and took them to the door and handed them to the relevant father waiting in a long queue to collect his own. And he took them off to queue for breakfast. Richard had regular trips to the ship's doctor for dressings to his arm, and of course half way through the three-week voyage I had a malaria bout and Sarah and I were put into the ship's hospital, and poor Richard was left alone in the dreadful dormitory. They wouldn't allow him to go with John or me, which was cruel.

I don't think we appreciated what a horrible time that whole journey must have been for Richard. And what a long effect it must have had. He must have felt that his whole world security had suddenly gone. And when we got to England he was quite a sad little boy, and had lost all his bounce. While we were looking out of the ship at Liverpool docks in the dark rain, waiting to go ashore he asked wonderingly 'Is that England?' and then quickly asked 'When are we going back?'

We went to my paternal grandparents at The Hoppet in Essex, which

in February was bitterly cold and spartan, and Richard's confidence ebbed and he took to having a stammer. He couldn't cope with his clothes. Having lived most of his life in very few garments, buttons and shoe-laces defeated him and the grand-parents thought he was backward: 'Can't he dress himself yet?' Neither we nor they having any child psychology books they set about getting some discipline going and he became quite a withdrawn little person and a bit paralysed. There was a ritual in the Hoppet that he must learn to say Good Morning when he came to breakfast and he was sent out of the room until it happened – His stammer got worse and none of us were happy. I am ashamed now that we allowed that grandparental behaviour.

Things were not much better with the maternal grandparents in Cornwall. I disliked almost all food, and would psych myself up to retch when grandparents made me eat it. Horrible, watery vegetable marrow was the worst, and I actually vomited into my plate. I think everyone was relieved when the time came for us to board the *Carnarvon Castle* at Southampton bound for Cape Town, and return to Nyasaland – not back to Makwapala in the south, but to the central district around Lilongwe. My father was posted first to the agricultural research station at Likuni, outside Lilongwe, and then to Lilongwe itself, now the capital of Malawi but then a small provincial town.

Both Likuni and Lilongwe are places of happy memory. I must have been interested in science by the age of six, because I can remember regaling my poor long-suffering little sister, in our shared bedroom at Likuni, with stories of Mars and Venus and the other planets, their distances from Earth and their respective likelihoods of harbouring life. I loved the stars in that most un-light-polluted place. Evening was a magically safe and secure time, which I associated with the Baring-Gould hymn:

> Now the day is over,
> Night is drawing nigh,
> Shadows of the evening
> Steal across the sky.
>
> Now the darkness gathers,
> Stars begin to peep;
> Birds, and beasts, and flowers
> Soon will be asleep.

I don't know how it came about that I knew any hymns at all, because we never went to church in Africa (although we did when staying with the grandparents in England). I suppose my parents must have taught me that hymn, along with 'There's a friend for littul chuldren, above the bright blue sky'.

Likuni was also where I first noticed, and was fascinated by, the long shadows of evening, which at the time had none of the foreboding evoked by T. S. Eliot's 'shadow at evening rising to meet you'. Today, whenever I hear Chopin's Nocturnes, I am transported back to Likuni and the secure, comforting feeling of evening when 'stars begin to peep'.

My father invented wonderful bedtime stories for Sarah and me, often featuring a 'Broncosaurus' which said 'Tiddly-widdly-widdly' in a high falsetto voice, and lived faaaaar away in a distant land called Gonwonkyland (I didn't finally take the allusion until undergraduate days when I learned about Gondwanaland, the great southern continent that broke up to form Africa, South America, Australia, New Zealand, Antarctica, India and Madagascar). We loved watching the luminous dial of his wristwatch in the darkness, and he would draw a watch on our wrists with his fountain pen, so we could keep track of the time under our mosquito nets during the comfortable night.

Lilongwe, too, was a place of precious childhood memory. The official house of the District Agricultural Officer was smothered in cascades of bougainvillea. The garden was filled with nasturtiums, and I loved to eat the leaves. Their unique, peppery taste, still encountered occasionally in salads, is the other candidate for my Proustian *madeleine*.

The identical house next door was the doctor's. Dr and Mrs Glynn had a son, David, of exactly my age, and we played together every day, in his house or mine or round about. There were dark blue-black grains in the sand, which must have been iron because we picked them up by dragging a magnet on a piece of string. On the verandah we made 'houses', with little rooms and corridors, by draping rugs and mats and blankets over upended chairs and tables.

We even equipped our verandah 'houses' with piped water, whose plumbing we made by sticking together hollow stems from a tree in the garden. Perhaps it was a *Cecropia*, but we called it a 'rhubarb tree', presumably deriving the name from a song that we liked to sing (to the tune of 'Little Brown Jug'):

> Ha ha ha. Hee hee hee.
> Elephant's nest in a rhubarb tree.

We collected butterflies, mostly yellow and black swallowtails, which I now realize were probably various species of the genus *Papilio*. David and I, however, didn't differentiate: we called them all 'Daddy Xmas', which he said was their proper name although it made no sense of their yellow and black colour scheme.

My butterfly habit was encouraged by my father, who made me a box for pinning them, using dried sisal instead of the cork favoured by professionals, and by my Dawkins grandfather – who was a collector himself – when he and my grandmother came to visit. They planned a grand tour of East Africa, calling on their sons in turn. They went first to Uganda to see Colyear, then made their way south to Nyasaland, through Tanganyika, as my mother recounted,

in a series of short-term local bus journeys, incredibly uncomfortably packed in with crowds of Africans and poor chickens with their legs tied, and enormous bundles of goods. But there was no transport further than Mbeya [in southern Tanganyika]. However, a young man with a little light aircraft offered to try to take them on. So they set off but got into bad weather and had to turn back. Meanwhile we had heard nothing from them at all. When their weather improved they tried again, flying low so that Tony [my grandfather, short for Clinton] could lean out and identify rivers and roads reading an old map as they went, and directing the pilot.

Grandfather would have been in his adventurous element. He loved maps. Also railway timetables, which he knew by heart and which came to constitute his only reading matter in extreme old age.

In Lilongwe everyone knew when a plane was coming about ten minutes before it arrived. This was because a local family kept pet crested cranes in their garden. These birds could hear an approaching plane long before people could and would start shrieking about it. Whether in fear or joy one didn't know! The regular weekly plane not being due, we wondered if it could be the grandparents when the cranes started shouting one day – so we went up to the air field, Richard and David on their tricycles, and we were in time to see the tiny plane arrive circling around the town twice before landing with enormous bumps and then Granny and Grandfather climbing out.

Nothing so obvious as Air Traffic Control, then. Just crested cranes.

It was in Lilongwe that we were struck by lightning. One evening a huge thunderstorm came. It was very dark and the children were having their suppers under their mosquito nets in the (wooden) beds. I was reading sitting on the floor and leaning against our so-called sofa (made of an old iron bedstead). Suddenly I felt as though a sledgehammer had landed on my head and I was completely flattened. It was a tremendous, carefully aimed blow. We saw that the wireless aerial and a curtain were on fire and we rushed into the children's bedroom to see if they were alright. They were totally unaffected and were chewing on their maize-cobs in a fairly bored sort of way!

History doesn't relate whether my parents extinguished the curtain fire before or after rushing into our bedroom to see if we were safe. My mother's memoir continues:

I had a long red burn all across my side where I had been leaning against the iron bed, and we discovered all sorts of other funny things later. Like a lump of concrete floor torn up and put onto the garage roof! The cook had a knife snatched out of his hand and was knocked over, a wire clothes-line was melted and the panes of glass in the sitting room were all splattered with molten wire from the radio aerial which totally disappeared, etc. etc. We now can't remember it all but it was dramatic.

My memory of that lightning strike is hazy, but I do wonder whether the cook's knife was really snatched out of his hand or whether he threw it in fright – as I would have. I do recall the multicoloured patterns made by some kind of residue all over the windows. And the actual moment of strike itself when the noise, instead of the usual boom boom de boom boom boom (which is mostly echoes) consisted of a single, prodigiously loud bang. There must have been a simultaneous very bright flash, but I have no memory of it.

Luckily it didn't leave us thunderstorm-shy because there were plenty of splendid ones in Africa. They were immensely beautiful, silhouetting mountain ranges black against brilliant-lit skies, all to the grand opera accompaniment of the sometimes almost non-stop thunder.

At Lilongwe we bought our first ever brand new car, a Willys Jeep station wagon called Creeping Jenny, to replace Betty Turner, the old Standard Twelve. I remember with nostalgic delight Creeping Jenny's exciting new-car smell. Our father explained to Sarah and me its advantages over all other cars, most memorable of which were the flat mudguards over the front wheels. He explained to us that these were especially designed to act as tables for us to put our picnic on.

At the age of five I was sent to Mrs Milne's school, a little one-room nursery school run by a neighbour. Mrs Milne couldn't really teach me anything because all the other children were learning to read, and my mother had already taught me to read; so Mrs Milne sent me off to one side with a 'grown-up' book to read to myself. It was too grown-up for me and, although I faithfully forced my eyes to travel over every word, I didn't understand most of them. I remember asking Mrs Milne what 'inquisitive' meant, but I couldn't muster enough of the stuff to keep asking her the meanings of words when she was busy teaching the other children. So I then

shared lessons with the doctor's son David Glynn taught by the doctor's wife. They were both bright, keen little boys and we think they probably learned a lot. Then he and David went on to the Eagle School together.

EAGLE IN THE MOUNTAINS

THE Eagle School was a brand new boarding school set high among the conifers of the Vumba Mountains, near the border with Mozambique, in Southern Rhodesia (now the sick joke dictatorship of Zimbabwe). I use the past tense because the school closed for ever during the conflicts that later beset that unhappy country. It was founded by Frank ('Tank') Cary, a former housemaster from the Dragon School in Oxford, I think the largest and arguably the best prep school in England, with a wonderful spirit of adventure and a remarkable list of distinguished alumni. Tank had come out to seek his fortune in Africa, and his school was a faithful scion of the Dragon. We had the same school motto (*Arduus ad solem*, a quotation from Virgil) and the same school song, to Sullivan's tune for 'Onward, Christian Soldiers': '*Arduus ad solem /* By strife up to the sun'. Tank had visited our family in Lilongwe when he was on a tour trying to drum up business from Nyasaland parents: mine liked him and decided that Eagle was the school for me, as did Dr and Mrs Glynn for David, and we went there together.

My memory of Eagle is hazy. I think I was there for only two terms, including the second term of the school's existence. I remember being there for the formal opening of the school, which was much talked about in advance as the 'Opening Day'. This mystified me because I took it to be an allusion to 'O God our help in ages past':

> Time like an ever-rolling stream,
> Bears all its sons away;
> They fly forgotten, as a dream
> Dies at the *opening day*.

Hymns made a big impression on me at Eagle, even 'Fight the good fight with all thy might', sung to a stupefyingly dreary tune more appropriate to dozing than fighting. All parents were told to equip their sons with a bible. My parents, for some reason, gave me *The Children's Bible*, which was not the same thing at all, and I felt rather left out and 'different'. In particular it was not divided up into chapters and verses, which I felt as a terrible deprivation. I was so intrigued by the biblical method of subdividing prose for easy reference that I went through some of my ordinary story books, writing in numbered 'verses' for them too. I have recently had occasion to look at the *Book of Mormon*, fabricated by a nineteenth-century charlatan called Smith, and it occurs to me that he must have had the same fascination with the King James Bible, laying his book out in verses and even imitating the sixteenth-century English style. Incidentally, it is a mystery to me why that last fact alone didn't instantly brand him a fake. Did his contemporaries think the Bible was originally written in the English of Tyndale and Cranmer? As Mark Twain cuttingly remarked, if you removed all occurrences of the phrase 'And it came to pass', the *Book of Mormon* would be reduced to a pamphlet.

My favourite book at Eagle was Hugh Lofting's *The Story of Doctor Dolittle*, which I discovered in the school library. It is now widely banned from libraries for its racism, and you can see why. Prince Bumpo of the Jolliginki tribe, steeped in fairy tales, desperately wanted to be the kind of prince that frogs magically turn into, or that falls in love with Cinderellas. Concerned that his black face might frighten any Sleeping Beauties he should chance to awaken with a princely kiss, he begged Doctor Dolittle to turn his face white. Well, it's easy enough to see now why this book, unremarkable and uncontroversial in 1920 when it was published, fell foul of the shifting *Zeitgeist* of the late twentieth century. But if we must talk moral lessons, the splendidly imaginative Doctor Dolittle books, of which I think the best is *Doctor Dolittle's Post Office*, are redeemed of their touch of racism by their much more prominent anti-speciesism.

In addition to its school song and motto, Eagle took over the Dragon School's tradition of calling the teachers by their nicknames or Christian names. We all called the headmaster Tank, even when being punished by him. At the time I thought the name meant the sort of tank that holds water in your roof, but I now realize that it almost certainly referred to the relentlessly unstoppable military vehicle. Probably Mr Cary acquired, during his years at the Dragon, a reputation for dogged persistence, moving forward in a straight line regardless of obstacles. Other masters were Claude (also a migrant from the Dragon), Dick (who had the popular duty of handing out a blessed ration of chocolate during our afternoon rest every Wednesday) and Paul, a darkly jovial Hungarian who taught French. Mrs Watson, who taught the most junior boys, was 'Wattie' and the matron, Miss Copplestone, was 'Coppers'.

I cannot pretend that I was happy at Eagle, but I was probably as happy as a seven-year-old sent away from home for three months can expect to be. Most poignant was the fantasy which I think I indulged almost daily when Coppers used to do her quiet morning rounds of the dormitories and we were still dozing: I imagined that she would somehow magically be transformed into my mother. I prayed incessantly for this – Coppers had dark curly hair like my mother, so in my childish naivety I reasoned that it wouldn't have taken a very big miracle to effect the transformation. And I was sure the other boys would like my mother just as much as we all liked Coppers.

Coppers was motherly and kind. I like to think that her report on me at the end of my first term was not entirely lacking in affection: I had, she wrote, 'only three speeds: slow, very slow and stop'. She did scare me once, without the slightest intention of doing so. I had a horror of going blind, having once seen an African with white blank-staring eyes like the ends of hard-boiled eggs. I used to fret that one day I would become either totally blind or totally deaf and I decided, after much painful deliberation, that it was a close-run thing but that going blind was the worst thing that could possibly happen. The Eagle School was modern enough to have electric light, driven by our own generator. One evening, as Coppers was talking to us in the dormitory, the generator engine must have died. As the light faded into total darkness, I quavered fearfully, 'Have the lights gone out?' 'Oh no,' said Coppers with breezy sarcasm, 'you must have gone blind.' Poor Coppers, she little knew what she said.

I was also terrified of ghosts, which I pictured as fully articulated, rattling skeletons with gaping eye sockets, sprinting towards me down long corridors at immense speed and armed with pickaxes, whose blows they would aim with devastating precision at my big toe. I also had weird fantasies of being cooked and eaten. I have no idea where these awful imaginings came from. Not from any books I had read, and certainly not from anything my parents had ever told me. Maybe tall stories recounted by other boys in the dormitory – of the type that I was to meet at my next school.

For Eagle was also my first exposure to the boundless cruelty of children. I wasn't bullied myself, thank goodness, but there was a boy called Aunty Peggy who was mercilessly teased, seemingly for no better reason than his nickname. As if in a scene from *Lord of the Flies*, he would be surrounded by dozens of boys, dancing around him in a circle and chanting 'Aunty *Peggy*, Aunty *Peggy*, Aunty *Peggy*' to a monotonous playground tune. The poor boy himself was driven demented by this, and would blindly rush at his tormentors in the circle, fists flying. On one occasion we all stood around and watched him in a serious and prolonged fight, rolling around the ground, with a boy called Roger, of whom we were in awe because he was twelve. The sympathy of the crowd was with the bully, who was good-looking and good at games, not the victim. A shameful episode, all too common among schoolchildren. Eventually, and not before time, Tank put a stop to this mass bullying, with a solemn warning at the morning assembly.

Every night in the dormitory we had to kneel on our beds, facing the wall at the head, and take turns on successive evenings to say the goodnight prayer:

Lighten our darkness, we beseech thee, O Lord; and by thy great mercy defend us from all perils and dangers of this night. Amen.

None of us had ever seen it written down, and we didn't know what it meant. We copied it parrot fashion from each other on successive evenings, and consequently the words evolved towards garbled meaninglessness. Quite an interesting test case in meme theory, if you happen to be interested in such things – if you are not, and don't know what I'm talking about, skip to the next paragraph. If we had understood the words of that prayer, we would not have garbled them, because their meaning would have had a 'normalizing' effect, similar to the 'proofreading' of DNA. It is such normalization that makes it possible for memes to survive through enough 'generations' to fulfil the analogy with genes. But because many of the words of the prayer were unfamiliar to us, all we could do was imitate their sound, phonetically, and the result was a very high 'mutation rate' as they passed down the 'generations' of boy-to-boy imitation. I think it would be interesting to investigate this effect experimentally, but have not so far got around to it.

One of the masters, probably Tank or Dick, used to lead us in community singing, including 'The Camptown Races' and:

> I have sixpence, jolly jolly sixpence,
> Sixpence to last me all my life
> I've tuppence to lend and tuppence to spend
> And tuppence to take home to my wife.

In this next one we were taught to sound the 'r' in 'birds', for reasons that I didn't understand at the time, but perhaps it was presumed to be an American song:

> Here we sits like brrrds in the wilderness
> Brrrds in the wilderness
> Brrrds in the wilderness
> Here we sits like brrrds in the wilderness
> Down in Demerara.

Some of the Dragon School's famous spirit of adventure had been exported to Eagle. I remember one exciting day when the masters organized the whole school into a large-scale game of Matabeles and Mashonas (a local version of Cowboys and Indians, using the names of the two dominant Rhodesian tribes) which had us roaming through the woods and meadows of the Vumba ('the mountains of the mist' in the Shona language). Goodness knows how we managed not to get lost for ever. And although the school had no swimming pool (one was built later, after I left) we were taken to swim (naked) in a lovely pool at the foot of a waterfall, which was far more exciting. What boy needs a swimming pool when you have a waterfall?

I made one journey to Eagle by plane, quite an adventure for a seven-year-old travelling alone. I flew in a Dragon Rapide biplane from Lilongwe to Salisbury (now Harare), from where I was to go on to Umtali (now Mutari). The parents of another Eagle boy, who lived in Salisbury, were supposed to meet me and set me on my onward journey, but they failed to show up. I spent what seemed like a whole day (with hindsight it cannot have been that long) wandering around Salisbury airport by myself. People were nice to me, somebody bought me lunch, and they let me wander into hangars and look at the planes. Weirdly, my memory is that it was quite a happy day and I wasn't at all frightened of being alone or of what might happen to me. The people who were supposed to meet me finally turned up and I got to Umtali where, I think, Tank met me in his Willys Jeep station wagon, which I liked because it reminded me of Creeping Jenny and home. I've told this story as I remember it. David Glynn has a different memory, and I'm guessing there were two journeys, one with him and one on my own.

FAREWELL TO
AFRICA

IN 1949, three years after their previous leave, my parents had another leave and we journeyed to England from Cape Town again, this time in a nice little ship called the *Umtali*, of which I don't remember much except for the lovely polished wood panelling and the light fittings, which I now think were probably art deco. The crew was too small to have a paid entertainments officer, so one of the passengers, a life-and-soul-of-the-party type called Mr Kimber, was elected to perform the role. Among other things, when we passed the Equator he organized a 'crossing the line' ceremony, in which Father Neptune appeared in costume complete with seaweed beard and trident. Mr Kimber also organized a fancy-dress dinner at which I was a pirate. I was jealous of another boy who came as a cowboy, but my parents explained that his admittedly superior costume was simply bought off the shelf, whereas mine was improvised and therefore really better. I understand the point now, but didn't then. One little boy came as Cupid, completely naked, with an arrow, and a bow which he threw at people. My mother came as one of the (male) Indian waiters, darkening her skin with potassium permanganate, which must have taken many days to wear off, and borrowing a waiter's uniform with its prominent sash and turban. The other waiters played along with the joke and none of the diners saw through her: not even me, not even the Captain when she deliberately brought him ice-cream instead of soup.

I learned to swim on my eighth birthday in the *Umtali*'s tiny swimming pool, made of canvas stretched between posts and erected on deck. I was so pleased with my new skill that I wanted to try it in the sea. So when the ship docked at Las Palmas in the Canary Islands in order to take on a large cargo of tomatoes and the passengers were put ashore for the day, we went to a beach where I proudly swam in the sea, with my mother vigilant on the shore. Suddenly she saw an abnormally huge wave about to break, as she thought, right over my diminutive, dog-paddling self. Gallantly she rushed, fully dressed, into the water to save me. In the event the wave lifted me harmlessly up – and then broke with full force on my mother, who was soaked from head to toe. The passengers were not allowed back on the *Umtali* until evening, so she spent the rest of the day in salt-wet clothes. Ungratefully, I have no memory of this act of maternal heroism, and the account I have given is hers.

The cargo of tomatoes must have been poorly loaded, because it shifted alarmingly at sea and the ship listed so far to starboard that our cabin porthole was permanently under water, causing my little sister Sarah to believe that we 'really *have* sunk now, Mummy'. Things became worse in the notorious Bay of Biscay, where the *Umtali* was seized by a spectacular gale, so strong it was hard to stand up. I excitedly rushed down to our cabin and pulled a sheet off my bunk to use as a 'sail' because I wanted to be blown along the deck like a yacht. My mother was furious, telling me – perhaps rightly – that I could have been blown overboard. Sarah's precious comfort blanket, 'the Bott', was indeed blown overboard, which would have been a serious tragedy but for our mother's prior foresight in cutting it in half so that she could keep a spare that had the right smell. I'm interested in the phenomenon of comfort blankets, though I never had one myself. They seem to be held in a position to be smelled while thumb- or finger-sucking. I suspect that there is a connection to the research of Harry Harlow on rhesus monkeys and cloth mother-substitutes.

We eventually docked in the Port of London and went to stay in a lovely old Tudor farmhouse called Cuckoos, opposite The Hoppet, which my paternal grandparents had bought to protect the land from developers. Living with us were my mother's sister Diana, her daughter Penny and her second husband, my father's brother Bill, on leave from Sierra Leone. Penny was born after her father, Bob Keddie, was killed in the war, as were both his gallant brothers – a terrible tragedy for old Mr and Mrs Keddie, who understandably then lavished their attention on little Penny, their only remaining descendant. They also were very generous to Sarah and me, her cousins, whom they treated as honorary grandchildren, and regularly gave us our most expensive Christmas presents and took us annually to a play or pantomime in London. They were rich – the family owned Keddie's Department Store in Southend – and possessed a grand house with a swimming pool and tennis court outside, and inside a lovely Broadwood baby grand piano and one of the first television sets. We children had never seen a television before, and we were enthralled to watch the blurry black-and-white image of Muffin the Mule on the tiny screen in the middle of the big, polished wood cabinet.

Those few months living as two families in one at Cuckoos provided the kind of magical memories that only childhood can. Beloved Uncle Bill made us giggle, calling us 'Treacle Trousers' (which Google now tells me is Australian slang for what the English call 'trousers at half mast') and singing his two songs, which we would frequently request.

Why has the cow got four legs? I must find out somehow.
I don't know and you don't know and neither does the cow.

And this one, to a sailor's hornpipe tune:

> Tiddlywinks old man, get a kettle if you can,
> If you can't get a kettle get a dirty old pan.

Penny's half-brother Thomas was born in Cuckoos while we were there. Thomas Dawkins is my double cousin, an unusual relationship. We share all four of our grandparents and hence all our ancestors except our immediate parents. Our proportion of shared genes is the same as for half-brothers, but we don't, as it happens, resemble each other. When Thomas was born the family hired a nurse, but she lasted only as long as it took her to see dear Uncle Bill in action making breakfast for the two families. He was on the stone-flagged kitchen floor, surrounded by a circle of plates into which he was throwing eggs and bacon in turn like dealing cards. This was before the days of 'health and safety' but it was more than the fastidious nurse could stand and she walked out, never to return.

Sarah, Penny and I went daily to St Anne's School in Chelmsford, the school that Jean and Diana had attended at the same age and under the same head teacher, Miss Martin. I don't remember much about it, except for the mincemeaty smell of school dinners, a boy called Giles who claimed that his father had lain down between the rails and let a train run over him, and the fact that the music master was called Mr Harp. Mr Harp had us singing 'Sweet Lass of Richmond Hill': 'I'd crowns resign to call her mine', but I didn't interpret it as 'I'd resign crowns to call her mine'. I heard 'crownsresign' as a single verb, which I guessed, from the context, must mean 'very much like'. I had made the same kind of misunderstanding with the hymn 'New every morning is the love / Our wakening and uprising prove'. I didn't know what 'our prisingprove' was, but evidently a prisingprove was an object anyone should be thankful to possess. The St Anne's school motto was quite admirable: 'I can, I ought, I must, I will' (not necessarily in that order, but it sounds about right). The adults in Cuckoos were reminded of Kipling's 'Song of the Commissariat Camels', and recited it with such a swing that I have never forgotten it:

Can't! Don't! Shan't! Won't!
Pass it along the line!

I was bullied at St Anne's by some big girls – not really badly bullied, but badly enough to provoke me to fantasize that, if I prayed hard enough, I could call down supernatural powers to give the bullies their come-uppance. I pictured a purplish black cloud with a scowling human face in profile, streaking across the sky above the playground to my rescue. All I had to do was *believe* it would happen; presumably the reason it didn't work was that I didn't pray hard enough – as when I prayed at the Eagle School for the metamorphosis of Miss Copplestone. Such is the naivety of the childhood view of prayer. Some adults, of course, never grow out of it and pray that God will save them a parking place or grant them victory in a tennis match.

I was expecting to return to Eagle after one term at St Anne's. While we were in England, however, my family's plans changed radically and I was never to see Eagle or Coppers or Tank again. Three years earlier, my father had received a telegram from England to say that he had inherited from a very distant cousin the Dawkins family property in Oxfordshire, including Over Norton House, Over Norton Park, and a number of cottages in the village of Over Norton. The estate had been much larger when it first came into the family, bought in 1726 by James Dawkins MP (1696–1766). He left it to his nephew, my great-great-great-great-grandfather Henry Dawkins MP (1728–1814), father of the Henry who eloped with the help of four hansom cabs galloping in different directions.

Thereafter it passed through many generations of Dawkinses, including the disastrous Colonel William Gregory Dawkins (1825–1914), a choleric Crimean War veteran who is said to have threatened tenants with eviction if they didn't vote his way, which was – oddly – liberal. Colonel William was irascible and litigious and squandered most of his inheritance suing senior army officers for insulting him: a drawn-out and futile process which benefited nobody except – as usual – the lawyers. Apparently a raving paranoid, he publicly insulted the Queen, assaulted his commanding officer Lord Rokeby in a London street, and sued the Commander-in-Chief, the Duke of Cambridge. Even more unfortunately, believing it to be haunted, he pulled down the beautiful Georgian Over Norton House and in 1874 built a Victorian replacement. His lawsuits drove him deeper and deeper into debt, forcing him to mortgage the Over Norton Estate to something more than the hilt, and he died in penury in a Brighton boarding house, living on the £2 per week allowed him by his creditors. The mortgage was eventually paid off by his unfortunate heirs in the early twentieth century, but only by dint of selling off most of the land, leaving the small nucleus that eventually passed to my father.

By 1945, the owner of what remained was Colonel William's great-nephew, Major Hereward Dawkins, who lived in London and seldom went near the place. Hereward, like William, was a bachelor, and he had no close relations bearing the name Dawkins. Evidently, when making his will, he looked up the family tree and lit upon my grandfather as the senior surviving Dawkins. His lawyer presumably advised him to skip a generation, and so he ended up naming my father, his much younger third cousin, as his heir. As things turned out, it was a brilliant choice, although he couldn't have known at the time that my father was ideally suited to preserve the land and make a go of it: the two of them had never met, and I don't think my father even knew of Hereward's existence when the telegram arrived in Africa, out of the blue.

In 1899 a long lease on Over Norton House had been given, as a wedding present, to a Mrs Daly. No doubt the rent vanished into the bottomless pit of Colonel William's debt repayments. Mrs Daly lived there in grand style with her family, a pillar of the local gentry and stalwart of the Heythrop Hunt, and my parents had no expectation that Hereward's legacy would change their lives. My father intended to rise through the ranks of the Nyasaland Department of Agriculture until he retired (or, as it would in fact have turned out, until the country became independent as Malawi).

When the *Umtali* docked in England in 1949, however, my parents received a piece of unexpected news: old Mrs Daly had died. Their immediate thought was that they should set about finding another tenant. But the possibility of leaving Africa and farming in England began to occur to them, and slowly gained favour in their minds. Jean's susceptibility to a dangerous strain of malaria was one reason, and I expect they were also attracted by the thought of English schools for Sarah and me. Their parents counselled against leaving Africa, as did the family lawyer. The Dawkins parents thought it was John's duty, in keeping with family tradition, to carry on serving the British Empire in Nyasaland, while Jean's mother was filled with dark forebodings that they would 'fail farming' as most people did. In the end, Jean and John went against all advice and decided to forsake Africa, live at Over Norton and take the estate in hand as a working farm – the first time after more than two centuries as parkland for the leisured gentry. John resigned from the Colonial Service, forfeiting his pension, and apprenticed himself to a series of English small farmers to learn the new skills he would need. He and my mother decided not to live in Over Norton House itself, but to divide it up into flats in the hope that it might pay for itself (lawyers' advice was to pull it down and cut their losses).

We ourselves would live in the cottage at the entrance to the drive, but it needed a lot of renovation, and while this was being done we did live – well, camp would be a better way of putting it – in a corner of Over Norton House.

I was still very keen on Doctor Dolittle, and my dominant fantasy during this brief interlude in Over Norton House was of learning, like him, to talk to non-human animals. But I would go one better than Doctor Dolittle. I would do it by telepathy. I wished and prayed and willed all the animals from miles around to converge on Over Norton Park, and me in particular, so that I could do good works for them. I did this kind of wishful praying so often, I must have been deeply influenced by preachers telling me that if you want something strongly enough you can make it happen; that all it takes is willpower, or the power of prayer. I even believed you could move mountains if your faith was strong enough. Some preacher must have said this in my hearing and, as is all too common with preachers, forgot to make the distinction between metaphor and reality clear to a gullible child. Actually, I sometimes wonder whether they even realize there is a distinction. Many of them don't seem to think it matters much.

My childhood games around the same time were imaginative in a science-fiction way. My friend Jill Jackson and I played spaceships in Over Norton House. Each of our beds was a spaceship, and we hammed it up for each other for hour after happy hour. It is interesting how two children can cobble together a storyboard for a joint fantasy, without ever sitting down together to work out the plot. One child suddenly says: 'Look out, Captain, Troon rockets are attacking on the left flank!' and the other instantly takes evasive action before announcing his side of the fantasy.

My parents had by now formally withdrawn me from Eagle and set about finding a school for me in England. They would probably have liked to send me to the Dragon, which was close by in Oxford, so that I could continue with something like the 'adventurous' Eagle experience. But such was the demand for places at the Dragon that you had to have your name down at birth to get in. So instead, they sent me to Chafyn Grove in Salisbury (the English Salisbury, after which the Rhodesian one was named), where my father and both his brothers had been, and not a bad school in its own right.

Chafyn Grove and Eagle were both – I should explain to those unfamiliar with such British arcana – 'preparatory schools': 'prep schools', for short. What did they 'prepare' us for? The answer is the even more confusing 'public schools', so called because they are in fact not public but private – open only to those whose parents can pay their fees. Close to where I live in Oxford there is a school called Wychwood, which for some years had a delightful notice outside the gates:

Wychwood School for Girls (preparatory for boys).

Anyway, Chafyn Grove was the prep school to which I was sent from eight to thirteen, to prepare me for public school from thirteen to eighteen. I don't, by the way, think it occurred to my parents to send me to anything other than the kind of boarding schools that Dawkinses normally attended. Expensive, but worth making sacrifices for – that would have been their attitude.

UNDER SALISBURY'S SPIRE

BEGINNING at any new school is bewildering. On the very first day I became aware that there were new words to learn. 'Puce' puzzled me. I saw it written on a wall and wrongly thought it must be pronounced 'pucky'. I eventually worked out that it was derogatory, synonymous with 'wet', also a favourite word, both meaning feeble. 'Muscle' meant the opposite: 'I was born in muscle India, Africa is puce' (in that era, many children who went to that kind of school were born in one or the other of those areas coloured Imperial pink on the map of the world). 'Wig', in the same school dialect, meant penis. 'Are you a roundhead or a cavalier? You know, your *wig*, is it a mushroom or a bootlace?' Such anatomical details were not confidential anyway, for we had to line up naked every morning for a cold bath. As soon as the rising bell sounded, we had to leap out of bed, take off our pyjamas, pick up our towels and stumble to the bathroom, where one of the three baths was filled with cold water. We plunged in and out as quickly as we could, supervised by the headmaster, Mr Galloway. From time to time the same bell was used to rouse us in the middle of the night for a fire practice. On one such occasion I was so dizzy with sleep that I went mechanically into morning getting-up routine, took my pyjamas off and had reached the bottom of the fire escape completely naked and carrying my towel before I noticed my mistake – everyone else was wearing pyjamas, dressing gown and slippers. Fortunately it was summer. The cold baths were not the only baths we had, of course. We had a proper hot bath in the evening (I forget how many times per week) in which we stood up to be washed by a matron, which we quite liked, especially when it was the pretty under-matron.

It was a time of austerity, close enough to the end of the war for many things still to be rationed. The food, with hindsight, was pretty horrible. Sweets were among the goods rationed by the government, and this had the paradoxical effect – presumably to the detriment of our teeth – that we actually had more sweets than we otherwise would have, because our sweet ration was scrupulously handed out after tea. I gave most of mine away. Now that I think about it, why was the wartime sweet ration anything other than zero? Couldn't what little sugar survived the U-boats have been put to better use?

My feet were frequently cold, and I suffered terribly from chilblains. Smells are notorious triggers of memory, and the eucalyptus smell of the chilblain liniment with which my mother supplied me is irrevocably associated with Chafyn Grove and the torment of itchy toes. We were often cold in bed at night, and we tried to stave it off by putting our dressing gowns on our beds. There was a chamber pot under each bed to obviate the need to go along the corridor during the night. I wish I had known at the time the North Country word for this object: gazunder (because it goes under).

Only one master was still at Chafyn Grove from my father's time: H. M. Letchworth, a kindly old Mr Chips-like figure who had fought in the First World War and had once been joint headmaster. We called him Slush, but not to his face, because Chafyn Grove didn't have the Dragon/Eagle convention about nicknames. The only exception was during the annual Scout Camp, when he liked to be called Chippi, an older nickname which I think dated from long before when he had known Baden-Powell. He didn't like the name

The Dawkins family have been members of the Chipping Norton set since the early eighteenth century, when my great-great-great-great-grandfather Henry Dawkins MP built a family mausoleum in St Mary's Church for, in the words of the inscription on the memorial tablet (*below*), 'himself and his heirs'. Brompton's 1774 portrait of Henry's family serves as backdrop to a family photograph taken in Over Norton house around 1958. My Grandfather Dawkins, with his pink Leander tie, sits between his wife Enid and his daughter-in-law Diana. My sister Sarah is in front of him; uncle Bill is behind him between Uncle Colyear and me. My father is on the far left. My mother is between Enid and Colyear's wife Barbara.

Is Zuleika Dobson among the spectators on the college barge as my grandfather Clinton G. E. Dawkins, leaning forward, prepares to row for Balliol?

My grandfather's education as an undergraduate (*right*) was supported by his uncle (later Sir) Clinton Edward Dawkins (*above*), whose freethinking views were celebrated in the Balliol Rhymes.

My father (*above*) and his rugby-playing brother Bill (*right*) followed their father and several other Dawkinses to Balliol after an idyllic childhood in the forests of Burma.

Above: The Smythies family at Dolton, Devon. *Top*: my paternal grandmother Enid, with dog and book, sits by her mother (in the very fine hat), brother Evelyn (with tennis racquet) and father (in panama hat), along with two unidentified guests. *Below*, Smythies cousins around 1923. Sitting on the ground, from right to left, are Bill, Yorick, John and Yorick's sister Belinda. Colyear is in his mother's arms.

Opposite: Evelyn Smythies' wife Olive was known as 'Tiger Lady' from her disagreeable hobby of shooting tigers. Her son, my father's first cousin Bertram Smythies, took a less destructive and more literary interest in the natural world.

THE AUTHOR ON AN ELEPHANT

RHINOCEROS HORNBILL

THE BIRDS OF BORNEO

BY

BERTRAM E. SMYTHIES
B.A., M.B.O.U.
Overseas Forest Service, Sarawak

with special chapters by TOM HARRISON, D.S.O., O.B.E.,
Curator of the Sarawak Museum, LORD MEDWAY, formerly
Technical Assistant, Sarawak Museum, and J. D. FREEMAN,
M.A., Reader in Social Anthropology in the National
University, Canberra, Australia

With 50 plates in colour by
COMMANDER A. M. HUGHES
O.B.E., R.N. (retd.)

and
48 photographic plates (in colour) by
LORD TOM WEL, A.R.P.S., ALFRED DOMINGO, A.R.P.S.,
T. G. H. ELGIE, E. N. WOODFORDE,
L. F. W. BROWN AND JOHN LINE

OLIVER AND BOYD
EDINBURGH: TWEEDDALE COURT
LONDON: 39a WELBECK STREET, W.1

1960

My maternal grandfather 'Bill' Ladner (seated third from left in the picture opposite), was among a group of naval officers sent to Ceylon to help build a wireless station during the First World War. Was the dog the station mascot? It seems to be the same dog my grandmother Connie is petting. The family returned to England when my mother Jean (*left*) was three. They lived in Essex (*above right*: my mother has her arms around a little friend) and spent their holidays at Mullion in Cornwall: here on the beach my Aunt Diana is holding hands with her mother and sister.

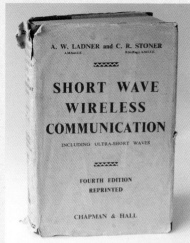

A. W. LADNER and C. R. STONER
A.M.Inst.C.E.   B.Sc.(Eng.), A.M.I.E.E.

~~~~~

**SHORT WAVE
WIRELESS
COMMUNICATION**

INCLUDING ULTRA-SHORT WAVES

~~~~~

**FOURTH EDITION
REPRINTED**

CHAPMAN & HALL

Above: my grandfather Ladner, a wireless engineer employed by Marconi and the author of the standard textbook on short wave wireless communication, demonstrates some equipment to visiting Arab royalty. He first met my grandmother in Cornwall while working at the Poldhu Wireless Station. Some of the thick slates used by the station as insulated instrument boards ended up as paving stones at our family house at the neighbouring Mullion Cove.

Slush. One Latin lesson the word *tabes* appeared in the vocabulary that we had to learn. Mr Letchworth was testing us and when the time came for a boy to translate *tabes* ('slush' in the context of the text we were reading) we all started sniggering. Mr Letchworth told us sadly that the name stemmed from that very passage of Livy ('All those years ago . . . that very sentence . . . all those years ago . . .'), though he never told us how it had come to stick to him.

The headmaster, Malcolm Galloway, was a formidable figure (maybe headmasters become formidable *ex officio*) whom we called Gallows. As befitted his nickname, he was not reluctant to use the extreme penalty, which in the case of Chafyn Grove was the cane. Unlike Eagle's 'bacon slice' beatings with a ruler, Gallows with the cane really hurt. He was reputed to have two canes, Slim Jim and Big Ben, and the punishment varied between three and six strokes depending on the severity of the misdemeanour. I never had Big Ben, thank goodness, but three strokes with Slim Jim was painful enough and caused bruises which we used to show off with pride, like battle scars, in the dormitory afterwards. They took several weeks to fade, turning from purple to blue to yellow on the way. Boys joked about stuffing an exercise book down the pants to soften the blows, but of course Gallows would have detected that instantly and I am sure it was never really tried.

Nowadays corporal punishment is illegal in England, and hindsight suspects teachers who employed it of cruelty or sadism. I am convinced that Gallows was guilty of neither. We have here an example of the speed with which customs and values change – an aspect of what I called, in *The God Delusion*, the 'shifting moral *Zeitgeist*'. Not under that name, the shifting moral *Zeitgeist* over a great span of history is massively documented by Steven Pinker in *The Better Angels of our Nature*.[1]

---

[1] Steven Pinker, *The Better Angels of our Nature: Why Violence has Declined* (New York, Viking, 2011).

Gallows was capable of great kindness. He would go around the dormitories before lights-out like a genial uncle, cheering us up, calling us by our Christian names (then only: never during the school day). One evening Gallows noticed the *Jeeves Omnibus* on a shelf in my dormitory, and he asked whether any of us knew P. G. Wodehouse. None of us did, so he sat down on one of the beds and read a story to us. It was 'The Great Sermon Handicap', and I suppose he must have spread it over several evenings. We *loved* it. It has remained one of my favourite Jeeves stories, and P. G. Wodehouse one of my favourite authors, read, reread, and even parodied to my own purpose.

Every Sunday evening, Mrs Galloway used to read to us in the family's private sitting room. We had to leave our shoes outside and we all sat on the floor, cross-legged amid a faint smell of damp socks. She would read a chapter or two each week, and would get through a book a term. They were usually stirring adventure stories like *Moonfleet* or *Maddon's Rock* or *The Cruel Sea* (the 'cadet edition' with the sex scenes removed). One Sunday, Mrs Galloway was away and Gallows read instead. We had reached the bit of *King Solomon's Mines* where the gallant pith-helmeted heroes were confronted with the twin mountains called Sheba's Breasts (fascinatingly, this name was censored in the film version starring Stewart Granger, a version which, bizarrely, included a woman on the expedition). Gallows paused to explain to us that these mountains were the Ngong Hills. (*I say, you fellows, that's utter rot. Gallows is just showing off that he's been to Kenya. King Solomon's Mines wasn't set in Kenya at all. Race you upstairs to the dorm.*)

When there was a severe thunderstorm at night, Gallows went up to the most junior dormitory, switched the light on and comforted the tinies (small enough to be allowed teddy bears) who might have been frightened. Halfway through each term, on 'Going Out Sunday' when parents came to take their sons out for the day, there were always one or two boys with no visitors, perhaps because their parents were abroad or ill. It happened to me once. Mr and Mrs Galloway took us out with their own children, in their big old 1930s touring car called Grey Goose and their little Morris 8 called James. We had a lovely picnic by a weir, and it makes me almost tearful remembering how kind they were to us, especially given that they might have preferred a day out with just their own children.

But as a teacher, Gallows was frightening. He would bellow at the top of his powerful voice and his stentorian scorn could be clearly heard in all the other classrooms throughout the school, provoking conspiratorial smiles between us boys and the other masters. 'What do you do when you meet "*ut* with the subjunctive"? . . . STOP AND THINK!' (Though, come to think of it, rules like that are not the way language really works.) Mr Mills, one of several who taught Latin, was more frightening still: too alarming even to have a nickname. He had a menacing presence and insisted on total accuracy and flawless handwriting: one mistake and we had to write the whole passage out again. Miss Mills – no relation – plump, sweet and motherly, with pigtails tied up in a sort of halo around the back of her head, taught the tinies and called us all 'dear'. Mr Dowson, the jovial, bespectacled maths master, was nicknamed Ernie Dow. We none of us knew where the 'Ernie' came from until one day he read us a poem and ended up telling us the author: Ernest Dowson, of course. I don't remember which poem it was, possibly 'They are not long, the weeping and the laughter', but it was certainly wasted on us anyway. Ernie Dow was a good teacher who, in his faintly northern accent, taught me most of the calculus I was ever to know. Mr Shaw didn't have a nickname, but his teenage daughter was called 'Pretty Shaw' for no better reason than to justify the puerile joke that inevitably followed when anybody said 'I'm pretty sure . . .' There was a continuous turnover of young masters, presumably students waiting to go up to university or just come down from it, whom we mostly liked, probably just because they were young. One of these was a Mr Howard, Anthony Howard, who later became a distinguished journalist and editor of the *New Statesman*.

My first term, in Form II, I was taught by Miss Long, a thin, angular lady of middle age with straight hair and rimless glasses, very kind like most of the teachers. Apart from Form II, she mainly taught the piano. Indeed, my first music lessons were with her, and I boasted to my parents that my progress was much faster than it was. Since the truth was sure to emerge eventually, what was the point of such boasting? I shall never know.

It became apparent that, if my parents had been pessimistic about the academic standards of Eagle in Southern Rhodesia, they were wrong. I had been only average among my contemporaries at Eagle, but found myself well ahead when I first arrived at Chafyn Grove. Embarrassingly so, to the extent that, since academic ability was not admired, I would pretend to know less than I did. Asked for the meaning of a Latin or French word, for example, I would pretend to um and er uncertainly, rather than instantly show what I knew and risk losing face with my peers. This tendency became positively illogical the following year in Form III, when I foolishly decided that, since the muscle boys who were good at games were mostly not very good academically, the only way for me to become good at games was to perform poorly in class. Actually, now that I think about it, that attitude was so stupid it's pretty self-evident that I didn't deserve to do well in class anyway.

I was evidently very confused about what it meant to be good at games. There were three brothers, Sampson *ma*, Sampson *mi* and Sampson *min* (*major*, *minor* and *minimus*), who were all good, especially Sampson *min* who was brilliant at all sports and once 'carried his bat' all the way from opening to when he ran out of partners, and then made a miraculous catch at silly-mid-on. It ludicrously occurred to me that the resemblance of the Sampson name to that of the famous muscleman of the Bible could be no accident. My naive mind conjectured that the Sampsons must surely have inherited their sporting prowess, if not from the biblical hero himself, then from some medieval strongman ancestor who had earned the name in the same kind of way as 'Smith' or 'Miller' – or, indeed, 'Armstrong', which really does derive from a nickname for a man with strong arms. Among many other things I got wrong here was my assumption that noticeable hereditary qualities go back more than a couple of generations – the same 'Tess of the D'Urbervilles' error I mentioned in the first chapter.

The Sampsons' father, who had only one eye – the other one having been pecked out by a heron (or so we were implausibly told) – owned a farm in Hampshire, on which the Chafyn Grove Scout Troop held its annual camp, supervised by Slush, with assistance from Gallows and a portly gentleman called Dumbo who was drafted in for the occasion. Scout camp was a high spot of the year for me. We put up our tents, dug our latrines, and made a hearth for our fire on which we cooked delicious dampers and twists (lumps of dough charred in the fire). We learned how to lash sticks together with elegantly turned sisal string, and lashed up all manner of useful camp furniture, from mug trees to clothes horses. We sang songs around the camp fire, special Scout songs like 'Dai's got a head like a ping pong ball', taught us by Slush/Chippi and not difficult to learn, most of them being very short:

> Gaily sings the donkey, as he goes to grass.
> Who knows why he does so, because he is an ass.
> Ee aw. Ee aw. Ee aw, ee aw, ee aw.

Some of them had no tune and were yells of solidarity rather than songs:

> There ain't no flies on us.
> There ain't no flies on us.
> There may be flies
> On some of you guys
> But there ain't no flies on us!

Pièce de résistance was an epic saga about a bad egg, sung by Chippi. I've reproduced it in the web appendix, in the sentimental hope that some of my readers might want to sing this otherwise forgotten song round a camp fire, and metaphorically stir the ashes of Henry Murray Letchworth MA (Oxon), Royal Dublin Fusiliers, alias Slush, alias Chippi, the benign and wistfully sad Goodbye-Mr-Chips patriarch of Chafyn Grove. In 2005, for my father's ninetieth birthday party in the Master's Lodgings, Balliol College, I faithfully wrote out the egg song for a bravura performance by the lovely soprano Ann Mackay and her piano accompanist, and my father jovially, if less tunefully, joined in.

At Scout camp we earned badges for accomplishments such as axemanship, knot-tying, semaphore and Morse code. I was good at Morse, using a technique perfected by my father in wartime Somaliland when signalling from his armoured car. For each letter, you learn a phrase beginning with that letter. Single-syllable words represent dots, longer words dashes. G, for instance, was 'Gordon Highlanders go' – dash dash dot. I could construct no such mnemonics for semaphore and that was perhaps why I was bad at it. Or it may have been because I have low spatial intelligence: I do well on IQ tests until I hit the spatial rotation questions at the end, and they pull my score right down.

The other high spot of the year was the annual school play, always an operetta, always produced by Slush, in a tradition that had been going at least since my father's time. My uncle Bill later explained to me that he was 'auditioned for the part of a bulb, but was found wanting'. The lead roles went to boys who could sing, and I was one. *The Willow Pattern Plate*, in which I played the female lead in my last year, was typical. The scenery backdrop consisted of a large painting of the famous blue plate design. The pagoda was the residence of the royal princess; she had died, and to avert the threat of a republic, the three little men on the bridge had long conspired to keep her death a secret. Their plot was threatened when a handsome Tartar prince sent word that he was galloping on his way as a suitor. At this point I, as the village maiden, appeared and sang my big number describing, with hammed-up histrionic gestures towards the scenery, the blue ceramic world in which we all lived:

> Blue is the sky above my aching head.
> The grass is blue beneath my weary feet.
> Blue are the trees that o'er the blue path shed
> A deeper shade of everlasting blue.
> And all the world is clothed in robes of blue.
> The restless sea is of the self-same hue.

That last line is quite witty (wasted on us schoolboys, of course), and I'd like to think it got a laugh from the adult audience, which consisted almost entirely of devotedly tolerant parents plus the reporter from the *Salisbury Chronicle* (who, by the way, gave me a very nice but undeserved notice).

> The royal pagoda glistens in the sun.
> The footballs grow on yon preposterous tree.
> (*The song has several verses more to run*
> *But that's the lot in my poor memory.*)

The three little men on the bridge seized their opportunity and bundled me into the pagoda to impersonate the dead princess, just in time before the Tartar prince bounded onto the stage with black moustache drawn on his face and sword drawn from its scabbard. I can't remember how the happy ending was achieved, but the prince ended up slinging me over his shoulder in a fireman's lift and carrying me off back to Tartary.

Moments of acute embarrassment linger in memory and wring an audible groan out of me when I recall them. At Chafyn Grove we had a sit-down tea every day, where we ate bread and butter. While we were lining up to file into the dining room the duty master would sometimes read out a list of names, supplied by a boy whose birthday it was. The named invitees would drop out of the line and go to a special table set aside for birthdays at the end of the dining hall, laden with birthday cake, jellies and other good things sent by the loving mother. I understood the principle, and I understood about supplying the duty master with the list of your friends' names. That was very clear. What slipped my attention was the small point that you had to arrange for your mother in advance to send the cake and jelly. On my birthday – perhaps my ninth – I wrote out the list of my friends and gave it to the duty master, who read it aloud. My chosen friends walked eagerly into the dining room, surveyed the empty table and . . . even after all these years embarrassment prevents me from describing the scene any further. What still baffles me is why it never occurred to me to wonder where the cake was supposed to come from. Perhaps I vaguely thought the school cook would provide it. But even so, shouldn't I have wondered how the cook was supposed to know it was my birthday? Perhaps I thought it materialized by supernatural magic, like sixpenny bits when you put a tooth under your pillow. As with my Zomba Mountain hide-and-seek story, this incident reveals a sad lack of anything remotely approaching critical or sceptical thinking in my childhood years. Even though I find these examples embarrassing, the lack of ability to think through the *plausibility* of things is a human trait common enough to be interesting. I'll return to the theme.

I was an exceptionally untidy and disorganized little boy in my early years at Chafyn Grove. My first school reports dwelt insistently on the theme of ink.

Headmaster's Report: He has produced some good work and well deserves his prize. A very inky little boy at present, which is apt to spoil his work.

Mathematics: He works very well but I am not always able to read his work. He must learn that ink is for writing, not washing, purposes.

Latin: He has made steady progress but unfortunately when using ink his written work becomes very untidy.

Miss Benson, my elderly French teacher, somehow managed to omit the ink *Leitmotiv*, but even her report had a sting in the tail.

French: Plenty of ability – a good pronunciation and a wonderful facility in escaping work.

Ink? Well, what do you expect if you equip every desk with an open inkwell and give the children dip pens that might have been designed to flick ink all over the room, or at least to deposit great round glistening drops of ink all over a page – drops which I would then draw out in spider shapes, or turn into Rorschach blots by folding the paper? No wonder the row of washbasins was strewn with pumice stones (we thought they were pummy stones) for cleaning ink stains off fingers. I'm afraid the ubiquitous ink somehow managed to spread itself over more than just my exercise books. I would desecrate printed textbooks too. I'm not talking about changing Kennedy's *Shorter Latin Primer* to *Shortbread Eating*

Primer: everyone did that automatically, of course. My ink habit went further than that. I doodled all over the school textbooks, filling in letters with ink or drawing little cartoon figures in the top right-hand corner of the pages so that they moved cinematically when you flicked through the book. The books didn't belong to us: we had to hand them back at the end of every term ready for the next cohort to inherit. And I knew I would be in trouble when the time came for me to hand in my ink-encrusted textbooks. Worrying about this kept me awake at nights, it made me seriously unhappy and put me off my (admittedly pretty nasty) food, yet I still went on doing it. I recognize that there is a sense in which the book-desecrating child was the same individual as my present bibliophile self, but this perverse childhood behaviour is beyond my understanding today. As is my erstwhile reaction, and that, I suspect, of just about all my contemporaries now, to bullying.

Much of the apparent bullying was pure *braggadocio*, futile threats whose emptiness was attested by their invocation of an indefinite future: 'Right! That does it. I'm putting you on my *beating-up list*' was about as nebulous a threat as 'You'll go to hell when you die' (though, alas, not everybody treats the latter threat as nebulous). But there was real bullying too, the especially unkind form of bullying where gangs of sycophantic henchmen rally around a bullying leader, courting his approval.

The 'Aunty Peggy' of Chafyn Grove was even more seriously bullied than the one at Eagle. He was a precociously brilliant scholar, large, clumsy and ungainly, with an unharmonious, prematurely breaking voice and few friends. I won't mention his name in case he should happen to read this and the memory is still painful. He was an unfortunate misfit, an ugly duckling doubtless destined for swanhood, who should have aroused compassion, and would have done in any decent environment – but not in the Goldingesque jungle of the playground. There was even a gang bearing his name, the 'anti- —— gang', the sole purpose of which was to make his life a misery. Yet his only crime was to be awkward and gangling, too uncoordinated to catch a ball, unable to run except with a graceless staggering gait – and very, very clever.

He was a day boy, which meant that he could escape to his home each evening – unlike today's victims of bullying, who are pursued beyond the school gates on Facebook and Twitter. But there came a term when, for some reason – perhaps his parents went abroad – he became a boarder. And then the fun really took off. His torment was exacerbated by the fact that he couldn't stand the cold baths. Whether it was the cold water or the nakedness I don't know, but what the rest of us took in our stride reduced him to a state of whimpering, abject horror, clutching his towel to him, trembling uncontrollably and refusing to let go. It was his Room 101. Finally, Gallows took pity on him and excused him the cold baths. Which, of course, did wonders for his already rock-bottom popularity among his peers.

I cannot even begin to imagine how human beings could be so cruel, but to a greater or lesser extent we were, if only through failing to stop it. How could we be so devoid of empathy? There's a scene in Aldous Huxley's *Eyeless in Gaza* where men recall with shame and bewilderment their bullying of a similar ugly duckling in the dormitory at their old school. Perhaps the guilt, which I and presumably all of my Chafyn Grove friends who remember the episode feel, goes some little way towards helping us understand how guards in concentration camps could possibly have done what they did. Could the Gestapo represent a sort of retention into adulthood of a psychology that is normal in children, giving rise to an adult psychopathy? That's probably too simplistic, but my adult self remains baffled. It's not as though I was without empathy. Doctor Dolittle had taught me to empathize with non-human animals to a degree that most people would consider excessive. At the age of about nine I was fishing from a boat out of Mullion harbour with my grandmother when I had the misfortune to catch a mackerel. I was immediately filled with tearful remorse and wanted to throw it back. I cried when I wasn't allowed to. My grandmother was kind and consoled me, but not kind enough to let me throw the poor creature back.

I also empathized, arguably again to an excessive extent, with schoolfellows who were in trouble with the authorities. I would go to ridiculous lengths – actually rather courageously foolhardy lengths – to try to exculpate them, and I have to regard this as evidence of empathy. Yet I didn't lift a finger to stop the grotesque bullying that I have just described. I think this was partly due to a desire to remain popular with dominant and popular individuals. It is a hallmark of the successful bully to have a posse of loyal lieutenants, and again we see this brutally manifested in the verbal cruelty and bullying that has become epidemic on internet forums, where the abusers have the additional protection of anonymity. But I don't recall feeling even secret pity for the victim of the bullying at Chafyn Grove. How is that possible? These contradictions trouble me to this day, together with a strong feeling of retrospective guilt.

Once again, as with the ink, I am struggling to reconcile the child with the adult that he became; and the same struggle, I suspect, arises with most people. The apparent contradiction arises because we buy into the idea that the child was the same 'person' as the adult: that 'the child is father of the man'. It is natural to do this because of the continuity of memory from day to day and, by extension, from decade to decade even though, we are told, no physical molecule of the child's body survives the decades. Since I kept no diaries, it is precisely that continuity that makes it possible for me to write this book. But some of our deepest-thinking philosophers, for example Derek Parfit and others cited by him in *Reasons and*

Persons,[1] have shown, with the aid of intriguing thought experiments, that it is by no means obvious what it means when we say that we are the same person through time. Psychologists such as Bruce Hood have approached the same problem from other directions. This is no place for a philosophical disquisition, so I will content myself with the observation that continuity of memory makes me *feel* as though my identity has remained continuous during my whole life, while I simultaneously feel incredulous that I am the same person as the young book-despoiler and the young empathy-failure.

I was also a games-failure, but the school had a squash court and I became obsessed with squash. I didn't really enjoy trying to win against an opponent. I just liked knocking the ball against the wall by myself, seeing how long I could keep going. I had squash-withdrawal symptoms during the school holidays – missed the echoing sound as ball hit wall, and the smell of black rubber – and I kept dreaming of ways in which I might improvise a squash court somewhere on the farm, perhaps in a deserted pig sty.

Back at Chafyn Grove I would watch games of squash from the gallery, waiting for the game to end so I could slip down and practise by myself. One day – I must have been about eleven – there was a master in the gallery with me. He pulled me onto his knee and put his hand inside my shorts. He did no more than have a little feel, but it was extremely disagreeable (the cremasteric reflex is not painful, but in a skin-crawling, creepy way it is almost worse than painful) as well as embarrassing. As soon as I could wriggle off his lap, I ran to tell my friends, many of whom had had the same experience with him. I don't think he did any of us any lasting damage, but some years later he killed himself. The atmosphere at morning prayers told us that something was up even before Gallows made his grim

---

[1] Derek Parfit, *Reasons and Persons* (Oxford, Oxford University Press, 1984).

announcement, and one of the woman teachers was crying. Many years later in Oxford, a large bishop sat next to me at high table in New College. I recognized his name. He had been the (ah me, much smaller then) curate at St Mark's church, to which Chafyn Grove marched in crocodile for Matins every Sunday, and he was evidently in touch with the gossip. He told me that the same woman teacher had been hopelessly in love with the paedophile master who had killed himself. None of us had ever guessed.

While Sunday morning service was in St Mark's, every weekday morning and every evening we had prayers in the school chapel. Gallows was extremely religious. I mean really religious, not token-religious: he truly believed all that stuff, unlike many educators (and even clergymen) who pretend to do so out of duty, and politicians who pretend to do so because they are under the (I suspect exaggerated) impression that it wins votes. Gallows usually referred to God as 'the King' (he pronounced it 'Keeng', surprisingly because his speech was otherwise standard English 'received pronunciation'). I think that, when I was very young, this led to a certain confusion in my mind. I must have been aware that King George VI was not really God, but there was a certain almost synaesthetic confusion in my mind between royalty and godhead. This persisted after George VI's death into the coronation of his daughter, when Gallows instilled in us a deep reverence for ceremonial nonsense such as anointing with holy oil. I can still conjure an echo of the same reverence when I see a 1953 coronation mug, or hear Handel's magnificent anthem 'Zadok the Priest', or Walton's 'Orb and Sceptre' march, or Elgar's 'Pomp and Circumstance'.

Every Sunday evening we had a sermon. Gallows and Slush took turns to preach it, Gallows in his Cambridge MA gown with white hood, Slush in his Oxford MA gown with red hood. One extraordinary sermon sticks in my memory. I can't remember who preached it. Whoever it was told the story of a squad of soldiers drilling by a railway line. At one point the drill sergeant's attention was distracted and he failed to shout 'About turn!' So the soldiers continued to march on – right into the path of an oncoming train. The story can't have been true, and I now think it also can't be true that – as I seem to remember from the sermon – we were supposed to admire the soldiers for their unquestioning obedience to military authority. Perhaps it is a failure of my memory. I certainly hope it is. Psychologists such as Elizabeth Loftus have shown that false memories can be indistinguishable from true ones, even when deliberately planted by unscrupulous therapists seeking, for example, to persuade distressed people that they *must* have been sexually abused as children.

One Sunday a junior master, a nice young man called Tom Stedman, was cajoled, obviously with the utmost reluctance, to do the preaching. He clearly hated it. I remember his frequent repetition of 'What's a heaven for?' It would have made more sense if I had realized – at the time, instead of years later – that it was a quotation from Browning. Another popular young master, Mr Jackson, had a fine tenor voice. He was persuaded one day to sing Handel's 'The Trumpet Shall Sound', which he also did with extreme reluctance, evidently realizing – correctly – that his art was wasted on us.

Also wasted on us were the occasional visiting lecturers and per-formers, although I suppose the fact that I have remembered them must say something. The ones that stick in my memory are R. Keith Jopp on 'It's there still' (archeology), Lady Hull playing the upright piano in the dining hall (Schumann's *Faschingsschwank*), somebody talking about Shackleton's Antarctic expeditions, somebody else showing flickering black and white films of athletes of the 1920s and 1930s, including Sydney Wooderson, and a trio of Irish troubadours who set up a little stage for themselves and sang 'I bought my fiddle for ninepence, and that is Irish too'. One lecturer talked about explosives. He fished out what he claimed was a stick of dynamite. Casually saying that if he dropped it the whole school would go sky high, he threw it up and caught it. Of course we believed him, gullible little naïfs that we were. How could we not believe him? He was an adult, and we were brought up to believe what we were told.

We didn't believe only adults. We were gullible too in the dormitory, where the resident yarn-spinner fooled us nightly. He told us King George VI was his uncle. The unfortunate king was held a prisoner in Buckingham Palace, from where he smuggled out desperate messages in code with a searchlight to our dormitory raconteur, his nephew. This young fantasist terrified us with stories of a horrible insect that would leap sideways from the wall onto your head, dig a neat round hole the size of a marble in your temple and bury a bag of poison in the hole to kill you. He told us, during a violent thunderstorm, that if you were struck by lightning you would be completely unaware of the fact for fifteen minutes. Your first inkling would be when blood started trickling out of both ears. Shortly after that you would be dead. We believed him and waited on tenterhooks after each lightning flash. Why? What reason did we have to think that he knew any more than we did? Was it even remotely plausible that you would feel nothing when you were struck by lightning, until fifteen minutes later? Once again, that sad lack of critical thinking. Shouldn't children be taught critical, sceptical thinking from an early age? Shouldn't we all be taught to doubt, to weigh up plausibility, to demand evidence?

Well, perhaps we should, but we weren't. On the contrary: if anything, gullibility was positively encouraged. Gallows was extremely keen that we should all be confirmed into the Church of England before we left the school, and almost all of us were. The only exceptions I can remember were the one boy who came from a Roman Catholic family (and went to a different church every Sunday in the envied company of the pretty Catholic under-matron), and the one precocious boy who struck awe into us by claiming to be an atheist – he called the Bible the Holy Drivel and we expected thunderbolts daily (his iconoclasm, if not his logic, carried over into his style of geometric proof: 'Triangle ABC *looks* isosceles, therefore . . .').

I signed up for confirmation with the rest of my cohort; and the vicar of St Mark's, Mr Higham, came to give us weekly confirmation classes in the school chapel. He was a handsome, silver-haired, avuncular figure and we went along with what he said. We didn't understand it, it didn't seem to make sense, and we thought this was because we were too young to understand. It is only with hindsight that I realize that it didn't make sense because, quite simply, there was no sense to make. It was all invented nonsense. I still have, and frequently have occasion to refer to, the bible I was given on my confirmation. This time it was the real thing, the King James translation, and I have some of the best bits in my head to this day, especially Ecclesiastes and the Song of Songs (not of Solomon, needless to say).

My mother has recently told me that Mr Galloway telephoned the parents individually to say how keen he was to have us confirmed. He said that thirteen was an impressionable age, and it was a good idea for children to be confirmed early, in order to give them a steady base in religion before they had to confront counter-influences at their public schools. Well, you can't say he wasn't honest in his designs on naive young minds.

I became intensely religious around the time I was confirmed. I priggishly upbraided my mother for not going to church. She took it very well and didn't tell me, as she should have, to take a running jump. I prayed every night, not kneeling at the bed but curled up in a foetal position inside it, in what I confided to myself was 'my own little corner with God'. I wanted (but never dared) to steal down into the chapel in the middle of the night and kneel at the altar where, I believed, an angel might appear to me in a vision. If I prayed hard enough, of course.

In my final term, when I was thirteen, Gallows made me a prefect. I don't know why this pleased me so much, but I walked on air for the whole term. Later in my life, when the head of my department at Oxford was knighted by the Queen, I attended his celebratory party. I asked a colleague whether our professor was pleased by the honour and received the memorable reply: 'Like a dog with three pricks, old boy.' That's pretty much how I felt on being made a prefect. Also on being accepted into the Railway Club.

The Railway Club was the main reason I had been pleased by my parents' decision to send me to Chafyn Grove. It was run by Mr K. O. Chetwood Aiken, not really a teacher except on the rare occasions when a boy would opt to learn German. A melancholy man with a long, sad face, his real love and apparently sole pastime was his Railway Room (although I recently learned, on Googling him, that he had been a known Cornish artist). One room in the school was set aside for him and he built there a magical simulacrum of the Great Western Railway, 0-gauge electric, with two terminuses called Paddington and Penzance and one station halfway between called Exeter. Each engine had a name, Susan or George for example, and the two dear little shunting engines were both called Boanerges (Bo One and Bo Two). Each station had a bank of switches, each switch activating its own portion of track, red switches for the Up line and blue for the Down. When a train arrived at Paddington, you had to uncouple it from the big engine that had pulled it, then drive one of the little shunting engines from its siding to move the train from the Up line to the Down, then send the engine to the turntable to turn it around, then couple it to the new front of the train and send it back along the Down line to Penzance, where the whole process would be reciprocated. I loved the smell of ozone that came from the electric sparks, and I adored working out which was the right combination of switches to flick on and off for each operation. I think the pleasure I got from it was similar to that which I later derived from programming computers, and also from soldering the connections in my one-valve radio set. Everybody wanted to get into the Railway Club, and all who did so doted on Mr Chetwood Aiken despite his lugubrious mien. With hindsight I think he may already have been very ill, for he died of cancer not long after I left. I don't know whether the Railway Room survived his death, but I think the school would have been mad to let it go.

Much as I enjoyed the Railway Club and being allowed to sashay uninvited through the door of the prefects' study, the time came when I had to move on to another school and start at the bottom again. When I was only three months old my father had put my name down for Marlborough, his old school, but was told that he was too late: I should have been put down at birth (how long before that sentence is quoted out of context?). Marlborough's snooty letter was quite hurtful to him as an old boy, but he put my name on the waiting list anyway, and when the time came I could have gone to Marlborough. Meanwhile, however, my father's thoughts had turned in another direction. He was impressed by the technical skills of the next-door gentleman farmer, Major Campbell, who had a well-equipped workshop and was an expert welder. My father naturally thought that I might become a farmer, and workshop skills give great advantages in that career (as I have recently learned from one of the most successful and certainly the most unconventionally enterprising of farmers I have ever met, the redoubtable and heroic George Scales).[1]

Major Campbell had acquired his expertise at his old school,

---

[1] http://old.richarddawkins.net/articles/2127-george-scales-war-hero-and-generous-friend-of-rdfrs.

Oundle in Northamptonshire. Oundle had the finest workshops of any school in the country, and its great headmaster from 1901 to 1922, F. W. Sanderson, had initiated a system whereby every boy spent one whole week of every term in the workshops, all normal school work suspended. Neither Marlborough nor any other school could boast anything like that. My parents therefore put my name down for Oundle, and I took the scholarship exam in my last term at Chafyn Grove. I didn't get a scholarship, but I did well enough to get a place, and Oundle was where I went, in 1954, aged thirteen.

I don't know, by the way, how much else Major Campbell had picked up during his time at Oundle. I presume his robust approach to recalcitrant underlings came rather from his days in the army. He caught one of his workers in petty theft, I think of a tool from his workshop, and fired him in somewhat literalistic terms: 'I'll give you fifty yards' start before you get both barrels.' Of course he wouldn't have carried out the threat, but it makes a good story and another fine illustration of the shifting moral *Zeitgeist*.

'AND YOUR ENGLISH SUMMER'S DONE'

OF COURSE, there was life beyond school. At Chafyn Grove, we longed for the end of every term, and our favourite hymn was the one we sang on the last day: 'God be with you till we meet again'. It ranked even higher than the stirringly martial missionary hymn that we also loved:

> Ho, my comrades! See the signal waving in the sky
> Reinforcements now appearing, victory is nigh.
> 'Hold the fort, for I am coming,' Jesus signals still.
> Wave the answer back to Heaven, 'By thy grace we will.'

We all went joyfully home for the holidays, some on the school train to London, some fetched by parents in their cars – in my case a battered old Land Rover which never caused me the embarrassment that snobbish boarding-school children are alleged to feel when their parents show up in anything less expensive than a Jaguar. I was proud of the ragged leaky-roofed old war-horse, in which my father had driven us crashing through the undergrowth on a dead straight compass course, on the child-delighting theory that there must have been a Roman road connecting two colinear stretches of dead straight highway on the well-thumbed Ordnance Survey map.

Very typical of my father, that kind of thing. Like his own father, he loved maps; and both loved keeping records. Weather records, for instance. Year after year my father filled notebooks with meticulously dated measurements of the daily maximum and minimum temperatures, and of rainfall – his enthusiasm only slightly dampened when we caught the dog peeing into the rain gauge. We had no way of knowing how many times dear Bunch had done this before and how many past rainfall records were similarly augmented.

My father always had an obsessive hobby on the go. It was usually one that would exercise his practical ingenuity, which was considerable, although he was of the scrap metal and red binder-twine rather than the Major Campbell lathe and welding-kit school of thought. The Royal Photographic Society elected him a Fellow for his beautiful 'dissolving' productions. These were carefully crafted sequences of colour slides, displayed by twin projectors working side by side in alternation, each slide artistically fading into the next, with musical and spoken accompaniment. Today it would all be done by computer, but in those days the fading in and fading out had to be achieved by iris diaphragms, inversely linked so that each opened as the other closed. My father fashioned cardboard iris diaphragms for the two projectors, coupled to each other by a fiendishly ingenious system of rubber bands and red string, activated by a wooden lever.

Family tradition changed 'dissolving' to 'drivelling', because that is how it had once been misread in a hastily scribbled note. We all became so used to calling his art-form 'drivelling' that we never thought of calling it anything else and the word lost its original meaning. On one occasion my father was giving a public presentation (one of many around that time) to a photographic club. It happened that this particular presentation was largely put together from earlier photographs, taken before he had begun his 'dissolving' hobby, and he began by explaining this to the audience. He had an endearingly halting and rambling style of delivery, and the audience warmed, in a somewhat bemused way, to his opening sentence: 'Er, I actually, I actually, er, these photographs mostly date, er, mostly date from before I started *drivelling . . .*'

His less than fluently accomplished style of speaking had earlier shown itself during his courtship of my mother, when he lovingly looked deep into her eyes and murmured, 'Your eyes are like . . . spongebags.' Bizarre as this sounds, I think I can make some sort of sense of it and it again has something to do with iris diaphragms. When seen end-on, a spongebag's drawstrings look a bit like the radiating lines which are an attractive feature of an eye's iris.

Another year, his hobby was making pendants for all his female relations, each one a sea-smoothed Cornish serpentine pebble bound with a leather thong. At yet another time in his life, his obsession was to design and build his own automated pasteurizer for the dairy, with flashing coloured signal lights and an overhead conveyance system for the churns, which provoked a lovely verse from one of his employees, Richard Adams (not the famous rabbit man), who managed the pigs:

> With clouds of steam and lights that flash, the scheme is most
> > giganto,
> While churns take wings on nylon slings like fairies at the panto.

My father had a ceaselessly creative mind. While cultivating a field on his little grey Ferguson tractor, wearing his battered old KAR hat and singing psalms at the top of his voice ('Moab was my washpot': by the way, the fact that he sang psalms emphatically didn't mean he was religious), he had plenty of time to think. He calculated that all the time spent doubling back at the end of each row was wasted. So he devised an ingenious scheme for zigzagging diagonally across and along the field with *shallow* turning angles, such that the whole field could be covered twice, in little more than the time it would normally take to cover it once.

Ingenious on the tractor he may have been, but always sensible he was not. On one occasion the clutch on the tractor stuck down. Unable to get it out of gear, he lay down on the ground beneath the clutch to see why it was stuck, and eventually succeeded in freeing it. Now, if you lie down under the clutch of a tractor you'll find that you are also lying directly in front of the large left rear wheel. The tractor enthusiastically leapt into action and ran him over, and all I can say is it was a good thing it was a Ferguson and not one of today's giant tractors. The little Fergie went bowling triumphantly across the field, and Norman, my father's employee who was standing there, was too dumbstruck with horror to do anything about it. My father had to sit up and tell him to chase after the tractor to stop it. Poor Norman was also too shaken to drive him to hospital, so my father had to do that himself. He spent some time in hospital with his leg in traction, but apparently suffered no lasting damage. His stay in hospital had the beneficial side-effect of prompting him to give up smoking his pipe. He never went back to it, and its only legacy was hundreds of empty baccy tins bearing the slogan 'And assuredly this is a grand old rich tobacco' which he was still using decades later for keeping assorted screws, nuts and washers and the miscellaneous dirty old metal scraps in which he took such delight.

Under the influence of an evangelical agricultural author called F. Newman Turner, and also perhaps of his eccentric friend from Marlborough and Oxford days, Hugh Corley, my father was an early convert to organic farming, long before it became fashionable or patronized by princes. He never used inorganic fertilizers or weed-killers. His organic farming mentors also disapproved of combine harvesters, and our farm was too small to justify one anyway, so in the early days we harvested with an old binder. It clattered noisily across the field behind the little grey tractor, scissoring the wheat or barley in front and spitting out neatly tied sheaves behind (I marvelled at the clever mechanism for tying the knots). And then the real work began, because the sheaves had to be stooked. An army of us walked behind the binder picking up sheaves two at a time and stacking them against each other to make little wigwams (stooks), six sheaves to a stook. It was hard work, leaving our forearms scratched and grazed and sometimes bleeding, but it was satisfying and we slept well that night. My mother would bring jugs of draught cider (scrumpy) out to the fields for the stookers, and a warm feeling of good fellowship suffused the Hardyesque scene.

The purpose of stooking was to dry the crop, after which the sheaves were carted and tossed up onto a rick. As a boy, I was not strong enough with a pitchfork to toss a sheaf right to the top of a high rick, but I tried hard and I envied my father's strong arms and horny hands, the equal of any of his employees. Weeks later, a threshing machine would be hired and parked next to the rick. The sheaves were fed in by hand, the grain threshed out and the straw baled. The farm workers all joined in with goodwill, regardless of what their real jobs might have been – cowman or pigman or general handyman or whatever. Later we moved with the times and hired a neighbour's combine.

In an earlier chapter I said I was a secret reader who used to escape to my bedroom with a book instead of rushing around outside in all weathers in true Dawkins tradition. Secret reader I may have been, but I can't honestly pretend that my reading in the school holidays had much to do with philosophy or the meaning of life or other such deep questions. It was pretty standard juvenile fiction: *Billy Bunter, Just William, Biggles, Bulldog Drummond,* Percy F. Westerman, *The Scarlet Pimpernel, Treasure Island.* For some reason my family disapproved of Enid Blyton and discouraged me from reading her. My Uncle Colyear gave me the Arthur Ransome books in succession, but I never really got on with them. I think I found them too girly, which was silly of me. Richmal Crompton's *William* has, I still think, genuine literary merit, with irony that can appeal to an adult as well as a child. And even the *Billy Bunter* books, though so formulaically written they might almost have been composed by computer, have pretensions to literary allusion in such phrases as: 'Like Moses of old, he looked this way and that, and saw no man' or 'Like a podgy Peri at the gates of Paradise'. *Bulldog Drummond* plumbs depths of jingoistic and racist bigotry which unmistakably label its era but passed over my naive young head. My maternal grandparents had a copy of *Gone with the Wind*, which I reread avidly on more than one summer holiday, never really noticing the paternalistic racism until I was older.

Family life at Over Norton was about as happy as family life gets. My parents were a united couple who celebrated their seventieth wedding anniversary together shortly before my father died in December 2010 aged 95. We were not a particularly rich family, but we weren't poor either. We had no central heating and no television, although the latter was from choice more than poverty. The family car was the dirty old Land Rover I mentioned or a cream van, neither of them luxurious but they did the job. Sarah's and my schools were expensive, and my parents surely had to skimp in other areas of life to send us there. Our childhood holidays were not in posh hotels on the Côte d'Azur but in army surplus tents in Wales, pelted with rain. On those camping trips we washed in an ex-Burma-Forest-Department canvas bath, warmed by the camp fire on which we also cooked our meals. Sarah and I, in our tent, heard our father sitting in his bath with his feet outside it, meditatively ruminating to himself, 'Well, I've never had me bath in me boots before.'

For three of my most formative years in my early teens I had the equivalent of an elder brother. Our great friends from Africa, Dick and Margaret Kettlewell, had stayed on in Nyasaland. Dick had at an unusually young age become Director of Agriculture, and distinguished himself in the job so resoundingly that he later became Minister of Lands and Mines in the provisional government on the way to full independence. When their son Michael, a playmate of mine in our very early days, turned thirteen he started as a boarder at Sherborne School in England; and, as with my father a generation before, the question arose of where he should go in the school holidays. I was delighted when he came to us. The age gap was only just over a year, and we did everything together: swimming in the freezing cold stream in the valley; indoor pursuits like chemistry sets, Meccano,[1] ping-pong, canasta, badminton, miniature snooker, various childish concoctions and recipes for making beetroot wine, or detergents, or vitamin pills. With Sarah, we had a junior farming enterprise called The Gaffers. My father gave us a litter of piglets, which we called The Barrels. We fed them daily and were wholly responsible for looking after them. Mike and I have remained life-long friends. Indeed, he is now my brother-in-law and the grandfather of most of my young relations.

There is a downside to having an elder brother in your formative years, however. It can mean that whenever you do anything, he actually does the operation and you pass him the instruments (since Mike later became a distinguished surgeon, the metaphor is not unsuitable). My Uncle Bill had a lifelong reputation for being 'no good with his hands', whereas my father had the opposite repu-tation, and it was probably for the same reason. The younger brother is apt to be the apprentice, never the master craftsman. The elder brother tends to be the decision-maker, the younger brother the follower, and early habits stick. Unlike my Uncle Bill, I didn't culti-vate a reputation for being no good with my hands. Nevertheless I was – and now am – no good. Mike did everything, with me as

[1] American: Erector Set.

superfluous assistant, and my father probably looked forward to my imminent exposure to the famous workshops of Oundle, which should set me belatedly in the footsteps of Major Campbell. But those workshops, as we shall see, proved to be a disappointment.

I was probably a disappointment as a naturalist, too, despite the rare privilege of spending a day with the young David Attenborough, when we were both guests of my Uncle Bill and Aunt Diana. Already famous but not yet a household name, he had been their guest on a filming expedition up-country in Sierra Leone, and they remained friends. When Bill and Diana moved to England and I happened to be staying with them, David brought his son Robert to visit, and had us children wading all day through ditches and ponds with fishing nets and jam jars on strings. I've forgotten what we were seeking – newts or tadpoles or dragonfly larvae, I expect – but the day itself was never to be forgotten. Even that experience with the world's most charismatic zoologist, however, wasn't enough to turn me into the child naturalist that both my parents had been. Oundle beckoned.

THE SPIRE BY
THE NENE

By the boys, *for* the boys. The boys know best.
Leave it to them to pick the rotters out
With that rough justice decent schoolboys know.

John Betjeman, *Summoned by Bells*

I GOT the English public school experience too late – thank goodness – for the real cruelties of the John Betjeman era. But it was
quite tough enough. There were ludicrous rules, invented 'by the
boys for the boys'. The number of buttons you were allowed to undo
on your jacket was strictly laid down according to seniority, and
strictly enforced. Below a certain seniority level, you had to carry
your books with a straight arm. Why? The masters must have known
this sort of thing was going on, yet they did nothing to stop it.

The fagging system was still going strong, although happily it no
longer is. (Note to American readers: this doesn't mean what you
think. In British English, a 'faggot' is not a homosexual but a bundle
of sticks or a rather nasty meatball. And 'fag' means cigarette or
boring task or – as in this case – schoolboy slave.) Each house
prefect at Oundle chose one of the new boys as his personal slave or
fag. I was chosen by the deputy Head of House, known as Jitters
because he had a tremor. He was kind to me, but I still had to do his

every bidding. I had to clean his shoes, polish the brasses of his Cadet Corps uniform and make toast for him at teatime every day on a paraffin pressure stove in his study. I had to be ready to run errands for him at any time.

Not that fags were totally immune to sexual importuning. On four separate occasions I had to fend off nocturnal visits to my bed from senior boys much larger and stronger than I was. I suspect that they were driven by neither homosexuality nor paedophilia in the normal sense of the outside world, but by the simple fact that there were no girls. Pre-pubescent boys can be pretty in a girlish sort of way, and I was. There was also folklore, rife throughout the school, of boys having 'crushes' on other boys with girlish appeal. Once again, I was the victim of many such rumours, whose only real damage was the – considerable – time they wasted in idle gossip.

Many things about Oundle were intimidating after Chafyn Grove. In the Great Hall for morning prayers on my first day, new boys were yet to be assigned places and we had to find empty chairs where we could. I found one and timidly asked the big boy next to it whether it was taken: 'Not as far as I can observe' was his icily polite reply, and I felt crushed very small. After the treble chorus and foot-pumped harmonium of Chafyn Grove, Oundle's deep bellowing of 'New every morning is the love', accompanied by the massive, thundering organ, was alarming. The stooping headmaster in his black MA gown, Gus Stainforth, was formidable in a different way from Gallows. In nasal tones he exhorted us to 'break the back of the term's work' by the third week: I wasn't sure how you set about breaking the back of anything, let alone a term's work.

My form master in 4B1, Snappy Priestman, was a gentle man, cultivated, kind and civilized except when he (very occasionally) lost his temper. Even then, there was something oddly gentlemanly about the way he did it. In one of his lessons he caught a boy misbehaving. After a lull when nothing happened, he began to give us verbal warning of his escalating internal fury, speaking quite calmly as an objective observer of his own internal state.

> Oh dear. I can't hold it. I'm going to lose my temper. Get down below your desks. I'm warning you. It's coming. Get down below your desks.

As his voice rose in a steady crescendo he was becoming increasingly red in the face, and he finally picked up everything within reach – chalk, inkpots, books, wood-backed blackboard erasers – and hurled them, with the utmost ferocity, towards the miscreant. Next day he was charm itself, apologizing briefly but graciously to the same boy. He was a kind gentleman provoked beyond endurance – as who would not be in his profession? Who would not be in mine, for that matter?

Snappy had us reading Shakespeare and assisted my first appreciation of that sublime genius. We did *Henry IV* (both parts) and *Henry V*, and he himself played the dying Henry IV, chiding Hal for having taken the crown prematurely: 'Oh my son. God put it in thy mind to take it hence, that thou mightst win the more thy father's love, pleading so wisely in excuse of it.' He asked for a volunteer who could do Welsh (Williams) and Irish (Rumary: 'Oh, Rumary, you are a treasure.') Snappy read us Kipling, putting on a creditable Scottish accent for the hymn of the Chief Engineer, M'Andrew (that really is Kipling's spelling). The hauntingly rhythmical opening verse of 'The Long Trail' put me sadly in mind of the ricks of Over Norton and the 'all is safely gathered in' satisfaction of early autumn (please read it aloud to get the Kipling rhythm).

> There's a whisper down the field where the year has shot her yield,
> And the ricks stand grey to the sun,
> Singing: 'Over then, come over, for the bee has quit the clover,
> And your English summer's done.'

And, right on cue for the mellow fruitfulness, Mr Priestman read us Keats.

Our mathematics master that same year, Frout, was prone to dizzy fits. Once, before he arrived in the classroom, I seem to remember that we set all the lights swinging from the ceiling. Then when he came in we swayed in unison with them. I don't recall what happened next. Maybe remorse has blocked out the memory. Or maybe it is a false memory based on a schoolboy folk legend about what others had earlier done to him. Either way, I now see it as yet another example of the lamentable cruelty of children – a recurring theme of my schoolday recollections.

We didn't always get it our own way. One time the 4B1 physics master, Bufty, was ill and the class was taken instead by the senior science master, Bunjy. Having ascertained that we had reached Boyle's Law in our curriculum, he proceeded to teach us, labelling us with numbers in place of our names, which he had no time to learn. Small, stooping, old, and more short-sighted than anybody I have ever encountered before or since, he was, we thought, easy meat for ragging. He seemed scarcely to notice our insolence. We were wrong. Hypermyopic he may have been, but he noticed. At the end of the lesson, Bunjy quietly announced that he was keeping us all in detention that very afternoon. Crestfallen, we returned in the afternoon and were instructed to write on a clean page in our notebooks: 'Extra Lesson for Form 4B1. Object of the Lesson: To teach 4B1 good manners and Boyle's Law'. I am confident that this is not a false memory and I, for one, have never forgotten Boyle's Law.

One of our masters – the only one we were allowed to call by his nickname – was prone to fall in love with the prettier boys. He never, as far as we knew, went any further than to put an arm around them in class and make suggestive remarks, but nowadays that would probably be enough to land him in terrible trouble with the police – and tabloid-inflamed vigilantes.

Like most schools of its type, Oundle was divided into houses. Each boy lived in, and dined in, one of eleven houses, and his house commanded his loyalty in all competitive fields of endeavour. Mine was Laundimer. I don't know what the others were like inside because we were discouraged from visiting other houses, but I suspect that they were all much the same. Interestingly, however, our minds tended to see each house as having a 'personality', and we unconsciously grafted that personality onto individual boys in the house concerned. These house 'personalities' were so nebulous that I cannot find it in me to attempt a description of any one of them. It was just something one 'felt', subjectively. I suspect that this observation represents, in a somewhat more innocent form than many prevalent in the wider world, that 'tribal' human impulse that lies behind much that is more sinister, such as racial prejudice and sectarian bigotry. I'm talking about the human tendency to identify individuals with a group to which they belong, rather than seeing them as individuals in their own right. Experimental psychologists have shown that this happens even when individuals are allocated to groups at random in the first place and labelled with badges as arbitrary as T-shirts of different colours.

As a particular illustration of the effect – actually rather an agreeable one in this case – there was a single boy of African ancestry at Oundle when I was there. It is my impression that he suffered no racial prejudice whatsoever at that time, possibly because, being the only black boy, he was not identified with a racial group within the school. But he was identified with the house to which he belonged. Along with his contemporaries at Laxton House, we saw him not as noticeably black at all but as 'one of the Laxton crowd', with a similar personality to others in Laxton. In hindsight, I doubt that there was any identifiable personality trait that could reasonably be associated with Laxton or any other house. My observation relates not at all to the reality of life at Oundle but to a general characteristic of human psychology, the tendency to see individuals as badged with a group label.

My reason for choosing Laundimer as my house was a rumour, which proved ill-founded, that it was one of the few houses that lacked the tradition of an initiation ceremony (what American college students call 'hazing'). As it turned out, we did have to stand on the table and sing a song. In my piping treble I sang one of my father's songs:

> Oh the sun was shining, shining brightly
> Shining as it never shone before – shone before.
> Oh the sun was shining so brightly,
> When we left the baby on the shore.
>
> Yes we left the baby on the shore.
> It's a thing that we've never done before – done before.
> When you see the mother, tell her gently
> That we left the baby on the shore.

Singing this was an ordeal, but not as bad, in the event, as I had feared.

I didn't see much individual bullying at Oundle, but there was a kind of formalized bullying which afflicted every new boy for one week in his first term or two, at least in Laundimer, and I think much the same happened in the other houses. This was the dread week when he was 'bell boy'. In your week as bell boy you were responsible for everything, and you were to blame if anything went wrong – which it usually did. You had to light the fire and make sure it didn't go out. On Saturday during your week of ordeal as all-purpose scapegoat, you had to go round all the studies taking orders for Sunday newspapers, and collecting money for them. Then, on Sunday morning, you had to get up very early, walk to the far end of the town to buy the newspapers, then carry them back and distribute them to all the studies. Your most publicly noticeable function was to ring the bell at exactly the right time to signal each of many deadlines throughout the day: getting up time, mealtimes, bedtime and so on. That meant you had to have a very accurate watch. By the end of my week as bell boy, I had got the hang of it, but the first day was a disaster. For some reason, I hadn't grasped that the five-minute warning bell had to be rung *exactly* five minutes before the breakfast gong. Many of the senior boys were in the habit of getting out of bed five minutes to the dot before the gong rang, and five minutes is not long to wash and dress so the timing was crucial. On my first day as bell boy I rang the five-minute bell, then strolled across to belabour the gong about half a minute later. Consternation was rife, and angry ridicule ensued.

The duties of bell boy and fag were such that it is a wonder we new boys got any work done at all, let alone succeeded in 'breaking the back of the term's work'. Fagging has now been abolished, I think in all English schools. But I remain at a loss as to why it was ever permitted in the first place, and why it lasted as long as it did. In the nineteenth century there was a weird belief that it had some kind of educational value. Perhaps its long persistence had something to do with the 'I went through it in my time so why shouldn't you?' mentality – a mentality that is still, incidentally, the bane of many a junior doctor's life in Britain.

Not entirely surprisingly, my stammer resurfaced in my early terms at Oundle. I had trouble with hard consonants like 'D' and 'T' and it was unfortunate that my surname begins with one of them, for it was often necessary to enunciate it. When we had tests in class, we had to tick our correct answers, count the ticks and then shout out the tally, out of ten, for the master to record in his book. When I got ten out of ten, I used to call out 'nine' because it was so much easier to say than 't-t-t-ten'. In the army Cadet Corps we were to be inspected by a visiting general. One by one we would have to march out from the ranks, stamp to attention in front of him, shout our name, salute, smart about turn and march back. 'Cadet Dawkins, sir!' I dreaded it. I had sleepless nights about it. It was fine to practise by myself, but when I had to shout it out in front of the whole parade? 'Cadet D-d-d-d-d . . .' In the event, it passed off all right, with just a long, hesitating pause before the D.

The Cadet Corps was not quite compulsory. You could get out of it if you joined the Boy Scouts. Or the other way out was to spend the time tilling the land with Boggy Cartwright. In a previous book I described Mr Cartwright as 'a remarkable, bushy-browed man, who called a spade a spade and was seldom seen without one'. Although paid to teach us German, what he actually taught us, in a slow, rural accent, was a kind of earthy, agricultural eco-wisdom. His blackboard permanently had the word 'Ecology' written on it and if anybody erased it when he wasn't looking he promptly rewrote it without saying a word. When writing German on the blackboard, if a sentence threatened to overwrite 'Ecology', he would cause the German sentence to flow around and over it. He once caught a boy reading P. G. Wodehouse and furiously tore the book clean in two. He had evidently bought into the calumny – assiduously fostered by Cassandra of the *Daily Mirror* – that Wodehouse had been a German collaborator during the war, on a par with Lord Haw-Haw or – the American equivalent – Tokyo Rose. But Mr Cartwright had the story even more garbled than Cassandra's slander. 'Wodehouse once had the opportunity to kick a German colonel downstairs, and he didn't take it.' That makes him sound like an angry man. He really wasn't, except under extreme provocation, which, bizarrely, P. G. Wodehouse (he said 'Woadhouse' instead of the correct 'Woodhouse') seems to have constituted. He was just a wonderfully original character, ahead of his time in his ecological eccentricity, slow-spoken and literally down to earth.

I was not enterprising enough to get out of the Cadet Corps by either of the two escape routes. I was probably too influenced by my peers – which actually was the story of my life at Oundle. Eventually I got out of the worst parts of army training by joining the band, playing first the clarinet and then the saxophone, conducted by a bandsman NCO: 'Right, we'll go from the *very commencement* of the 'ole march.' Of course, being in the band didn't get us out of the weekly duty of polishing our army boots, blanco-ing our belts and shining our brasses with Duraglit or Brasso. And we had to go to army camp once a year, living in the barracks of some regiment or other, going on long route marches and fighting mock battles with blank ammunition in our antiquated Lee-Enfield rifles. We also fired live rounds at targets, and one boy in my platoon accidentally shot the adjutant in the fleshy part of the leg. He fell to the ground and immediately lit a cigarette, while we witnesses, still on the ground with our Bren guns, felt very queasy.

On one expedition to the Leicester barracks we were exposed to a real sergeant major, the genuine article complete with huge, waxed ginger moustache. He would bellow, 'Seeerloooooope ARMS' or 'Ordeeeeeer ARMS', the first word in each case being a bass and pro-longed bellow, while the second word was a staccato – and absurdly high – soprano shriek. We suppressed our laughter into terrified snorts, in the manner of Pontius Pilate's soldiers in the Monty Python 'Biggus Dickus' scene.

We had to pass an examination called Certificate A, which involved rote learning of army knowledge: an exercise clearly designed to suppress anything remotely resembling intelligence or initiative – commodities not valued in the ranks of general infantry. 'How many kinds of trees do we have in the army?' The correct answer was three: Fir, Poplar and Bushy Top (the poet Henry Reed picked up on this point, but our drill sergeants would not have appreciated his satire).

Peer pressure among schoolchildren is notoriously strong. I and many of my companions were abject victims of it. Our dominant motivation for doing anything was peer pressure. We wanted to be accepted by our fellows, especially the influential natural leaders among us; and the ethos of my peers was – until my last year at Oundle – anti-intellectual. You had to pretend to be working less hard than you actually were. Native ability was respected; hard work was not. It was the same on the sports field. Sportsmen were admired more than scholars in any case. But if you could achieve sporting brilliance without training, so much the better. Why is native ability more admired than hard graft? Shouldn't it be the other way around? Evolutionary psychologists might have interesting things to say on the question.

But such missed opportunities! There were all sorts of exciting clubs and societies, any of which I could have joined with benefit. There was an observatory with a telescope – perhaps the gift of an old boy – and I never went near it. Why not? I would be enthralled to do so now, to be instructed by a knowledgeable astronomer with a real telescope that I didn't have to set up myself. I sometimes think schooldays are too good to be wasted on teenagers. Perhaps devoted teachers, instead of casting their pearls before piglets, should be given the opportunity to teach pupils old enough to appreciate their beauty.

For me at Oundle, the biggest missed opportunity of all lay in the workshops, which were my father's main reason for sending me to the school in the first place. It wasn't entirely my fault. Sanderson's unique innovation of a compulsory week in the workshops was still in full swing, and the workshops were superbly well equipped. We learned how to use lathes, milling machines and other advanced machine tools which we would be unlikely to meet in the big world outside. What we did not learn was precisely what my father was so good at: improvising, designing, making do and coping, knocking things up from what was available – in his case, mostly red binder twine and dirty old bits of iron.

The first thing we made in the Oundle workshops was a 'marking gauge'. We weren't even told what a marking gauge was. We copied exactly what the instructors told us to do. We made a wooden pattern for the metal object we were trying to make. We took it into the foundry and made a mould of our wooden pattern by pounding sticky sand around it. We donned protective goggles and assisted in pouring molten aluminium from a glowing crucible into the mould. We disinterred the cooled metal from the sand and took it to the metalwork shop to file it, drill it and finish it. And we took home our finished marking gauge, still with no idea what a marking gauge was and having used no initiative or creativity of any kind. We might as well have been workers in a mass production factory.

And part of the problem may indeed have been that the instructors were not teachers but were recruited – I'm guessing – from the ranks of factory floor foremen. They taught us not how to develop skills in general, but how to do particular things. I met the problem again when I took professional driving lessons in the town of Banbury. I was taught how to reverse round a particular corner in Banbury, which happened to be the favourite corner the examiner headed for when testing that particular skill: 'Wait till that lamp-post is level with the back window, then swing hard around.'

The one exception in the Oundle workshops, the one partial upholder of the Sanderson tradition for me, was an old retired blacksmith who manned a little forge in a corner of the metal shop. I hived myself off from the 'factory floor' and apprenticed myself to this kindly, bespectacled little old man. He taught me the traditional arts of the smith, plus acetylene welding, and my mother still has the poker I made, sitting in its scrolled stand. Even with the old smith, however, I pretty much did exactly what I was told, rather than exercising much creative resourcefulness.

A bad workman blames his tools – and his instructors. What was definitely my own fault is that I never went near the workshops except during the prescribed week. I didn't seize the opportunity to go in the evenings and make things to my own design. Just as I didn't go to the observatory to look at the stars. Mostly I wasted my spare time in the same way my colleagues did, lazing around, making toast on a Primus stove and listening to Elvis Presley. Plus, in my case, tootling on musical instruments rather than playing real music. Such a waste of first-class, expensively bought opportunities is little short of tragic. Once again, is school too good for teenagers?

I did, however, join the beekeeping club, run by Ioan Thomas, Oundle's inspiring young zoology master, and the smell of beeswax and smoke still evokes happy memories. Happy in spite of the fact that I was quite frequently stung. On one such occasion (I am mildly proud to report) I didn't brush the bee off my hand but carefully watched as she slowly waltzed round and round on my hand, 'unscrewing' her sting from my skin. The stings of bees, unlike those of wasps, are barbed. When a bee stings a mammal, the barbs cause the sting to stick in the skin. When you brush the bee off, the sting stays behind and tears out some of the bee's vital organs. From an evolutionary point of view, the individual worker bee is behaving altruistically, sacrificing her life as a kamikaze fighter for the benefit of the hive (strictly speaking, for the benefit of the genes that pro- grammed her to do it, in the form of copies in queens and males). While she goes off to die, her sting remains in the victim, the poison gland still pumping venom and therefore acting as a more effective deterrent to the putative hive-raider. This makes perfect evolution- ary sense, and I'll return to the theme in the chapter on *The Selfish Gene*. Given that she is sterile, the worker bee has no chance of pass- ing on copies of her genes via offspring, so instead she works to pass them on via the queen and other non-sterile members of the hive. When I let my worker unscrew herself from my hand I was behaving altruistically towards her – but my motivation was mostly curiosity: I wanted to watch at first hand the procedure I had heard about from Mr Thomas.

I've mentioned Ioan Thomas in previous publications. My very first lesson with him, at the age of fourteen, was inspirational. I don't remember the details, but it conveyed the kind of atmosphere I was later to strive for in *Unweaving the Rainbow*: what I would now call 'science as the poetry of reality'. He had come to Oundle as a very young teacher because of his admiration for Sanderson, although he was too young to have met that old headmaster. He did meet Sanderson's successor, Kenneth Fisher, and told a story showing that something of the spirit of Sanderson had lived on. I retold the tale in my inaugural Oundle Lecture, given in 2002.

Kenneth Fisher was chairing a staff meeting when there was a timid knock on the door and a small boy came in: 'Please, sir, there are Black Terns down by the river.' 'This can wait,' said Fisher decisively to the assembled committee. He rose from the Chair, seized his binoculars from the door and cycled off in the company of the small ornithologist, and – one can't help imagining – with the benign, ruddy-faced ghost of Sanderson beaming in their wake. Now that's education – and to hell with your league table statistics, your fact-stuffed syllabuses and your endless roster of exams . . .

Some 35 years after Sanderson's death, I recall a lesson about *Hydra*, a small denizen of still freshwater. Mr Thomas asked one of us: 'What animal eats *Hydra*?' The boy made a guess. Non-committally, Mr Thomas turned to the next boy, asking him the same question. He went right round the entire class, with increasing excitement asking each one of us by name, 'What animal eats *Hydra*? What animal eats *Hydra*?' And one by one we guessed. By the time he had reached the last boy, we were agog for the true answer. 'Sir, sir, what animal does eat *Hydra*?' Mr Thomas waited until there was a pin-dropping silence. Then he spoke, slowly and distinctly, pausing between each word.

'I don't know . . .' *(Crescendo)* 'I don't know . . .' *(Molto crescendo)* 'And I don't think Mr Coulson knows either.' *(Fortissimo)* 'Mr Coulson! Mr Coulson!'

He flung open the door to the next classroom and dramatically interrupted his senior colleague's lesson, bringing him into our room. 'Mr Coulson, do you know what animal eats *Hydra*?' Whether some wink passed between them I couldn't say, but Mr Coulson played his part well: he didn't know. Again the fatherly shade of Sanderson chuckled in the corner, and none of us will have forgotten that lesson. What matters is not the facts but how you discover and think about them: education in the true sense, very different from today's assessment-mad exam culture.

Those two occasions, when I fancifully invoked the ghost of a long-dead headmaster, have been held up as showing that I must be in some sense a supernaturalist. Of course they show nothing of the kind. Such imagery should perhaps be called poetic. It is legitimate so long as it clearly is understood to be non-literal. I hope the context of those two quotations is sufficiently clear to obviate misunderstanding. Problems arise when (especially) theologians use such metaphorical language without realizing that that is what they are doing, and without even realizing that there is a distinction between metaphor and reality – saying something like: 'It is not important whether Jesus really fed the five thousand. What matters is what the *idea* of the story *means* to us.' Actually it is important, because millions of devout people do believe the Bible is literally true. I hope and trust that no reader thinks I believe Sanderson really was standing in the corner beaming at Mr Thomas's lesson.

Our lesson on *Hydra* was the scene of a slightly embarrassing story, but I should tell it as it might be revealing. Mr Thomas asked us whether any of us had seen *Hydra* before. I think I was the only boy to put his hand up. My father had an old brass microscope, and we had spent a lovely day a few years earlier looking at hugely magnified pond life: mostly crustaceans such as *Cyclops, Daphnia* and *Cypris*, but also *Hydra*. I had regarded the slowly waving, almost plant-like *Hydra* as rather dull compared with the crustaceans, leggy and vigorously kicking. *Hydra* was the least exciting memory of that memorable day, and I think I snobbishly looked down upon all the attention that Mr Thomas was giving to it in that lesson. So, when he asked me for more details of my previous encounter with *Hydra*, I said: 'I've seen all those sorts of animals.' To Mr Thomas, of course, *Cyclops, Daphnia* and *Cypris* were not at all the same sort of animal as *Hydra*, but to me they were because I had seen them all on the same day with my father, and so lumped them together. Mr Thomas probably suspected that I hadn't seen *Hydra* at all, and he cross-examined me closely. I am sorry to say that this had exactly the wrong effect on me. Perhaps I took his cross-questioning as some sort of slur on my father, who had introduced me to 'all those sorts of animals' and told me their Latin names. I obstinately dug in my toes and, instead of saying, clearly and unequivocally (and truthfully), that I had indeed seen *Hydra*, I persisted in my refusal to separate it from 'all those sorts of animals'. Embarrassing to recall. Revealing? Maybe, but I don't know of what. Perhaps it was connected with the fierce loyalty that I felt towards all things associated with my parents, whether it was Ferguson tractors ('Dirty old Fordson!') or Jersey cows ('Friesians don't give milk, they give water').

Mr Thomas having introduced me to beekeeping, I was able to carry on with the hobby in the school holidays when my father's eccentric old schoolfriend Hugh Corley gave me a hive. They were a wonderfully docile strain which literally never stung, and I used to work them without veil or gloves. Unfortunately they were later poisoned by insecticide wafting over from a neighbour's field. Mr

Corley, passionate organic farmer and early eco-warrior, was out-raged and gave me another hive. Unfortunately these went to the opposite extreme – undoubtedly a genetic difference – and stung everything that moved. I didn't react badly to stings in those days. But I wonder whether those many stings in my boyhood sensitized me to stings in later life. I have been stung only twice as an adult, once in my forties and once in my fifties, and on both occasions I reacted strangely and in a way that never happened when I was an active beekeeper. The region around one eye swelled up hugely, so that I could scarcely see. Why the eye, given that the stings were respectively on hand and foot? And, especially, why only one eye?

Apart from beekeeping with Mr Thomas, I suppose my other mildly constructive spare-time occupation at Oundle was playing music. I spent many hours in the Music School, but even there I have to confess to massive wasting of opportunities. From my earliest childhood, musical instruments of any kind would draw me like a magnet, and I had to be dragged away from shops that had violins or trumpets or oboes in the window. Even today, if a string quartet or a jazz band has been engaged to play at a garden party or a wedding, I will neglect my social duties and hover around the musicians, watching their fingers and talking to them during the intervals about their instruments. I don't have perfect pitch like my first wife Marian, and my harmonic sense is poor, unlike that of my present wife Lalla, who can effortlessly improvise harmonious descants to any melody. But I do have a natural melodic ability, meaning that I can play a tune about as easily as I can sing it or whistle it. I'm sorry to say that one of my pastimes in the Music School was illicitly to pick up instruments that didn't belong to me and teach myself to play tunes on them. On one occasion I was caught playing 'When the Saints Go Marching In' on a rather expensive trombone belonging to a senior boy, and got into trouble because the trombone was later found to be damaged. I genuinely believe I didn't do the damage, but I was blamed (not by the owner himself, who was rather nice about it).

My facile melodic gift turned out to be a curse rather than a blessing, at least in a child as lazy as I was. Playing by ear was so easy for me that I neglected other important skills such as reading music or creative improvisation. It was worse than laziness. For a while I even snobbishly looked down upon musicians who 'needed' to read music. I thought improvisation was a superior skill. But it turned out that I was no good at improvising either. Invited to join the school jazz band, I soon discovered that, although I could play any tune faultlessly, I had absolutely no capacity to improvise upon it. I was very slovenly about practising scales. I have a very slight, partial excuse, which is that nobody ever explained to me what scales are for. With hindsight, as an adult scientist, I can piece the reason together. You play scales in order to become totally at home with every key, so that, once you've read the key signature at the beginning of the line, your fingers automatically and effortlessly feel their way into that key.

The hours I spent in the Music School are best described as tootling rather than playing. I did learn to read a score adequately with the clarinet and saxophone. But on the piano – where you are expected to play more than one note at a time – I was unbearably slow, like a child learning to read and laboriously spelling his way through the words letter by letter, rather than fluently reading whole sentences at a time. My kind piano teacher, Mr Davison, recognized my innate melodic ability and taught me some rudimentary rules for accompanying myself with left-hand chords. But though I quickly learned these, I could do them only in the keys of C major and A minor (minimizing the black notes), and my style of left-hand chord-thumping was pretty monotonous – although inexpert listeners were impressed by my ability to play instant requests.

I had a true and pure, though not very loud, singing voice as a treble, and was early recruited into the rather small and select Chancel Choir in the Oundle school chapel. I hugely enjoyed this; the regular rehearsal, under the Director of Music Mr Miller, was the high spot of my week. I think it was rather a good choir, up there with a typical English cathedral choir. And I can't resist adding that we sang without the affectation of the half-rolled 'r' – sounding more like a 'd' – which, at least to my prejudiced ear, spoils much choral singing: 'Maady was that mother mild / Jesus Cdist, her little child.' 'The dising of the sun / And the dunning of the deer / The playing of the meddy organ . . . ' By the way, while I'm doing my grumpy act, the fake Italian 'r' of John McCormack-vintage tenors is even worse: 'Seated one day at the Oregon . . .'

We performed an anthem every Sunday: Stanford or Brahms or Mozart or Parry or John Ireland, or earlier composers such as Tallis or Byrd or Boyce. We had no conductor, but two of the basses, facing each other in the back rows on the two sides of the chancel, performed the role by their miming head movements. One of these basses, C. E. S. Patrick, had a spellbindingly beautiful voice – probably the better for not being trained. I never spoke to him (one didn't meet senior boys in other houses), but I hero-worshipped him as the star of the Male Voice Choir, which performed under the direction of another gifted music master, Donald Payne, at school concerts. Unfortunately I was never invited to join the Male Voice Choir. When my voice broke, it dropped in quality as well as in pitch.

Oundle had a tradition – again founded by Sanderson – of involving the entire school in an annual oratorio. The choice of music was staggered in such a way that every boy would experience Handel's *Messiah* and Bach's B Minor Mass during his five years at the school. The intervening years offered a variety of works. My first term we did Bach's *Sleepers Wake* cantata and Haydn's Imperial

Mass, and I *loved* them, especially the Bach, with its slow chorale for the voices cunningly set against the leaping counterpoint melody in the orchestra. This was a magical experience, of a kind I had never known before. Every morning, for five minutes after prayers, the tall, thin figure of Mr Miller would stride briskly forward and rehearse the entire school, just a few pages at a time, until the big day came for the performance. Professional soloists arrived from London: glamorous soprano and contralto in long dresses, tenor and bass in immaculate tailcoats. Mr Miller treated them with great deference. Goodness knows what they thought of the throaty roar of the 'non-choir'. But none of the soloists, in my youthfully amateur opinion, could hold a candle to C. E. S. Patrick of the Male Voice Choir.

It is hard to convey the atmosphere of the English public school during the era that I experienced it. Lindsay Anderson captured it well in his film *If*. I'm not referring to the massacre at the end of the film, of course, and he exaggerated the beating. Maybe prefects with swagger sticks and embroidered waistcoats took a run at it in earlier, crueller eras, but I'm sure it didn't happen in my time. Actually, I never knew of anyone being caned at all while I was at Oundle and only recently heard (from a victim) that it did happen.

If also beautifully captured the burgeoning sexuality that surrounds pretty boys in a school that has no girls. The flashlight inspection of groins by the matron in the enormous starched hat was only slightly exaggerated in the film. Our inspection was done by the school doctor, who didn't peer as pruriently as the *If* matron. Nor did our mild doctor stalk the touchline of the rugby field like she did, screaming 'Fight! Fight! Fight!' But what Lindsay Anderson caught to perfection was the squalid conviviality of the studies where we mostly lived, worked, burnt toast, listened to jazz and Elvis, and fooled around. He caught the hysterical laughter that bonded teenage friends like wrestling puppies – not physical wrestling but verbal wrestling with odd, private languages and weird nicknames that grew and evolved term by term.

As an illustration of the weirdness of nickname evolution (and maybe of memetic mutation generally), one friend of mine was called 'Colonel', although there was nothing remotely military about his personality. *'Seen the Colonel anywhere?'* Here's the evolutionary history. Years earlier, an older boy, who had by now left the school, was said to have had a crush on my friend. That older boy's nickname was Shkin (corruption of Skin, and who knows where that came from – maybe some connection with foreskin, but that name would have evolved before I arrived). So my friend inherited the name Shkin from his erstwhile admirer. Shkin rhymes with Thynne, and at this point something akin to Cockney rhyming slang stepped in. There was a character in the BBC radio *Goon Show* called Colonel Grytte Pyppe Thynne. Hence my friend became Colonel Grytte Pyppe Shkin, later contracted to 'Colonel'. We loved the *Goon Show*, and would vie with each other to mimic (as did Prince Charles, who went to a similar school around the same time) the voices of the characters: Bluebottle, Eccles, Major Denis Bloodnok, Henry Crun, Count Jim Moriarty. And we gave each other Goon nicknames like 'Colonel' or 'Count'.

Some of the squalor would positively not be allowed by a health inspector today. After playing rugby we would have a 'shower'. My hypothesis is that at some time in the past it really had been a shower, and other houses in the school probably had proper showers still. But in Laundimer House, all that was left of the shower was the porcelain rectangular base, which we would fill with hot water. It was just big enough for two boys to sit in, face to face, with their knees up under their chins. We queued up to enter the 'shower' and by the time all fifteen rugby players had been through it the 'water' was not so much water as dilute mud. The odd thing is that I don't think we minded being in the last pair. It had the advantage that you could linger on in the warmth instead of rushing to let the queue go through. I don't remember minding the fact that I was bathing in the muddy bathwater of fourteen other people, any more than I minded getting in a very small bath with another naked male – both things that I would dislike intensely today. Another indication, I suppose, that we are not the same people we once were.

Oundle didn't really live up to my parents' expectations. The vaunted workshops were a failure, at least where I was concerned. There was too much adulation of the rugby team and too little prestige attached to intelligence or scholarship, or indeed any of the qualities that Sanderson fostered. But in my last year at least, my set of peers finally started valuing the mind. A bright young history master started a club called Colloquium for intellectual discussion among sixth-formers. I can't remember what happened at the meetings: maybe we even used to 'read a paper', like earnest under-graduates. Equally earnestly, outside the meetings we would evaluate each other's intelligence, in an atmosphere of po-faced snobbery not unlike that conjured by John Betjeman's couplet:

Objectively our common room is like a small Athenian state . . .
Except for Lewis: he's all right, but do you think he's quite first
    rate?

I and two friends in my house became militantly anti-religious in our last year, when we were seventeen. We refused to kneel down in chapel and sat with folded arms and closed lips, defiantly upright like proud, volcanic islands in the sea of bowed and mumbling heads. As you'd expect of Anglicans, the school authorities were very decent and never complained, even when I took to skipping chapel altogether. But here I need to go back and trace my loss of religious faith.

I had arrived at Oundle a confirmed Anglican, and I even went to Holy Communion a few times in my first year. I enjoyed getting up early and walking through the sunlit churchyard listening to the blackbirds and thrushes, and I basked in righteous hunger for breakfast afterwards. The poet Alfred Noyes (1880–1958) wrote: 'If ever I had any doubts about the fundamental realities of religion, they could always be dispelled by one memory – the light upon my father's face as he came back from early communion.' It's a spectacularly silly piece of reasoning for an adult, but it sums me up at the age of fourteen.

I'm happy to say it wasn't long before I reverted to earlier doubts, first planted at the age of about nine when I learned from my mother that Christianity was one of many religions and they contradicted each other. They couldn't all be right, so why believe the one in which, by sheer accident of birth, I happened to be brought up? At Oundle, after my brief phase of going to Communion, I gave up believing in everything that was particular about Christianity, and even became quite contemptuous of all particular religions. I was especially incensed by the hypocrisy of the 'General Confession' in which we mumbled in chorus that we were 'miserable offenders'. The very fact that the exact words were written down to be repeated the following week, and the week after and for the rest of our lives (and had been so repeated ever since 1662), sent a clear signal that we had no intention of being anything other than miserable offenders in the future. Indeed, the obsession with 'sin' and the Pauline belief that everybody is born in sin, inherited from Adam (whose embarrassing non-existence was unknown to St Paul), is one of the very nastiest aspects of Christianity.

But I retained a strong belief in some sort of unspecified creator, almost entirely because I was impressed by the beauty and apparent design of the living world, and – like so many others – I bamboozled myself into believing that the appearance of design demanded a designer. I blush to admit that I had not at that stage worked out the elementary fallacy of this argument, which is that any god capable of designing the universe would have needed a fair bit of designing himself. If you are going to allow yourself to conjure a designer out of thin air, why not apply the same indulgence to that which he is supposed to have designed, and cut out, so to speak, the middle man? In any case, of course, Darwin provided the magnificently powerful alternative to biological design which we now know to be true. Darwin's explanation had the huge advantage of starting from primeval simplicity and working up, by slow, gradual degrees, to the stunning complexity that pervades every living body.

But at the time the 'it's all so beautiful, there must have been a designer' argument swayed me. My faith was reinforced by, of all people, Elvis Presley, of whom I was a dizzily enthusiastic fan, like most of my friends. I bought his records as soon as they were released: 'Heartbreak Hotel', 'Hound Dog', 'Blue Moon', 'All Shook Up', 'Don't be Cruel', 'Baby I Don't Care' and many others. Their sound is irrevocably – it seems now so appropriate – linked in my mind with the faintly sulphurous smell of the ointment with which many of us battled our adolescent spots. I once embarrassed myself by singing 'Blue Suede Shoes' loudly at home, thinking I was alone in the house and not knowing that my father was in earshot. 'You can knock me down / Step on my face / Slander my name / All over the place.' To imitate Elvis properly in this song you have to rasp the words with a kind of venom, like a modern rap performer. It took my chagrined self a while to convince my father that I was not having some kind of fit, or suffering from Tourette's Syndrome.

So, I worshipped Elvis and I was a strong believer in a non-denominational creator god. And it all came together when I passed a shop window in my home town of Chipping Norton and saw an album called *Peace in the Valley* featuring a song called 'I Believe'. I was transfixed. Elvis was religious! In a frenzy of excitement I dived into the shop and bought it. Hurrying home, I slipped the record out of the sleeve and on to the turntable. I listened with delight – for my hero sang that every time he saw the wonders of the natural world around him, he felt his religious faith reinforced. My own sentiments exactly! This was surely a sign from heaven. Why I was surprised that Elvis was religious is now beyond me. He came from an uneducated working-class family in the American South. How could he *not* have been religious? Nevertheless I was surprised at the time, and I sort of half-believed that in this unexpected record Elvis was speaking personally to me, calling me to devote my life to telling people about the creator god – which I should be especially well qualified to do if I became a biologist like my father. This seemed to be my vocation, and the call came from none other than the semi-divine Elvis.

I am not proud of this period of religious frenzy, and I'm happy to say that it didn't last long. I became increasingly aware that Darwinian evolution was a powerfully available alternative to my creator god as an explanation of the beauty and apparent design of life. It was my father who first explained it to me but, to begin with, although I understood the principle, I didn't think it was a big enough theory to do the job. I was biased against it by reading Bernard Shaw's preface to *Back to Methuselah* in the school library. Shaw, in his eloquently muddled way, favoured Lamarckian (more purpose-driven) and hated Darwinian (more mechanistic) evolution, and I was swayed towards the muddle by the eloquence. I went through a period of doubting the power of natural selection to do the job required of it. But eventually a friend – one of the two, neither of them biologists, in whose company I later refused to kneel in chapel – persuaded me of the full force of Darwin's brilliant idea and I shed my last vestige of theistic credulity, probably at the age of about sixteen. It wasn't long then before I became strongly and militantly atheistic.

I said that the school authorities were decently Anglican about my refusal to kneel in chapel, and turned a blind eye. But that may not be quite true, at least not of two of them. The first was my English teacher at the time, Flossie Payne, familiar as an erect figure on his sit-up-and-beg bicycle with raised umbrella. Flossie publicly challenged me in class to explain why I was leading a rebellion against kneeling in chapel. I'm afraid I didn't give a good account of myself. Far from seizing the opportunity to lead my classmates in the same direction, I miserably stammered something about an English lesson not being the appropriate place to have the discussion, and retreated into my shell.

Second, I have only recently learned that my housemaster, Peter Ling (actually a nice man, if rather too conformist and conventional), telephoned Ioan Thomas, my zoology master, to voice his concern about me. In a recent letter to me, Mr Thomas reported that he warned Mr Ling that 'requiring someone like you to attend chapel twice a day on Sunday was doing you positive harm. The phone went down without comment.'

Mr Ling also summoned my parents for a heart-to-heart talk, over tea, about my rebellious behaviour in chapel. I knew nothing of this at the time and my mother has only just told me of the incident. Mr Ling asked my parents to try to persuade me to change my ways. My father said (approximately, by my mother's recollection): 'It is not our business to control him in that sort of way, that kind of thing is your problem, and I'm afraid I must decline your request.' My parents' attitude to the whole affair was that it wasn't important.

Mr Ling, as I said, was in his way a decent man. A contemporary and friend of mine in the same house recently told me the following nice story. He was illicitly up in a dormitory during the day, kissing one of the housemaids. The pair panicked when they heard a heavy tread on the stairs, and my friend hastily bundled the young woman up onto a window sill and drew the curtains to hide her standing shape. Mr Ling came into the room, and must have noticed that only one of the three windows had the curtains drawn. Even worse, my friend noticed, to his horror, that the girl's feet were clearly visible protruding under the curtain. He firmly believes that Mr Ling must have realized what was going on but pretended not to, perhaps on 'boys will be boys' grounds: 'What are you doing up in the dormitory at this hour?' 'Just came up to change my socks, sir.' 'Oh, well, hurry on down.' Good call on Mr Ling's part! The boy went on to become probably the most successful Old Oundelian of his generation, the knighted chief executive officer of one of the largest international corporations in the world, and a generous benefactor of the school, endowing, among other things, the Peter Ling Fellowship.

The headmaster of a large school is a remote and formidable figure. The stooping Gus Stainforth only taught me for one term – Divinity – and we were terrified of him. We read *The Pilgrim's Progress*, and then had to produce our own artist's impression of that rather unpleasant book. Halfway through his expected time at Oundle, Gus left to head his own old school, Wellington, and was succeeded at Oundle by Dick Knight, a large, athletic man who won our respect by his ability to hit a ball out of the ground (he had played cricket for Wiltshire) and by the way he sang with the 'non-choir' in the annual oratorio. He drove a big Rolls-Royce, 1920s vintage I would guess from its imposingly upright style – very different from the sleek purrers of later decades. He happened to be visiting Oxford on business at the same time as I and another boy were taking the Oxford entrance exam and being interviewed in our respective colleges of choice. When they heard this, Mr and Mrs Knight kindly offered us a lift back to Oundle in their ancient Rolls, and on the journey he discreetly raised the subject of my rebellion against Christianity. It was a revelation to talk to a decent, humane, intelligent Christian, embodying Anglicanism at its tolerant best. He seemed genuinely interested in my motives and not at all inclined to condemn. Years later, I was not surprised to learn from his obituary that, an outstanding classical scholar in his youth as well as a noted athlete, in retirement he took a degree in mathematics from the Open University. Sanderson would have loved him.

My father and grandfather had never contemplated any destination for me after Oundle other than Balliol College, Oxford. At the time, Balliol still retained its reputation as the foremost Oxford college, top of the examination league table and *alma mater* of a glittering list of distinguished old members: writers, scholars, statesmen, prime ministers and presidents all around the world. My parents went to see Ioan Thomas about my prospects. Mr Thomas was realistically frank: 'Well, he might just scrape into Oxford, but Balliol is probably aiming too high.'

Mr Thomas might doubt that I was good enough for Balliol but – great teacher that he was – he was determined that I should give it my best shot. He had me round regularly at his home in the evenings for extra tuition (unpaid, of course; he was that sort of teacher), and by some miracle he got me into Balliol. More importantly, that meant I got into Oxford. And insofar as anything was the making of me, Oxford was.

DREAMING SPIRES

'Mr Dawkins? Sign here, sir. I remember your three brothers, very fine winger one of them was. I don't suppose you play rugby, sir?'

'No, I'm afraid not, and, er, actually I never had any brothers. You must be thinking of my father and my two uncles.'

'Yes, sir, very fine young gentlemen, sign here please. You are on Staircase 11, Room 3, sharing with Mr Jones. Who's next?'

Well, that's approximately how the conversation went. I didn't write it down at the time. The Balliol College porter took the timeless view characteristic of his bowler-hatted profession. Young gentlemen might come and go, but the college goes on for ever. Indeed, it was to celebrate its 700th anniversary during my time there. Talking of that loyal and ancient bowler-hatted profession, I can't resist an anecdote more recently told me by the Head Porter of my present college, New College (well, it was new in 1379). An inexperienced new porter hadn't yet got the hang of the porters' incident book and what it was for. His entries in the log for his first night duty, at hourly intervals, consisted of (approximately; the details will be wrong):

8 p.m. Raining.

9 p.m. Still raining.

10 p.m. Raining harder.

11 p.m. Raining harder still. I could hear it banging on me bowler as I did me rounds.

Oxford, I should explain, is a federal university: a federation of thirty or so colleges, of which Balliol is one of three claiming to be the oldest. Except for the newer colleges, each one is built around a series of quadrangles. These beautiful old buildings mostly don't have horizontally running corridors like hotels or halls of residence, with rooms along a passageway: instead, there are lots of staircases leading off doors from the quadrangle, each staircase giving access to a number of rooms on three or four floors. Thus each room is known by a staircase number and a room number within its staircase. In order to visit a near neighbour, you'd probably have to go out into the quadrangle and then in at another staircase entrance. In my time there was a bathroom on every staircase, so we no longer had to go out into the cold in our dressing gowns. Nowadays, the rooms are more likely to have their own en-suite bathrooms, which my father would have called 'terribly molly' (soft, namby-pamby). I suspect that a large part of the motivation for installing them is to cater for the lucrative conference trade, which all the Oxford and Cambridge colleges ply out of termtime.

The colleges at both Oxford and Cambridge are financially autonomous self-governing institutions, some of them, such as St John's, Oxford, and Trinity, Cambridge, very wealthy. Trinity, by the way, is outstandingly rich in achievement as well as money. This one Cambridge college can boast more Nobel Prizes than any single *country* in the world except the USA, Britain (obviously), Germany and France. The University of Oxford can make the same proud

My grandmother Enid with her dog Susan (*left*) in the garden of The Hoppet, where my parents first met. On the eve of war they were married (*above*) from Water Hall, seen below with my mother's younger sister Diana in the garden.

Discovering on her arrival in Africa in May 1940 that my father (*above*) had been called up, my mother accompanied him (illegally) to Kenya in the station wagon Lucy Lockett, seen here on a makeshift bridge where my mother is washing her face in the river (*above right*), and at breakfast-time in one of their many camps (*below*).

One of my father's training locations coincided with Baden-Powell's funeral and he was invited, as a former Scout, to be a pall-bearer (*above*). I think he looks very dashing in his KAR uniform, marching next to Lord Erroll (out of step) who was murdered soon after. For my mother, wartime domestic life in Kenya had its surprises: here (*below*) is her painting of the incident involving the lioness described on page 36.

To signal landmarks in family life, my mother had the custom of painting big tableaux representing scenes and events. This is a small part of one called 'The Ways that We Went', which she did for her Golden Wedding in 1989. Alongside generic African scenes are my father's armoured car in Somaliland, my mother and me striding into my life together, a sandy Lake Nyasa beach, Hookariah my pet chameleon, Percy our pet bushbaby, and our house at Makwapala with me pushing Sarah in the lorry towards Tui the dachshund.

NCHEMA
CHENA

ren

bagatki

LILONGWE   CHITALA

TIWI

NCHISI

NCHEU

I evidently looked up to my father from an early age, and accompanied him in climbing the lower slopes of Kilimanjaro. Baraza kindly tolerated my dogged help in pushing my pram. Later, we moved to Makwapala in Nyasaland (*lower right*) where I seem to have grown bored with the sewing class my mother was conducting in the garden. In 1946, on a brief leave, we stayed with my grandparents in England. During this time my uncle Bill and aunt Diana (middle row left, next to my parents) married at Mullion, and the whole family picnicked at Kynance Cove.

On our return to Nyasaland we lived in Lilongwe, where my parents bought Creeping Jenny, our first new car. I was sent to board at the Eagle School in Southern Rhodesia. In the picture here, Tank (the headmaster) is in the centre with Coppers (matron) and Dick (another teacher) on his right. I'm the very small boy third from left in the same row and David Glynn, also small, is in the mirror position the other side, next to Wattie who is next to Paul. David and I collected the beautiful swallowtail butterflies which he mysteriously called 'Daddy Xmas'.

claim, but no single Oxford college comes close to Trinity Cambridge, not even Balliol, which tops the table of Nobel Prizes for Oxford colleges. My father, I have just realized, is one of few people to have studied at both Balliol, Oxford, and Trinity, Cambridge.

At both Oxford and Cambridge, the relationship between the colleges and the university bears the same uneasy tension as that between the federal and state governments in the USA. The rise of science has increased the power and importance of the 'federal government' (university), because science is too big an enterprise to be handled by each college separately (though one or two of them tried to go it alone in the nineteenth century). The science departments belong to the university, and it was the Zoology Department rather than the college that was to dominate my life at Oxford.

That porter must have been one of the first people to call me 'Mr Dawkins' (let alone 'Sir') – treat me as an adult – and I wasn't used to it. I think it was characteristic of my generation of undergraduates that we worked rather self-consciously at appearing to be more adult than we were. Later generations of undergraduates have tended towards the opposite, dressing scruffily with hoods or baseball caps, loosely slung rucksacks and sometimes even more loosely slung jeans. But my generation favoured tweed jackets with leather elbow patches, smart waistcoats, corduroy trousers, trilby hats, moustaches, ties, even bow ties. Some (not I, despite the example of my father) put the finishing touches to this image by smoking a pipe. These affectations may have been prompted by the fact that many of my fellow freshmen really were two years older; for my cohort was almost the first of the post-war generations not to be called up for military service. Those of us who came straight from school in 1959 were boys, sharing lectures, quadrangles and a dining hall with militarily trained *men*, and this perhaps raised our aspiration to grow up and be taken seriously as adults. We left Elvis behind and listened to Bach or the Modern Jazz Quartet. We solemnly intoned Keats and Auden and Marvell to each other. Chiang Yee captured the mood in his charming book from a slightly earlier era, *The Silent Traveller in Oxford*,[1] when he drew, in his elegant Chinese style, a pair of freshmen bounding, two steps at a time, up their college staircase. His deliciously perceptive caption read: 'I could tell that they were freshmen because I heard one say to the other, "Do you read much Shelley?"'

The claim that army service turns boys into men is the basis of a lovely story about Maurice Bowra, legendary Warden of Wadham College (anecdotes about Bowra are so numerous as to be best avoided, but this is an especially charming one). Immediately after the war, he was interviewing a young man for a place at the college.

'Sir, I have been away at the war, and I have to confess that I have forgotten all my Latin. I cannot pass the Latin exam to qualify for entrance.'

'Oh, don't you worry about that, dear boy, war counts as Latin, war counts as Latin.'

My older colleagues back from National Service in 1959 were not literally 'battle-hardened' like Bowra's entrance candidate, but they had an unmistakable air of being worldly-wise and grown up, in a way that I was not. As I said, I think that those of my generation who affected pipe-smoking, bow ties and neatly trimmed moustaches may have been struggling to keep up with the military veterans. Am I right in suspecting that today's undergraduates aspire in the opposite direction, towards juvenilization? On the first day of a new university year, a modern college noticeboard is likely to have notices saying things like this: 'Freshers! Feeling lonely? Lost? Missing Mum? Do drop in for coffee and a chat. We love you.' Such cosseting invitations would have been inconceivable on the notice-board of my first term, which was more likely to carry

---

[1] Chiang Yee, *The Silent Traveller in Oxford* (London, Methuen, 1944).

announcements calculated to make me feel I had arrived in the adult world: 'Would the "gentleman" who "borrowed" my umbrella . . .'

I had applied to read biochemistry. The tutor who interviewed me, the kindly Sandy Ogston, who later became Master of Trinity, declined – thank goodness – to let me in as a biochemist (perhaps because he was one himself and would have had to teach me) but offered me a place to read zoology instead. I accepted gratefully, and it turned out to be the perfect course for me. Biochemistry could not have captured my enthusiastic interest the way zoology did: Dr Ogston was as wise as his venerable grey beard suggested.

Balliol had no tutorial fellow in zoology, so I was sent out of college to the wonderfully convivial Peter Brunet in the Department of Zoology. He would be responsible for tutoring me or for arranging tutorials with others. One incident in an early tutorial with Dr Brunet may have marked the beginning of my weaning away from a school attitude to learning in favour of a university one. I asked Dr Brunet a question about embryology. 'I don't know,' he mused, sucking on his pipe. 'Interesting question. I'll ask Fischberg and report back.' Dr Fischberg was the department's senior embryologist, so this was an entirely reasonable response. At the time, however, I was so impressed by Dr Brunet's attitude that I wrote to my parents about it. My tutor didn't know the answer to a question and was going to ask an expert colleague and report back to me! I felt that I'd joined the big boys.

Michael Fischberg was from Switzerland, with a very strong Swiss German accent. His lectures made frequent mention of things called 'tonk bars' and I think most of us wrote 'tonk bars' in our lecture notes before we finally saw the phrase written down: 'tongue bars', a feature of embryos at a certain stage of development. Endearingly, while at Oxford Dr Fischberg developed a great enthusiasm for our English national game, founding and captaining the departmental team. He had a most unusual bowling action.

Unlike a baseball pitcher, a cricket bowler has to keep his arm straight. Throwing is strictly forbidden: you must not bend your arm. Given this constraint, the only way to propel the ball at any speed is to run and then bowl while still running. The fastest bowlers in the world, such as the terrifying Jeff Thomson ('Tommo') of Australia, have achieved ball velocities of 100 mph (comparable to a baseball pitcher with his bent arm), and they do it by running very fast before releasing the ball with a straight overarm action in graceful rhythm with their running. Not Dr Fischberg. He stood rigidly to attention facing the batsman, raised his straight bowling arm horizontally to take careful aim at the wicket, then swung it over in a single arc and let go of the ball at the top.

I was hopelessly bad at cricket, but was sometimes cajoled into playing for Zoology when they couldn't find anybody better and were really desperate. I do, though, quite enjoy watching cricket, fascinated by the strategy of a captain placing his fielders around the batsman – like a chess master deploying his pieces to encircle the king. The best cricketer I ever saw playing in the Oxford University Parks was the Nawab of Pataudi ('Tiger'), the Oxford captain and my exact contemporary at Balliol. As a batsman, the effortless way he steered the ball to outwit the fielders was sublime. But it was as a fielder himself that he especially impressed me. On one occasion a batsman hit the ball and called for what must have seemed like an easy run. Then he noticed that the fielder charging down upon the ball was Tiger Pataudi, and he frantically shrieked to his partner to go back to his crease. Sadly, Tiger later lost one eye in a car accident and had to change his stance to bat monocularly, but he was still good enough to captain India.

I said that Oxford was the making of me, but really it was the tutorial system, which happens to be characteristic of Oxford and Cambridge. The Oxford zoology course also had lectures and laboratory classes, of course, but these were no more remarkable than those at any other university. Some lectures were good, some were bad, but it scarcely made any difference to me because I hadn't yet worked out the point of going to a lecture. It is not to imbibe information, and there is therefore no point in doing what I did (and what virtually all undergraduates do), which was take notes so slavishly that there was no attention left over for thinking. The only time I departed from this habit was once when I had forgotten to bring a pen. I was much too shy to borrow a pen from the girl sitting next to me (having been to a single-sex school, and shy to boot, I was in boyish awe of all girls at the time, and if I was too timorous to borrow a pen you can imagine how often I dared approach them for anything more interesting than that). So, for that one lecture I took no notes and just listened – and thought. It was not an unusually good lecture, but I got more out of it than from other lectures – some of them much better ones – because my lack of pen freed me to listen and think. But I didn't have the sense to learn my lesson and refrain from taking notes at subsequent lectures.

Theoretically the idea was to use your lecture notes in revision, but I never looked at mine ever again and I suspect that most of my colleagues didn't either. The purpose of a lecture should not be to impart information. There are books, libraries, nowadays the internet, for that. A lecture should inspire and provoke thought. You watch a good lecturer thinking aloud in front of you, reaching for a thought, sometimes grabbing it out of the air like the celebrated historian A. J. P. Taylor. A good lecturer thinking aloud, reflecting, musing, rephrasing for clarity, hesitating and then grasping, varying the pace, pausing for thought, can be a role model in how to think about a subject and how to transmit a passion for it. If a lecturer drones information as though reading it, the audience might as well read it – possibly in the lecturer's own book.

I exaggerate a little when I advise never to take notes. If a lecturer produces an original thought, something striking that makes you think, then by all means write yourself a memo to think again about it later, or look something up. But struggling to record a piece of every sentence the lecturer utters – which is what I tried to do – is pointless for the student and demoralizing for the lecturer. When lecturing to a student audience today, all I notice is a sea of tops of heads, bowed over notebooks. I prefer lay audiences, literary festivals, memorial lectures, guest lectures at universities where if the students come it is because they want to and not because it is on their syllabus. At such public lectures, the lecturer sees not bowed heads and scribbling hands but alert faces, smiling, registering comprehension – or the reverse. When lecturing in America, I get quite cross if I hear that some professor has *required* students to attend my lecture for 'credit'. I'm not keen on the idea of 'credit' at the best of times, and I actively hate the idea that students are getting credit for listening to me.

Niko Tinbergen, my later mentor, entered my life as the lecturer on molluscs. He announced no special affinity for that group save a fondness for oysters, but he played along with the department's tradition of handing out a phylum to each lecturer, more or less at random. From those lectures, I recall Niko's swift blackboard drawings; his deep voice (surprisingly deep for a small man), accented but not obviously Dutch; and his kindly smile (avuncular, as I thought it then, although he must have been much younger than I am now). In the following year he again lectured to us, this time on animal behaviour, and the avuncular smile broadened with enthusiasm for his own subject. In that heyday of his research group in the gull colony at Ravenglass in Cumberland, I was enchanted by his film on eggshell removal by black-headed gulls. I especially liked his method of plotting graphs – laying out tent-poles on the sand for axes, with strategically placed eggshells for data points. How very Niko. How very un-PowerPoint.

After each lecture there was a practical class in the laboratory. I had no aptitude for practical work, and – so young and immature was I – the opposite sex was even more of a distraction in the laboratory than in lectures. It was really only the tutorial system that educated me, and I shall forever be grateful to Oxford for this unique gift – unique because, at least where science subjects were concerned, I think even Cambridge was not equal in this respect. The Cambridge Natural Science Tripos Part I, which occupies the first two years of the undergraduate course, is commendably broad but in consequence it cannot give the student the exhilarating experience, as Oxford does, of becoming a world authority – I mean it only slightly short of literally – on a set of (admittedly very narrow) subjects. I explained this in an essay which was published in various places and definitively in a book called *The Oxford Tutorial: 'Thanks, you taught me how to think'*.[1] Parts of the following paragraphs are derived from this article.

I made the point there that our Oxford course was not 'lecture-driven' in the way that many undergraduates like their studies to be, feeling that they should be examined on, and only on, topics directly covered in lectures. On the contrary, when I was an undergraduate the entire subject of zoology was fair game for the examiners. The only constraint was an unwritten convention that the exam in any one year should not depart unfairly from the general precedent of previous years. And tutorials, too, were not 'lecture-driven' (as I fear they may be today); they were zoology-driven.

In my penultimate term Peter Brunet managed to secure for me the rare privilege of tutorials with Niko Tinbergen himself. Since he was solely responsible for all the lectures in animal behaviour, Dr Tinbergen would have been well placed to give 'lecture-driven'

---

[1] 'Evolution in biology tutoring?', in David Palfreyman, ed., *The Oxford Tutorial: 'Thanks, you taught me how to think'* (Oxford Centre for Higher Education Policy Studies, 2001; 2nd edn 2008). When the essay first appeared (in *The Oxford Magazine*, No. 112, Eighth Week, Michaelmas Term 1994), it bore the 'deliberately graceless' title 'Tutorial-Driven', in reflection of the 'lecture-driven' teaching I was criticizing.

tutorials. I need hardly say that he didn't. Each week my tutorial assignment was to read one DPhil (Oxford-speak for PhD) thesis. My essay was to be a combination of DPhil examiner's report, review of the history of the subject in which the thesis fell, proposal for follow-up research, and theoretical and philosophical discussion of the issues that the thesis raised. Never for one moment did it occur to either tutor or pupil to wonder whether this assignment would be directly useful for answering some exam question.

Another term Peter Brunet, recognizing that my bias in biology was more philosophical than his own, arranged for me to have tutorials with Arthur Cain, effervescently brilliant rising star of the department, who went on to become Professor of Zoology at Manchester and later Liverpool. Far from these tutorials being driven by any lectures on our course, Dr Cain had me reading nothing but books on history and philosophy. It was up to me to work out the connections between zoology and the books that I was reading. I did, and I loved it. I'm not saying that my juvenile essays on the philosophy of biology were any good – with hindsight I know they weren't – but I can say that I have never forgotten the exhilaration of writing them, or the feeling of being a real scholar as I read in the library.

The same is true of my more mainline essays on standard zoological topics. I have no memory of whether we had a lecture on the water-vascular system of starfish. Probably we did, but that fact had no bearing upon my tutor's decision to assign an essay on the topic. The starfish water-vascular system is one of many highly specialized topics in zoology that I now recall for the same reason – that I once wrote an essay on them. Starfish don't have red blood; instead, they have piped sea water, constantly circulated through an intricately plumbed system of tubes which form a ring around the centre of the star and lead off in branches down each of the five arms. The piped sea water is used in a unique hydraulic pressure system, operating the many hundreds of tiny tube feet arrayed along the five arms. Each tube foot ends in a little gripping sucker, and these shuttle back and forth in collusion to pull the starfish along in a particular direction. The tube feet don't move in unison but are semi-autonomous and, if the circum-oral nerve ring that gives them their orders should chance to become severed, the tube feet in different arms can pull in opposite directions and tear the starfish in half.

I remember the bare facts about starfish plumbing, but it is not the facts that matter. What matters is the way in which we were encouraged to discover them. We didn't just mug up a textbook: we went into the library and looked up books old and new; we followed trails of original research papers until we had made ourselves as nearly world authorities on the topic as it is possible to become in one week (nowadays one would do much of this work on the internet). The encouragement provided by the weekly tutorial meant that one didn't just *read* about starfish hydraulics, or whatever the topic was: for that one week, I remember that I slept, ate and dreamed starfish hydraulics. Tube feet marched behind my eyelids, hydraulic *pedicellariae* quested and sea water pulsed through my dozing brain. Writing my essay was the catharsis, and the tutorial was the justification for the entire week. And then the next week there would be a new topic and a new feast of images to be conjured up in the library. We were being educated . . . And I believe it is largely to this week-by-week training that I owe such writing ability as I may be judged to possess.

The tutor for whom I wrote the starfish essay was David Nichols, who went on to become Professor of Zoology at Exeter. Another notable tutor who shaped me as a young zoologist was John Currey, later Professor of Zoology at York University. He introduced me to, among other things, his – and now my – favourite example of revealingly bad 'design' in animals: the recurrent laryngeal nerve. As

I explained in *The Greatest Show on Earth*, instead of going directly from the brain to its end organ the larynx, this nerve makes a detour (in the case of the giraffe, a spectacularly long detour) down into the chest, where it loops around a large artery before proceeding back up the neck to the larynx. This is eloquent of terribly bad design, but is completely explicable the moment you forget design and start thinking in terms of evolutionary history instead. In our fishy ancestors the shortest route for the nerve was posterior to the then equivalent of that artery, which in those early days supplied one of the gills. Fish don't have necks. When necks started to lengthen on land, the artery gradually moved backwards relative to the head, step by tiny step through evolutionary time further away from brain and larynx. The nerve kept abreast – kind of literally – making at first only a small detour but then, as evolution progressed, a longer and longer detour until, in a modern giraffe, its diverted route is a matter of several metres. Just a few years ago, as part of a television documentary, I was privileged to assist in a dissection of this remarkable nerve in a giraffe that had unfortunately died a few days earlier.

My genetics tutor was Robert Creed, pupil of the eccentric and misogynistic aesthete E. B. Ford, himself heavily influenced by the great R. A. Fisher, whom we were all taught by Ford to revere. I learned from those tutorials, and from Dr Ford's own lectures, that genes are not atomistically separate from each other, where their effects on bodies are concerned. Rather, a gene's effect is conditioned by the 'background' of the other genes in the genome. Genes modify one another's effects. Later, when I became a tutor myself, I devised an analogy to try to explain this to my pupils. The body is represented by the shape of a bed-sheet, hanging approximately horizontally by thousands of strings attached to an array of hooks in the ceiling. Each string represents one gene. A mutation in the gene is represented by a change in the tension in that string's attachment to the ceiling. But – here is the important part of the analogy – each string is not isolated in its attachment to the sheet hanging below it. Rather, it is tangled up with lots of other strings, in a complicated cat's cradle. This means that when a mutation occurs in any one 'gene' (change of tension in its attachment to the ceiling hook), the tensions in all the other strings with which it is entangled change at the same time, in a series of knock-on effects throughout the cat's cradle. And the shape of the sheet (the body) is consequently influenced by the interaction of all the genes, not by each gene working separately on its 'own' little part of the sheet. In fact, no gene does 'own' any single part of the sheet. The body is not like a butcher's diagram, with 'cuts' of the body corresponding to particular genes. Rather, a gene may influence the whole body in interaction with other genes. An elaboration of the parable introduces environmental – non-genetic – influences tugging on the cat's cradle from the side.

From Arthur Cain, whom I mentioned above, I learned to dissent from the still fashionable trashing of numerical systems for classifying animals by mathematical measurement of the similarities and differences between them. Quite separately, I also learned from Dr Cain to be impressed by the power of natural selection to produce adaptations of extreme perfection – notwithstanding important and interesting exceptions such as the recurrent laryngeal nerve, just mentioned. Both these lessons set me somewhat at odds with certain orthodoxies, which still dominate the world of zoology. Arthur also taught me to be sparing in my use of the word 'mere' – an exercise in consciousness-raising that has stayed with me ever since. 'Humans are not *mere* bags of chemicals . . .' Well, of course they are not, but when you have said that you have said nothing interesting, and the word 'mere' is supererogatory. 'Humans are not *mere* animals . . .' What have you just said that is more than trite? What weight does the word 'mere' carry in that sentence? What is 'mere' about an animal? You haven't said anything meaningful. If you intend to mean something, say it.

Arthur also told me a never-forgotten story about Galileo, which summarizes what was new about Renaissance science. Galileo was showing a learned man an astronomical phenomenon through his telescope. This gentleman said, approximately: 'Sir, your demonstration with your telescope is so convincing that, were it not that Aristotle positively states the contrary, I would believe you.' Today it amazes us – or ought to – that anybody could possibly reject real observational or experimental evidence in favour of what some supposed authority had simply asserted. But that's the point. That is what has changed.

For us zoologists, unlike undergraduates reading history or English or law, tutorials almost never happened in our college, or indeed in any college. Nearly all were in the Department of Zoology, a rambling up-stairs and down-dale appendage to the University Museum. It was this warren of rooms and corridors which, as I have already mentioned, was the centre of my being. This was very different from the typical experience of an Oxford undergraduate reading a non-scientific subject, for whom the college was the centre of existence. Old-style college tutors think that tutoring outside the college walls is a sort of second-best. My experience suggests exactly the opposite. It was refreshing to have a different tutor every term, for reasons that seem to me almost too obvious to specify.

I did have friends in Balliol, most of whom were reading non-scientific subjects. Nicholas Tyacke (with whom I later shared lodgings, and who became a professor of history at University College, London) and Alan Ryan (who became a distinguished political philosopher and Warden of New College) were on my staircase. As it happened, several of my friends were in the college's acting fraternity, which led me to see some amateur dramatic productions. One of the most moving theatrical evenings I ever experienced was a Balliol College Dramatic Society production of Robert Ardrey's *Shadow of Heroes*, about the Hungarian revolution of 1956. More light-hearted were the Balliol Players, a travelling company who each year would put on a pastiche production of an Aristophanes play. I think that when they started in the 1920s the Players did Aristophanes straight, even in Greek. But the tradition changed, and by my time they were rewriting Aristophanes into revues satirizing modern politics. The leading lights of the Players in my time were Peter Snow, who became a familiar face on television, and John Albery, a witty and talented member of the famous theatrical dynasty, who later became Master of University College, Oxford. John Albery did a splendid General Montgomery ('Now God said – and I agwee with him . . .'), and Peter Snow an equally memorable General de Gaulle: '*La gloire . . . la victoire . . . l'histoire . . . et . . . la plume . . . de ma tante.*' Jeremy Gould scarcely had to act at all to do Harold Macmillan singing 'My birthday honours list is certain to contain . . . And plenty of OBEs . . .' It was the time of the twilight of empire, and the Players did a lovely valedictory song, presumably written by John Albery, of which I remember only five lines:

> Sunset and the evening star
> From Aden to Zanzibar.
> The bonds of the Empire sundering
> And final salutes are thundering
> And man will not cease his wondering . . .

The same theatrical set introduced me to the Victorian Society, in whose company I spent some of my happiest times in Balliol. We met once or twice a term to sing music-hall songs to piano accompaniment, while sipping port. A master of ceremonies would call up soloists one by one to sing their special songs, and we'd all join in the chorus. Mostly they were cheerful, cheeky songs ('Where did you get that hat?' 'Don't have any more, Mrs Moore'; 'You can't do that there 'ere'; 'I'm 'Enery the Eighth I am'; 'My old man said follow the van') interspersed with some sentimental weepies, for which tissues would be handed out ('She's only a bird in a gilded cage'; 'Silver threads among the gold'), and the evening would end with jingoistic patriotism ('Soldiers of the Queen'; 'We don't want to fight, but by jingo if we do . . . The Russkies shall not have Constantinople'). If there's one experience from Balliol days that I would dearly love to relive, it would be an evening with the Victorian Society.

It was much later in my life, but the nearest approach to such a reliving took place at the regular Friday evening sing-song at the Killingworth Castle pub in Wootton, a village just outside Oxford, to which I was introduced by my second wife Eve, mother of my beloved daughter Juliet. The music was British 'folk', not music hall, and the drink was beer, not port, but here I relived something of the atmosphere of the Victorian Society: a warm conviviality fuelled by music and community, more than by drink. The soloists and instrumentalists (guitar, squeezebox, penny whistle) on these Friday nights rotated between four or five regular performers or groups, all of them good in their different ways, all with their particular repertoires of songs, which were known to the regular chorus including Eve and me. For some songs quite stylish canons and descants would be produced, and – as with the Victorian Society – the chorus was always disciplined and up to a brisk tempo, very different from the usual 'Just a song at twilight' drunken dirge. We knew the more prominent members by private nicknames given them by Eve: 'Two Pints' (a large, bearded young man with a huge bass voice as muscular as the arms that raised his pints and took the collection for the musicians); 'Big Daddy' (a grandfatherly figure with an agreeable tenor, who sometimes volunteered 'Cock Robin' as a solo after the main soloists had finished); 'Maynard Smith' (a cheery, bespectacled fellow, named for his facial resemblance to the great scientist); 'the Incredible Hulk' (one of few who sang out of tune) and others.

Back in undergraduate days, my Balliol friends and I often went to the cinema, usually to the Scala in Walton Street: intellectual films by Ingmar Bergman, or Jean Cocteau, or Andrzej Wajda or other continental directors. I was especially affected by Ingmar Bergman's dark monochrome images in *Wild Strawberries* and *The Seventh Seal*, and the lyrical love scenes of *Summer Interlude* before it turned tragic. Films of that kind, and poetry to which my father introduced me – Rupert Brooke, A. E. Housman and above all the early W. B. Yeats – turned my young self into unrealistic, indeed deluded, byways of romantic fantasy. Like many a naive nineteen-year-old I fell in love – not with any particular girl, but with the idea of being in love. Well, there was a girl, and she happened to be Swedish, which chimed with my Bergman-led fantasies, but it was the idea of love itself, with me in the role of a tragic Romeo, that I loved. I moped over her for a ludicrously long time after she had returned to Sweden and – no doubt – had long forgotten her brief Summer Interlude with me.

I didn't finally lose my virginity until much later, at the rather advanced age of twenty-two, to a sweet cellist in London, who removed her skirt in order to play to me in her bedsitter (you can't play the cello in a tight skirt) – and then removed everything else. It is fashionable to decry one's first such experience but I shall not. It was wonderful, and what I chiefly remember is the feeling of atavistic fulfilment: 'Yes, of course, *this* is what it was always going to feel like. This is the way it was going to be from the beginning of time.' It isn't difficult for a biologist to explain why nervous systems evolved in such a way as to make sexual congress one of the consistently greatest experiences life has to offer. But explaining it doesn't make it any the less wonderful – just as Newton's spectral unweaving never diminishes the glory of the rainbow. And it doesn't matter how many rainbows you see throughout your life. The glory is reinvented afresh, and the heart leaps up every time. But I'll say no more on the subject, and will betray no confidences. It isn't that kind of autobiography.

Wordsworth, as it happened, was never a favourite of mine, but I would like to quote here a few fragments of some of the poems that did move me as a young man. These verses were an important part of making me what I am, and they were all (in some cases still are) word-perfect in my memory.

> Breathless, we flung us on the windy hill,
> Laughed in the sun, and kissed the lovely grass.
> You said, 'Through glory and ecstasy we pass;
> Wind, sun, and earth remain, the birds sing still,
> When we are old, are old . . .' 'And when we die
> All's over that is ours; and life burns on
> Through other lovers, other lips,' said I,
> 'Heart of my heart, our heaven is now, is won!'
> 'We are Earth's best, that learnt her lesson here.
> Life is our cry. We have kept the faith!' we said;
> 'We shall go down with unreluctant tread
> Rose-crowned into the darkness!' . . . Proud we were,
> And laughed, that had such brave true things to say.
> —And then you suddenly cried, and turned away.
>
> Rupert Brooke

> Tell me not here, it needs not saying,
> What tune the enchantress plays
> In aftermaths of soft September
> Or under blanching mays,
> For she and I were long acquainted
> And I knew all her ways.
>
> A. E. Housman

I dreamed that I stood in a valley, and amid sighs,
For happy lovers passed two by two where I stood;
And I dreamed my lost love came stealthily out of the wood
With her cloud-pale eyelids falling on dream-dimmed eyes:
I cried in my dream, *O women, bid the young men lay*
Their heads on your knees, and drown their eyes with your hair,
Or remembering hers they will find no other face fair
Till all the valleys of the world have been withered away.

W. B. Yeats

Heart handfast in heart as they stood, 'Look thither,'
Did he whisper? 'look forth from the flowers to the sea;
For the foam-flowers endure when the rose-blossoms wither,
And men that love lightly may die–but we?'
And the same wind sang and the same waves whitened,
And or ever the garden's last petals were shed,
In the lips that had whispered, the eyes that had lightened,
Love was dead.

A. C. Swinburne

My father kept a loose-leaf folder in which he bound a large
number of his favourite poems, all copied out in his own hand. My
own taste in poetry was strongly influenced by this private
anthology, which my mother still possesses. I was touched to learn
that it originated in letters to her in their early twenties, sent from
Cambridge where he was doing postgraduate studies, each poem
enclosed with a letter and preserved by her.

But, to my own undergraduate days and my thoughts of what
should come next: I don't think I ever seriously contemplated joining my father in farming. Increasingly, I wanted to stay on at Oxford
to do a research degree. I didn't have any very clear idea of what
might follow after that, or of what kind of research I wanted to do.
Peter Brunet offered me a biochemical project, and I gratefully

signed up to it and studied the relevant research literature, though without much enthusiasm. But then I went for tutorials with Niko Tinbergen on animal behaviour – and my life changed. Here was a subject I could really think about: a subject with philosophical implications. Niko was apparently impressed by me: his end-of-term report to my college said I was the best undergraduate he had ever tutored – although that verdict must be tempered by the fact that he didn't do much undergraduate tutoring. Anyway, it raised my courage to the point of asking him whether he would like to take me on as his research student, and to my enduring delight he said yes. My future was assured, for the next three years at least. And for the rest of my life, now that I think about it.

LEARNING
THE TRADE

PERHAPS all scientists recall their graduate student years as an idyll. But surely some research environments are more idyllic than others, and I think there was something special about the Tinbergen group at Oxford in the early 1960s. Hans Kruuk has captured the atmosphere in his affectionate but not hagiographic biography, *Niko's Nature*.[1] He and I arrived too late for the heroic 'hard core' period described by Desmond Morris, Aubrey Manning and others, but I think our time resembled it – though we saw less of Niko himself, because his room was in the main Zoology Department while all the rest of us were housed in the annexe at 13 Bevington Road, a tall, narrow house in north Oxford, about half a mile from the main Zoology building tacked onto the University Museum in Parks Road.

The senior figure in 13 Bevington Road was Mike Cullen, probably the most important mentor in my life – and I believe most of my contemporaries in the Animal Behaviour Research Group (ABRG) would agree. To try to explain the debt that all of us owe to this remarkable man, I can do no better than quote the closing

---

[1] Hans Kruuk, *Niko's Nature: The Life of Niko Tinbergen and his Science of Animal Behaviour* (Oxford, Oxford University Press, 2003).

words of the eulogy that I spoke at his memorial service in Wadham College, Oxford, in 2001.

He did not publish many papers himself, yet he worked prodigiously hard, both in teaching and research. He was probably the most sought-after tutor in the entire Zoology Department. The rest of his time – he was always in a hurry and worked a hugely long day – was devoted to research. But seldom his own research. Everybody who knew him has the same story to tell. All the obituaries told it, in revealingly similar terms.

You would have a problem with your research. You knew exactly where to go for help, and there he would be for you. I see the scene as yesterday. The lunchtime conversation in the crowded little kitchen at Bevington Road, the wiry, boyish figure in the red sweater, slightly hunched like a spring wound up with intense intellectual energy, sometimes rocking back and forth with concentration. The deeply intelligent eyes, understanding what you meant even before the words came out. The back of the envelope to aid explanation, the occasionally sceptical, quizzical tilt of the eyebrows, under the untidy hair. Then he would have to rush off – he always rushed everywhere – perhaps for a tutorial, and he would seize his biscuit tin by its wire handles, and disappear. But next morning the answer to your problem would arrive, in Mike's small, distinctive handwriting, two pages, often some algebra, diagrams, a key reference to the literature, sometimes an apt verse of his own composition, or a fragment of Latin or classical Greek. Always encouragement.

We were grateful, but not grateful enough. If we had thought about it we would have realized, he must have been working on that mathematical model of my research all evening. And it isn't only me for whom he does this. Everybody in Bevington Road gets the same treatment. And not just his own students. I was officially Niko's student, not Mike's. Mike took me on, without payment and without official recognition, when my research became more mathematical than Niko could handle. When the time came for me to write my thesis, it was Mike Cullen who read it, criticized it, helped me polish every line. And all this, while he was doing the same thing for his own official students.

When (we all should have wondered) does he get time for ordinary family life? When does he get time for his own research? No wonder he so seldom published anything. No wonder he never wrote his long-awaited book on animal communication. In truth, he should have been joint author of just about every one of the hundreds of papers that came out of 13 Bevington Road during that golden period. In fact, his name appears on virtually none of them – except in the Acknowledgements section . . .

The worldly success of scientists is judged – for promotion or honours – by their published papers. Mike did not rate highly on this index. But if he had consented to add his name to his students' publications, as readily as modern supervisors insist on putting their names on papers to which they contribute much less, Mike would have been a conventionally successful scientist, lauded with conventional honours. As it is, he was a brilliantly successful scientist in a far deeper and truer sense. And I think we know which kind of scientist we really admire.

Oxford sadly lost him to Australia. Years later, in Melbourne, at a party for me as visiting lecturer, I was standing, probably rather stiffly, with a drink in my hand. Suddenly, a familiar figure shot into the room, in a hurry as ever. The rest of us were in suits, but not this familiar figure. The years vanished away. Everything was the same – though he must have been well into his sixties, he seemed still to be in his thirties – the glow of boyish enthusiasm, even the red sweater. Next day he drove me to the coast to see his beloved penguins, stopping on the way to look at giant Australian earthworms, many feet long. We tired the sun with talking – not, I think, about old times and old friends, and certainly not about ambition, grant-getting and papers in *Nature*, but about new science and new ideas. It was a perfect day, the last day I saw him.

> We may know other scientists as intelligent as Mike Cullen –
> though not many. We may know other scientists who were as
> generous in support – though vanishingly few. But I declare, we
> have known nobody who had so much to give, combined with so
> much generosity in giving it.

I almost wept when I spoke that eulogy in Wadham chapel, and I
almost wept again just now when rereading it twelve years later.

I don't know whether the camaraderie of 13 Bevington Road
was exceptional, or whether all groups of graduate students nurture
a similar *esprit de corps*. I suspect that being housed in a separate
annexe rather than in a large university building improves the social
dynamics. When the ABRG (and other outliers such as David Lack's
Edward Grey Institute of Field Ornithology and Charles Elton's
Bureau of Animal Populations) eventually moved into the present
concrete monster on South Parks Road, something, I believe, was
lost. But it may just be that by then I was older and more weighed
down by responsibilities. Whatever the reason, I retain a loyal
affection for 13 Bevington Road and my comrades of those times
who foregathered at the Friday evening seminars, or in the lunch
room, or over the bar billiards table in the Rose and Crown: Robert
Mash, whose epidemic sense of humour I later recalled in my foreword to his book *How to Keep Dinosaurs*;[1] Dick Brown, chain
smoking, hard drinking and implausibly rumoured to be religious;
Juan Delius, whose deliriously eccentric brilliance never ceased to
amuse; Juan's supernormally delightful wife Uta who gave me
German lessons; the tall, blond Dutchman Hans Kruuk, who later
wrote Niko's biography; the Scotsman Ian Patterson; Bryan Nelson
the gannet man, known to me in my first six months only from the
enigmatic notice on his door, 'Nelson is on the Bass Rock'; bearded
Cliff Henty; David McFarland, Niko's eventual successor who,

---

[1] Robert Mash, *How to Keep Dinosaurs* (London, Orion, 2005).

although based in the Psychology Department, was a sort of honorary member of our group because his vivacious wife Jill was Juan's research assistant, and the couple had lunch in Bevington Road every day; Vivienne Benzie, who introduced the sunny New Zealand girls Lyn McKechie and Ann Jamieson as yet other honorary members of the lunch group; Lou Gurr, another smiling New Zealander; Robin Liley; the jovial naturalist Michael Robinson; Michael Hansell, who later shared a flat with me; Monica Impekoven, with whom I wrote a paper later; Marian Stamp, whom I was to marry; Heather McLannahan, Robert Martin, Ken Wilz; Michael Norton-Griffiths and Harvey Croze, who later formed a consulting partnership in Kenya; John Krebs, who later collaborated with me in writing three papers; the daredevil Iain Douglas-Hamilton, unwilling exile from Africa while he wrote his thesis on elephants; Jamie Smith, with whom I wrote a paper on optimal foraging in tits; Tim Halliday the newt man, Sean Neill with his lovingly restored Lagonda and gift for drawing cartoons, Lary Shaffer, master photographer, and other friends whom I apologize for omitting.

The Friday evening seminars were the highlight of the week for the Tinbergen group. They lasted two hours and frequently spilled over into the following Friday, but the time flashed by because, instead of the soporific formula of an hour spent listening to one speaker's voice followed by questions at the end, our two hours were enlivened by argument throughout. Niko set the tone by interrupting almost before the speaker could complete his first sentence: '*Ja, ja*, but what do you mean by . . . ?' This wasn't as irritating as it sounds, because Niko's interventions always aimed at clarification and it was usually necessary. Mike Cullen's questions were more penetrating, better informed and more feared. Other notable contributors – brilliant in their idiosyncratic ways – were Juan Delius and David McFarland, but the rest of us chipped in without inhibition too, almost from the first day we were there. Niko encouraged that. He insisted on absolute clarity about the question we were asking in our research. I recall how shocked I was, on visiting our sister research group at Madingley in Cambridge, to hear one of the graduate students beginning to describe his research with the words: 'What I do is . . .' I had to restrain myself from imitating Niko's voice: '*Ja, ja*, but what is your *question*? Years later, I related this story when I gave a research seminar at Madingley. I refused to identify the culprit to a mock-scandalized Robert Hinde, the formidably intelligent and charismatic leader of the Madingley group who later became Master of St John's College, Cambridge, and my lips are sealed to this day.

The question Niko set for me was a version of the question often labelled with the 'nature or nurture?' cliché derived from *The Tempest*:

> A devil, a born devil, on whose nature
> Nurture can never stick . . .

Philosophers down the centuries have pondered the question. How much of what we know is natively built in, and to what extent is the young mind a blank slate, waiting to be written over, as John Locke believed?

Niko himself, like Konrad Lorenz (with whom he is credited with co-founding the science of ethology), was early associated with the 'nature' school of thought. His most famous book, *The Study of Instinct*,[1] which he later pretty much disowned, used 'instinct' as a synonym for 'innate behaviour', defined as 'behaviour that has not been changed by learning processes'. Ethology is the biological study of animal behaviour. Various schools of psychology also study animal behaviour, but with different emphasis. Psychologists

---

[1] N. Tinbergen, *The Study of Instinct* (Oxford, Clarendon Press, 1951).

historically tended to study animals like rats or pigeons or monkeys as substitutes for humans. Ethologists historically were interested in the animals in their own right, not as proxies for anything. Consequently they have always studied a much wider range of species, and they tend to emphasize the role of behaviour in the natural environment of the species. Ethologists also, as I have said, historically emphasized 'innate' behaviour, whereas psychologists were more interested in learning.

In the 1950s, a group of American psychologists started to take an interest in the works of the ethologists. Prominent among them was Daniel S. Lehrman, a big man with a deep knowledge of natural history as well as of psychology. He also spoke adequate German, which made him an effective bridge between the two approaches to animal behaviour.

In 1953 Lehrman wrote a very influential critique of the traditional ethological approach. He strongly criticized the whole notion of innate behaviour, not because he thought everything was learned (although some psychologists whom he quoted did), but because he thought it was in principle impossible to define innate behaviour: impossible to devise an experiment to demonstrate that any particular piece of behaviour is innate. Theoretically, the obvious method was the 'deprivation experiment'. Imagine if humans were given no verbal instruction in how to copulate and no opportunity to observe other species – not even the smallest inkling. Would they know how to do it when the opportunity finally presented itself? It's an intriguing question, and there might be telling anecdotes, perhaps about over-sheltered and naive Victorian couples. But in non-human animals we can do experiments. Deprivation experiments.

If you rear a young animal in deprived conditions without the opportunity for experience, and it still knows how to behave properly, that must mean the behaviour is innate, inborn, instinctive. Mustn't it? But Lehrman objected that you couldn't deprive the young animal of everything – light, food, air, etc. – and that it is never obvious how much deprivation is needed in order to satisfy the criterion of innateness.

The dispute between Lehrman and Lorenz got personal. Lehrman, whose family background was Jewish, caught Lorenz out in some suspiciously Nazi-inflected writings from the war years and did not shrink from mentioning this in his famous critique. Lorenz, on first meeting Lehrman after the critique was published, said (approximately): 'I thought from your writings that you must be a small, mean, wizened little man. But now that I see you are a BIG man [and Lehrman was indeed a very big man] we can be friends.' This avowal of friendship didn't stop Lorenz trying – Desmond Morris tells the story as an eye-witness from inside the car – to intimidate Lehrman by almost mowing him down with an enormous American car that he was driving in Paris.

But back to the controversy over nurture or nature. Male sedge warblers (to take just one example) have a complex and elaborate song, and they can perform it even when reared in isolation, never having heard another sedge warbler. The Lorenz–Tinbergen school would therefore have said it must be 'innate'. But Lehrman empha- sized the complexity of developmental processes and always wondered whether learning was involved in some less obvious way. For Lehrman, it wasn't good enough to say that the young animal had been reared under deprived conditions. For him, the question was: 'Deprived of what?'

Since Lehrman's critique was published, ethologists have indeed discovered that many young songbirds, including sedge warblers, even when reared in isolation, *learn* to sing their correct species song by listening to their own fumbling efforts, repeating the good fumbles and discarding the bad. So that looks like nurture after all. But in that case, Lorenz and Tinbergen might reply, how do the young birds know which of their fumbles are good and which bad? Surely that 'knowledge' – a template for what their species song ought to sound like – has to be innate? All learning does is transfer the song pattern from the sensory part of the brain (the built-in template) to the motor side (the actual skill of singing the song).

Other species, by the way, such as the American white-crowned sparrow, also teach themselves to sing in this 'fumbling' way, but do need to have heard the species song earlier in life. It is as though the young bird takes a 'tape recording' before it can sing, and uses it as a template for teaching itself how to sing. And there are intermediates between the 'learned tape recording' and the 'innate tape recording' as templates for later learning.

This was the philosophical minefield into which Niko Tinbergen released me in 1962. I think he wanted to back away from his perceived association with Lorenz and saw me as a bridge towards the Lehrman camp. My experimental subject was to be not singing birds but baby chicks pecking. I did a series of experiments of which I'll mention only one here.

Baby chicks straight out of the egg start pecking at small objects, presumably looking for food. But how do they know what to peck at? How do they know what's good for them? One extreme would be for nature to endow them, before they have any experience at all, with a template picture of a grain of wheat in the brain. That's un-realistic, especially in an omnivore. Do wheat grains and mealworms and barleycorns and millet seeds and beetle larvae have anything in common, as opposed to boring and inedible marks and stains? Yes, they do. For one thing, they are solid.

How do you recognize something as solid? One way is by surface shading. Look at the photographs of moon craters overleaf. They are the same photograph, but one is rotated through 180° relative to the other. My guess is that on the left you will see hollow craters and on the right solid flat-topped hills – and the other way around if you

swivel the book upside down. The illusion has been known for a long time. It depends upon a preconception about where the light is coming from: in effect, a preconception about the location of the sun. Solid objects tend to be brighter on the side nearest the sun, which will usually be approximately above. A photograph of a solid object can therefore look hollow if you turn it upside-down, and vice versa.

The sun is seldom *directly* overhead, but the general direction of its light is more likely to be down than up. Therefore any predator seeking solid objects as possible prey can use surface shading cues based upon that assumption. And on the other side of the predator–prey arms race, natural selection might well favour prey animals that manage to disguise their solidity by 'countershading'. Many species of fish are darker on top, lighter below, which tends to neutralize the natural tendency for sunlight to come from above, and thereby makes the fish look flatter. One fish, the 'upside-down catfish', is a genuine 'exception that proves the rule'. It habitually swims upside-down and, sure enough, it is *reverse countershaded*: darker on its belly than its back.

A Dutch student of Tinbergen called Leen De Ruiter did some

neat experiments on reverse countershaded caterpillars, who habitually rest upside-down. The upper picture on this page shows *Cerura vinula* in its normal position. It looks flat and inconspicuous. The lower picture shows what it looked like when De Ruiter turned its twig upside-down: much more conspicuous to my eyes and – more significantly – to the eyes of jays, when De Ruiter used them as experimental predators.

But none of this says anything about whether – in jays or

humans – the knowledge that the sun is normally overhead is innate or learned. The solid shading illusion seemed to me to provide a good opportunity to test the question, using baby chicks in deprivation experiments.

First, did chicks see the illusion? Apparently, yes. I photographed a half ping-pong ball lit asymmetrically, and printed the image to be about the size of a tempting grain or seed. When I viewed the photograph with the illuminated side at the top, the hemisphere looked solid. When I inverted the photograph, it didn't. When chicks were offered a choice between the two orientations, they strongly chose to peck at the apparently solid picture, the one lit from above. This suggested that chicks possess the same 'preconception' we do, that the sun is normally overhead.

So far, so good; but these chicks, though young, were not completely naive. They were three days old and had been feeding in normal overhead light during this time. They might have had time to learn the appearance of solid objects illuminated from above.

To test this I did a crucial experiment. I reared chicks with light coming from *below* and tested them under the same conditions. So, at the time of testing, they had never had any experience of overhead light. As far as they were concerned, the world into which they had hatched was a world with a sun underneath them. Every solid object they had ever seen, whether food objects or parts of other chicks, was lighter underneath than on top. I expected that, when tested with the two ping-pong ball photographs, they would prefer to peck at the one illuminated from below.

But I was delighted to be proved wrong. The chicks overwhelmingly pecked at the photograph illuminated from above. If you accept my interpretation, this means that the chicks are genetically equipped by ancestral natural selection with something equivalent to 'advance information': in the world in which they are to live, the sun will normally shine from above. My experiment had pinpointed a true example of innate information which is not reversed by a positive attempt to teach the contrary.

I can't think of any group of humans who habitually live with underfloor lighting. If they exist, it would be interesting to test them in the same way I tested my chicks. I thought about offering an intuitive guess as to what the result would be, but I honestly prefer not to place a bet. Wouldn't it be fascinating if we too saw the illusion innately? Having been surprised by the chicks, I'd be only slightly more surprised if humans did the same. We may never know, but there could be ways to do the experiment on very young babies. They don't peck, but they do fixate their eyes on objects that interest them, and you can measure that. Could a developmental psychologist offer babies a version of my ping-pong ball experiment and measure the time they spend staring at each of the two photographs? Would it be considered unethical to use underfloor lighting for a baby's room for the first few days of life? I can't see why, but who knows what the verdict of a modern 'ethical committee' might be?

In the end, my work on 'nature or nurture' constituted only a small part of my doctoral research,[1] and it was relegated to an appendix in my thesis. The main part of my thesis had little in common with it, except that it also involved pecking in chicks. And it was also an attempt to illustrate a point of philosophical interest – although taken from a different part of philosophy. It became possible through an improved technique for recording pecks.

Bevington Road, and especially its satellite research stations in the great gull colonies of the north, ran a system of 'slaves' – young unpaid volunteers who wanted a brief taste of the Tinbergen experience before going to university. Among them were Fritz

---

[1] R. Dawkins, 'The ontogeny of a pecking preference in domestic chicks', *Zeitschrift für Tierpsychologie*, 25 (1968), pp. 170–86.

Vollrath (who later returned to Oxford to head a flourishing group working on spider behaviour, and remains a close friend) and (also from Germany) Jan Adam. Jan and I found an immediate affinity, and we worked together. He had remarkable workshop skills – combining the very different virtues of my father and Major Campbell – and, fortunately, these were the days before health and safety regulations interfered to protect us from ourselves and sap our initiative. Jan and I had the freedom of the departmental workshops: lathes, milling machines, bandsaws and all. We (that is to say Jan, with me as willing apprentice – the younger brother syndrome again, I suppose) built an apparatus to automate the counting of chick pecks, using delicately hinged little pecking keys, elegantly made from scratch by Jan, with sensitive micro-switches. Previously, when working on the surface shading illusion, I had counted pecks by hand. Suddenly, I was in a position to collect huge quantities of data automatically. And this opened the door to a completely different kind of research, motivated by a different philosophy, Karl Popper's philosophy of science, which I learned from Peter Medawar.

As I have already explained, I had come to know of Medawar early on through my father, who was a schoolfriend of his. As British biology's star intellectual, Medawar came to give a visiting lecture at his old Oxford department when I was an undergraduate there, and I remember the excited buzz in the standing-room-only audience waiting for this tall, handsome, gracious figure to arrive ('This lecturer has never been thought ungracious in his life,' as a later critic said of him). The lecture prompted me to read Medawar's essays, later anthologized in *The Art of the Soluble* and *Pluto's Republic*,[1] and it was from them that I learned about Karl Popper.

I became intrigued by Popper's vision of science as a two-stage

---

[1] Peter Medawar, *The Art of the Soluble: Creativity and Originality in Science* (London, Methuen, 1967); *Pluto's Republic: Incorporating The Art of the Soluble and Induction and Intuition in Scientific Thought* (Oxford, Oxford University Press, 1982).

process: first the creative – almost artistic – dreaming up of a hypothesis or 'model', followed by attempts to *falsify* predictions deduced from it. I wanted to do a textbook Popperian study: dream up a hypothesis that might or might not be true, deduce precise mathematical predictions from it, and then try to falsify those predictions in the lab. It was important to me that the predictions should be mathematically precise. It was not enough to predict that a measurement X should be larger than Y. I wanted a model that would predict the exact value of X. And this kind of exact prediction demanded large quantities of data. Jan's apparatus for counting massive numbers of pecks gave me the opportunity. Instead of pecking at photographs of ping-pong balls, my birds pecked at little coloured hemispheres mounted on Jan's hinged windows, which triggered micro-switches. They preferred blue over red over green, but that wasn't what interested me. I wanted to know what governed each individual pecking decision, whichever colour it was directed towards. And this, of course, was only a specimen of a more general question about how decisions are made at any time by any animal.

Medawar elsewhere made the point that scientific research doesn't develop in the same orderly sequence as the final published 'story'. Real life is messier than that. In my own case it was so messy that I can't remember what gave me the idea for my 'Popperian' experiments. I remember only the finished story which, as Medawar would have expected, gives an implausibly tidy impression.

The finished story is that I dreamed up an imaginary 'model' of what might be going on inside a chick's head when it makes a decision between alternative targets, did some algebra to deduce precise, quantitative predictions from the model, then tested them in the lab. The model itself was a 'drive/threshold' model. I postulated that there was a variable ('drive' to peck) in the bird's head, whose graph was continuously wiggling up and down as the drive strengthened or weakened (perhaps at random; it

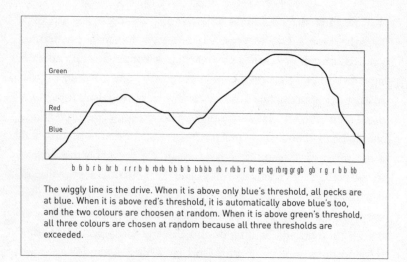

b b b r b br b r r r b b rbrb bb b b bbbb rb r rbb r br gr bg rbrg gr gb gb r g r bb bb

The wiggly line is the drive. When it is above only blue's threshold, all pecks are at blue. When it is above red's threshold, it is automatically above blue's too, and the two colours are choosen at random. When it is above green's threshold, all three colours are chosen at random because all three thresholds are exceeded.

didn't matter). Every time the drive happened to rise above the threshold for a colour, the bird was capable of pecking at that colour (something else, for which I developed and tested another model which I'll mention later, determined the *timing* of pecks). Blue, being a preferred colour, had a lower threshold than green. But if the drive rose above green's threshold, it automatically had to be above blue's threshold as well. What would the bird do then? I postulated that it would be indifferent between the two colours, since both thresholds were exceeded: it would 'toss a coin' to decide between them. So the model predicted that a bird's choices over a long period would consist of periods of pecking at only the preferred colour, interspersed with periods of choosing at random between the two. There would be no periods of consistently positive choice of the less preferred colour.

I didn't at first look directly at sequences of pecks. That was to come later, after I moved to California. I think the reason I didn't test sequences at first was as unambitious as the fact that Jan's apparatus could count pecks but not record the exact order in which they happened; and Jan himself had by now gone back to Germany, so wasn't there to modify his apparatus. I think, too, that I was simply seduced by the Popperian elegance of deducing a mathematical formula which would predict some measured quantity from some other measured quantities.

The chicks happened to prefer blue over red over green. I imagined an experiment in which I would present Blue versus Green, Blue versus Red, and Red versus Green, counting the proportion P of pecks to the preferred colour in each case. This would give me three numbers ($P_{BestWorst}$, $P_{BestMedium}$, $P_{MediumWorst}$). It's only to be expected that $P_{BestWorst}$ would be larger than either of the other two. But could the model predict precisely how much bigger? Could I deduce from the model a formula to predict exactly what $P_{BestWorst}$ should be, if I fed in $P_{BestMedium}$ and $P_{MediumWorst}$? Yes, that is exactly what I succeeded in doing. I defined algebraic symbols to stand for the time spent by the drive between various thresholds, did some school algebra (simultaneous equations as taught by Ernie Dow) to eliminate the unknown variables, and was pretty pleased when, at the end of pages of algebra, a simple, precise, quantitative prediction dropped out. The drive/threshold model predicts that

$$P_{BestWorst} = 2(P_{BestMedium} + P_{MediumWorst} - P_{BestMedium} \cdot P_{MediumWorst}) - 1.$$

I called this Prediction 1. The thing that interested me about Prediction 1 was that it is quantitatively precise.

So now to test it. Would the chicks obey the prediction? Yes: to my delight and amazement, in seven out of eight repeats of the experiment they did, very closely. The eighth experiment was way off, so much so that, to my acute embarrassment, when one of my papers was published in the journal *Animal Behaviour*,[1] the printer

---

[1] R. Dawkins, 'A threshold model of choice behaviour', *Animal Behaviour*, 17 (1969), pp. 120–33.

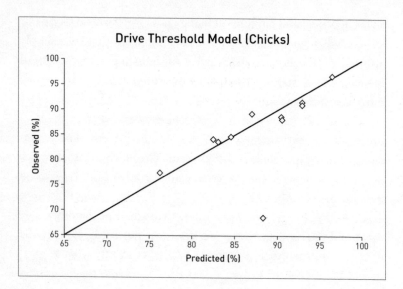

removed the relevant point from the graph, thinking it must be a blemish on the block! Fortunately the offending datum was clearly present in the accompanying table, otherwise I might have been accused of dishonesty. I did another set of experiments on chicks, involving not pecking but walking into chambers illuminated by light of different colours. The graph shown here combines the two sets of experiments and plots the observed against predicted percentages for all 11 chick experiments.

If the model's predictions were perfect, the points should all lie exactly along the diagonal line. With the exception of Experiment 8, as already mentioned, the Drive Threshold Model does a far better job than we ever dare to expect in animal behaviour experiments (physicists expect higher precision because there is usually less statistical error in their measurements).

I also used all the same data to test the predictions of an alternative model, one which simply assumed that each colour has a 'value' for the animal, and that the animal allocates its choices in proportion to the colour's value. The two models gave similar predictions, so that if one is right the other one can't help being nearly right. But the Drive Threshold Model was consistently more accurate in predicting the observed result. The 'colour value' model consistently overestimated $P_{BestWorst}$. The 'colour value' model was falsified. The Drive Threshold Model triumphantly survived the attempt to falsify it, and indeed its predictions were (with the exception of the one experiment) remarkably accurate.

Does this good performance of the model really mean that there is something equivalent to a fluctuating 'drive' in the chick's head, crossing 'thresholds', and that something equivalent to tossing a coin happens when the drive is above more than one threshold? Well, Popper would say that the model survived a strong attempt to disprove it; but that says nothing about what the 'drive' and the 'thresholds' actually correspond to in the language of nerves and synapses. It is at least an interesting thought that you can make inferences about what is going on inside the head without cutting it open.

The same method of imagining a model and testing its predictions has proved enormously productive in many branches of science. In genetics, for example, you can infer the existence of chromosomes as one-dimensional linear sequences of genetic code without ever looking down a microscope, using only the data from breeding experiments. You can even work out the order in which the genes are arrayed along the chromosomes, and how far apart genes are from each other, entirely by imagining what might be the case and testing predictions in breeding experiments. As with my experiments on solidity and shading, I think of my Drive Threshold Model as an illustrative example of the *kind* of thing that can be done with a model, rather than as a conclusive discovery of what is really going on inside a chick's head.

I elaborated the Drive Threshold Model in various directions (that's also something that is supposed to happen according to Popperian philosophy) and tested nine predictions in all, with good success. One of these elaborations of the model, as I mentioned above, was an attempt to explain the exact timings of pecks ('samplings' of the position of the 'drive' relative to the 'thresholds'). The predictions of this model stood up well against data on black-headed gull chicks from my colleague and close friend Dr Monica Impekoven, a visitor to Bevington Road from Switzerland. We published a joint paper on this work.[1]

Another elaboration of the model, which I published as the 'Attention Threshold Model',[2] was an attempt to probe more deeply the 'penny tossing' of the original Drive Threshold Model: the indiscriminate choice of target when more than one threshold is exceeded. Briefly, I suggested that chicks attend to one dimension at a time – colour, shape, size, texture etc. – and in a definite order. Each of these attention systems has its own version of the Drive Threshold Model. The chick attends to the first dimension – say, colour. If the colour system's drive/threshold delivers a definite choice, the chick goes for the preferred colour, say Blue. But if the colour system's verdict is a 'penny toss' the chick switches its attention to something else, say shape, and ignores colour. From the colour system's point of view, choosing by shape is equivalent to choosing at random. But of course it is not random from the shape system's point of view. This trickle-down process continues through all the attention systems. If all else fails, the equivalent of a 'penny toss' is something like 'choose the nearest one'. The Attention Threshold Model yielded a series of further predictions (making nine predictions in all), which I tested with success.

Again, as with the solid shading experiments: is it possible that a

[1] R. Dawkins and M. Impekoven, 'The peck/no-peck decision-maker in the black-headed gull chick', *Animal Behaviour*, 17 (1969), pp. 243–51.
[2] R. Dawkins, 'The attention threshold model', *Animal Behaviour*, 17 (1969), pp. 134–41.

version of the Drive Threshold Model might apply to humans? I searched the scientific literature and found that several psychologists had done pairwise preference tests on humans. Their motivation was different from mine, but I could use their published results. There are various reasons why a psychologist might present a range of choices in all possible pair combinations: to test an idea in Voting Theory, for instance. Instead of offering a three-way choice between Conservative, Liberal and Socialist, with either winner-take-all or rank-order voting, a pollster might investigate the benefits of pairwise testing: 'How would you vote between Conservative versus Liberal (if there were no other choice), between Liberal versus Socialist (given no other choice), and finally between Conservative versus Socialist?' Anyway, for whatever reason, psychologists have presented humans with choices in all possible pairwise combinations. I was therefore able to feed their measurements of Best versus Medium, and Medium versus Worst, into my formula, and test my model's prediction of Best versus Worst. The data came from a diverse set of studies: American students choosing handwriting samples, American students choosing vegetables, American students choosing bitter/sweet tastes, and Chinese students choosing colours. In addition, I was especially delighted to be able to use a big study of preferences for composers shown by the members of the Boston Symphony Orchestra, the Philadelphia Orchestra, the Minneapolis Symphony Orchestra and the New York Philharmonic. Overleaf is a graph pooling all the results from humans. Once again, if the Drive Threshold Model's predictions were perfect, the points would fall along the diagonal line. I must say I was pretty excited when I saw how closely the prediction was fulfilled. Predictions in behavioural biology just aren't usually fulfilled that precisely!

The orchestra study was large, and processing the data laborious. I discussed the problem with my Uncle Colyear, who was by then in the Oxford Forestry Department, lecturing and advising

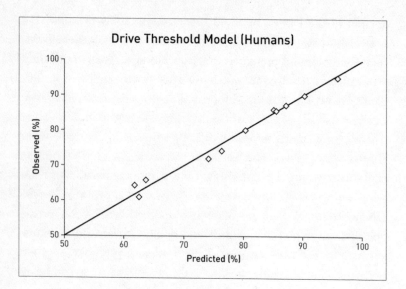

on statistical methods. He suggested that I should learn to program the university computer. He and his wife Barbara got me started and helped me to write a program for the composer preferences. And so began my forty-year time-wasting and soul-consuming love affair with computer programming, a love affair that is now happily over: I am still an intensive computer user, but I now leave the programming to professionals.

Back then in the mid-1960s there was only one computer in Oxford University: a brand new English Electric KDF9, less powerful than an iPad today, but state-of-the-art in its time and filling a large room. The programming language favoured by my uncle and aunt was K-Autocode, a British alternative to Fortran with a similar structure and grammar, and a similar tendency to foster bad programming practice (absolute jumps, for instance). At that time American computers used great stacks of punched cards (vulnerable to being dropped and irretrievably shuffled) and British ones used punched paper tape (spewed out into great spaghetti mounds on the floor, which then had to be rolled up and were vulnerable to tearing). Thank goodness those days are over. Thank goodness, too, that computers now communicate with us via a screen or loudspeaker, not on reams and reams of paper – and immediately, instead of after a 24-hour delay.

But in those days we knew no better, and I was captivated. I was entranced by the idea of pre programming a sequence of operations and chugging through them step by step with a pencil to check them, then unleashing them in the computer to be rerun at very high speed and thousands upon thousands of times. I had a terrible night when I dreamed that I was a computer running my program and I spent – as it seemed – the whole night going through repeat loop after repeat loop in my fevered brain. To be fair, conditions that night were not ideal for sleeping anyway. Along with several of the Bevington Road crew, I had been persuaded by my friend Robert Mash to spend the weekend hunting the Surrey Puma.

Ever since 1959, sightings of a mysterious large carnivore had been reported in the woods of Surrey in southern England. Dubbed the Surrey Puma, it had achieved the status of a minor Yeti-style myth, and a group of us joined together to spend a weekend in May 1966 trying to find it. The newspapers heard about the scheme and, with news material getting thinner as the summer Silly Season approached, the *Observer* printed a photograph of me in a British Empire pith helmet (solar topee) of the kind that I had worn as a child. I have forgotten where my companions pitched their tents, but my assigned role was to spend the night in a sleeping bag, out in the open under the stars, surrounded by great haunches of raw meat. I was allocated a camera with a flash, and my marching orders were to photograph the puma if it should come to take the meat – or me, I suppose. I didn't sleep peacefully, to say the least, so perhaps it was not surprising that my computer nightmare should have come on that particular night. Dawn came as a great relief to me and my companions, and a dreamily misty dawn it was (as you can see from the photograph reproduced in the picture section). We never did find the Surrey Puma, and it may be significant that sightings went on being reported as late as 2005, which would seem to suggest that the Surrey Puma lived more than twice as long as the maximum recorded lifespan of its species, even in captivity.

My computer habit moved on from the KDF9 to a smaller but more accessible computer. Oxford Zoology had acquired a dynamic new professor (head of department in the Oxford-speak of the time) in succession to the genial and slightly batty-looking Sir Alister Hardy. The stiff new broom from Cambridge was 'Laughing John' Pringle (one of those ironic nicknames, like 'Lofty' for a very short man), and the department was thrown into a whirlwind of modernization. In one way after another, dear old Alister Hardy's dear old department was 'upgepringled', no doubt for the better. Among the more exciting instances of upgepringleheit was the arrival from London of an equally dynamic group of X-ray crystallographers (think Watson/Crick, but with protein molecules instead of DNA). And, most exciting for me, they brought with them their own computer, which I was allowed, by its friendly custodian Dr Tony North, to use at night when it wasn't needed for number-crunching the patterns of X-ray scattering by crystals. An Elliott 803, it was even more primitive by modern standards than the KDF9, but it had the enormous advantage that I was allowed to get my hands on it.

This was the time when I became fully aware of the addictive lure of computers. I really did literally – and frequently – spend all night in the warm, glowing computer room, entangled in a spaghetti of punched paper tape, which must have resembled my insomnia-tousled hair. The Elliott had the charming habit of beeping an acoustic rendering of its inner processing. You could listen to the progress of your computation through a small loudspeaker which hummed and hooted a rhythmic serenade, doubtless meaningful to Dr North's expert ear but merely companionable to my nocturnal solitude. Nocturnal dalliance with computers in youth is character- istic of those – now called geeks – whose love affair with computers lasted longer (and more profitably) than mine: Bill Gates, to name one. With hindsight I cannot say that my affair with the Elliott was productive. No doubt I obtained some valuable practice in the art of programming. But Elliott Autocode was not a language that could be used on any other computers, and my nocturnal geekery, though diligent and very hard work, bore the same relation to serious pro- gramming as my tootling in the Oundle music school bore to real music.

I gave a talk on my Drive Threshold Model at the 1965 International Ethological Conference in Zurich. For the talk, I built a physical model of my theory, incorporating a rubber tube filled with mercury which I jiggled up and down to represent fluctuating 'drive'. The rubber tube was attached to the bottom of a vertical glass tube, into which were let three electrical contacts at different depths, representing 'thresholds'. Mercury is an electrical conductor, so when the jiggling column hit any of these contacts (the 'drive' exceeded the 'threshold') a circuit was completed. Obviously, if mercury was in contact with any electrode, it was automatically in contact with all lower electrodes too, which captured the major assumption of the model. I implemented the rules of the model by means of a noisy system of clattering electromechanical relays switching on coloured lights to represent pecks at different colours. The whole Heath Robinson[1] affair was calculated to bring the house down, just as, at an earlier Ethological Conference in Oxford, a spoof hydraulic simulation devised by Desmond Morris, Aubrey Manning and friends reputedly had. How I managed to transport it from Oxford to Zurich evades my

---

[1] American: Rube Goldberg.

memory, and indeed my comprehension. There's not a chance that today's airport security would allow anything remotely like it through, bristling as it was with amateurishly soldered wires, relays, batteries and mercury.

Alas, just as I was about to go on the big stage for my first ever conference performance, something went wrong and my contraption wouldn't work. In a sweat of panic and unable to think straight, I was kneeling on the floor outside the theatre, frantically tinkering, when I suddenly became aware of an amused Austrian accent barking out peremptory orders at great speed behind me. The rapid-fire, stuttering voice told me exactly what to do. As in a dream I obeyed – and it worked. I turned to look at my saviour, and beheld Wolfgang Schleidt, whom I hadn't previously met although I knew who he was. Without any prior knowledge of what my infernal machine was supposed to do, this rising star of continental ethology had come upon my panic, instantly sized up the problem and dictated the solution to me. I have been grateful ever since to Dr Schleidt, who had, as I later learned without surprise, a reputation for technical ingenuity. I bore my strange device up into the theatre and at the end of my talk its spluttering coloured lights and air of Heath Robinson amateurishness received a reception little short of an ovation. Thank you, Wolfgang Schleidt: and for more than saving my blushes. For in the audience that day was the handsome figure of George Barlow, rising star of American ethology, and he was sufficiently impressed by my talk to get me invited to become an assistant professor at the University of California at Berkeley, without an interview or submission of a curriculum vitae: my first proper job.

But that was to come later. Meanwhile, back in Oxford, Niko Tinbergen had a sabbatical leave in 1966 and he invited me to take over the undergraduate lectures on animal behaviour for that year. He offered me his lecture notes, but I decided instead to develop my own lectures from scratch. Because this was the first course of lectures I ever gave, I wrote them out pretty fully. I thought I had lost them long ago but, to my surprise, they turned up in a cardboard box in the basement of my house while I was writing this memoir, and it was quite interesting for me to read them forty-six years on – especially the lecture on social behaviour, because it so clearly demonstrates both the central message and the style of *The Selfish Gene*, despite having been written a full ten years before that book.

In 1964, the *Journal of Theoretical Biology* had published two long and rather difficult mathematical papers by W. D. Hamilton, a young graduate student at the University of London whom none of us knew at the time, although he was later to become a close colleague. Mike Cullen characteristically recognized the importance of Hamilton's papers before almost anyone else in the world except John Maynard Smith, and he expounded them one evening to the Bevington Road group. Mike's enthusiasm was infectious, and I was immediately fired up, to the extent that I wanted to explain Hamilton's ideas to the undergraduates in my stand-in lectures on animal behaviour.

Hamilton's theory, now often labelled 'kin selection' (Maynard Smith's name, not Hamilton's own), follows directly from the neo-Darwinian 'Modern Synthesis' – directly, in the sense that kin selection is not an extra, not an addition bolted on to the neo-Darwinian synthesis: it is a necessary part of the synthesis. You cannot divorce kin selection from neo-Darwinism, any more than you can divorce the Pythagorean theorem from Euclidean geometry. A field biologist attempting to 'test' kin selection is in the same position as Pythagoras setting forth with a ruler seeking triangles to measure.

The neo-Darwinian synthesis, as opposed to Darwin's own version of his theory, is centred on the gene as the unit of natural selection. Genes are discrete entities that can be *counted* in a population, more or less ignoring the fact that they are in fact sitting in the cells of organisms. Each gene has a *frequency* in the 'gene pool', which is approximated by the number of reproductive individuals possessing it. Successful genes are those whose frequency increases at the expense of unsuccessful alternatives, which decrease in frequency. Genes that cause an animal to be good at caring for its offspring tend to increase because they are carried in the bodies of the offspring cared for. Hamilton realized (as Fisher and Haldane had too, sort of, but they didn't make much of it) that offspring are not the only category of relatives who share genes, and who therefore might be beneficiaries of evolved caring.

Hamilton derived a simple rule (now called Hamilton's Rule): any gene 'for' altruism towards kin will tend to spread through the population, if the cost to the altruist C is less than the benefit B to the recipient devalued by the degree of relatedness r between the two. The degree of relatedness r is a proportion (i.e. a number between 0 and 1) which Hamilton showed how to calculate (its exact meaning is hard, though not impossible, to explain intuitively).[1] Between full siblings, r is 0.5. Between uncles and nephews it is 0.25, between first cousins 0.125. Hamilton had a special interest in the social insects, and he made brilliant use of his theory of kin selection in explaining how ants, bees, wasps and (in a rather different way) termites evolved their remarkable habits of social altruism.

A typical underground nest of ants is a factory for propagating genes and spreading them around the countryside. The genes are churned out from the factory packaged up inside the winged bodies of young queens and males. These flying ants (which you might not realize are ants because of their unfamiliar wings) erupt from holes in the ground and fly up to mate on the wing. During her mating flight, each female (young queen) collects a lifetime's supply of sperm, which she will store inside her body and eke out over the course of a long life. Laden with sperm, the mated female flies off and settles down to dig a

---

[1] The clearest explanation is given by my Oxford colleague and sometime graduate student Professor Alan Grafen, 'A geometric view of relatedness', in R. Dawkins and M. Ridley, eds, *Oxford Surveys in Evolutionary Biology*, vol. 2 (Oxford, Oxford University Press, 1985), pp. 28–89.

hole and found a new nest. In some species, she bites or breaks off her wings, as she will need them no more in her role as subterranean queen.

Most of her offspring will be sterile workers, but the important children from the point of view of gene propagation are the young (winged) queens and males. Workers (all female in the case of ants, bees and wasps; male and female in the case of termites) normally have no prospect of passing on their genes through offspring, and devote their efforts to feeding and caring for their fertile collateral relatives, young queens and males, their siblings or nieces, for instance. A gene that makes a sterile worker care for a sister who is destined to become a queen can pass into future gene pools, carried there in the body of the young queen. The caring behaviour may never be expressed by the young queen herself, but the gene for that behaviour will be passed on to her worker daughters, who will consequently care for young queens and males who can pass it on.

The social insects are just a special case. Hamilton's Rule applies to all animals and plants, whether or not they in practice care for relatives. If they don't care for relatives, the reason will be that the economic costs and benefits in Hamilton's Rule (the *Bs* and *Cs*) don't pan out in such a way as to favour it, in spite of the fact that the coefficient of relatedness *r* may be high. And – a point that is often misunderstood, even by professional biologists – individuals care for their offspring for the same reason that elder siblings care for younger siblings (when they do): in both cases they share the genes for caring.

As I said, I was fired with enthusiasm when Mike Cullen introduced us to Hamilton's brilliant ideas, and I very much wanted to try my hand at explaining them in my own way, in the lectures that I was to give as understudy for Niko Tinbergen. I was diffident about departing so far from Niko's message and substituting my own rhetoric – about 'selfish genes' inhabiting a succession of mortal bodies, to be thrown away in the genes' relentless march into the future. Seeking

Natural selection acts directly on phenotypes, but it will affect
evolution only insofar as phenotypic xxxxxxxxxxxxxx differences
are correlated with genetic differences. The important effect of
natural selection is therefore on genes.

Genes are in a sense immortal. They pass through the generations,
xxxxxxxx reshuffling themselves each time they pass from parent to
offspring. The body of an animal is but a temporary resting place
for the genes; the further survival of the genes depends on the
survival of that body at least until it reproduces, and the genes
pass into another body. The structure and behaviour of the body
are to a large extent determined by the genes - the genes build
themselves a temporary house, mortal, but efficient for as long as
it needs to be. Natural selection will favour those genes which
build themselves a body which is most likely to succeed in passing

lovely stuff!

handing themxdown safely to the next generation, xxx a large number
or replicas of those genes.

To use the terms "selfish" and "altruistic" then, our basic
expectation on the basis of the orthodox neo-Darwinian theory of
evolution, is that Genes will be "selfish".

~~But this mean that individuals will be selfish? Not necessarily,
though it does mean that we must be very suspicious of expressions
like "the good of the species." There are two main ways in which
individual altruism~~

(This gives us the most important difference between individuals and
social groups. If an individual body is a colony of cells, it is
a very special kind of colony, because all those cells are genetically
identical. Every xxxxxxxxxxxxx somatic cell, muscle, bone, skin,
brain etc., contains the same complement of genes. Furthermore the
reproduction of all the genes in these somatic cells is limited to
the life-span of the body. Only the genes in the germ cells will *may*
survive. The other cells are built by the genes simply to ensure the
survival of the xxxxxxxxxx identical genes in the germ cells.
In say a fxxx colony of gulls, the individual birds all contain
different xxxxx sets of genes (except identical twins), and because
of the arguments given above, we shall have to think very carefully
about whether we should expect altruism between individuals. Only
in the social insects where the workers are sterile and very closely
related, do we have a social group that is really comparable with
the many-celled body. We will return to this later.)

If genes are selfish then, how can individuals evolve altruism?

reassurance, I showed my typed-out lecture to Mike Cullen, and seeing again his handwritten *marginalis* today reminds me that I was hugely encouraged by it at the time (see the facsimile opposite). It was Mike's 'lovely stuff' that emboldened me to persist with my plan to lecture on this topic and in this style. And I suppose this could be said to be the moment of conception of *The Selfish Gene*, born ten years later. My lecture notes even contained the phrase 'genes will be selfish'. I'll return to this later when I come to the book itself.

In the summer of 1967, in the tiny Protestant church of Annestown on the south coast of Ireland, where her parents had a holiday cottage, I married Marian Stamp, a member of Niko Tinbergen's group of graduate students, who later became his successor but one as Oxford's Professor of Animal Behaviour and the world's leading authority on the experimental science of animal welfare. I was by now committed to taking up the offer of an assistant professorship at the University of California at Berkeley. Niko was confident of Marian's ability to continue her doctoral research there, with a minimum of long-distance supervision from him, and indeed his confidence was well justified. We had a brief honeymoon driving around Ireland in a hired car. Marian had to drive because I had forgotten my licence, and we had an awkward moment when the car hire clerk discovered she was a 'graduate' (it seems that graduates had a poor track record). Almost immediately after the honeymoon we set off for San Francisco, where we were met at the airport by the ever-kind George Barlow. A new life in the New World had begun.

WEST COAST
DREAMTIME

B ERKELEY in the late 1960s was politically seething, and the politics of Telegraph Avenue, and Haight-Ashbury across the bay in San Francisco, were to dominate our two years there. Lyndon Johnson, who might otherwise have been remembered as a great reforming president, was mired in the disaster that was the Vietnam War – inherited from Kennedy. Just about everyone in Berkeley was against the war, and we joined them – in marches in San Francisco, in tear-gassed parades in Berkeley, in demonstrations, classroom disruptions and sit-ins.

I am proud of my part in protesting against American involvement in Vietnam, proud of having worked hard in the anti-war campaign of Senator Eugene McCarthy, less proud of some of the other political movements in which I was involved. The most memorable of these concerned the surreal episode of the 'People's Park' (fictionalized by David Lodge as the 'People's Garden' in his campus novel *Changing Places*). The People's Park campaign was an attempt (ultimately successful, as I discovered when I revisited Berkeley on a filming trip recently) to take over for public recreation a piece of waste ground owned by the university and intended for building. With hindsight it was a trumped-up excuse for radical political activism for its own sake, trumped up by anarchist student leaders cynically manipulating the gentle 'flower-power' 'street people'. The radical student leaders and the infamous Governor Ronald Reagan ('Ronald Duck' in David Lodge's novel) gleefully played into each other's hands, each mining the situation to enhance their following among their respective constituents, and each probably knowing exactly what they were doing. And I, together with most of the younger faculty of the university, played right into their hands. We demonstrated, sat in, ran from the tear gas, wrote outraged letters to the newspapers (my first letter to *The Times* was on the subject) and cheered as the street people stuck flowers down the rifle barrels of the bewildered and rather scared young National Guardsmen. Honesty compels me to admit a frisson of exhilaration – of which I am now quietly ashamed – at having been tear-gassed and (very slightly) endangered.

I try to peer into my own state of mind in my twenties in Berkeley as honestly as I can. I think what I see there is a kind of youthful excitement at the very idea of rebellion: a Wordsworthian 'Bliss was it in that dawn to be alive / But to be young was very heaven'. A student called James Rector was shot dead by an Oakland policeman. It was right to march in protest against that, and with hindsight that seems to have justified, in our minds, our decision to march for the People's Park in the first place. But of course it didn't justify it at all, not in itself. The decision to march for the People's Park required completely separate justification.

We, the younger faculty, convened meetings where we tried to bully our colleagues into cancelling their lectures in solidarity with the activists – and I use the word 'bully' advisedly, for I have seen the same thing more recently on the internet in the form of 'cyberbullying' by radical activists powerful enough to act as a kind of thought police, just as I saw the same thing at school when willing accomplices would rally around a playground bully. I remember with particular regret a faculty meeting at Berkeley when a decent older professor was reluctant to cancel his lecture and we tried to vote to force him to do so. With remorse I salute his courage, and that of an even older professor whose hand was the sole one raised in support of his colleague's right to fulfil what he perceived as his duty to give his scheduled lecture. As with Aunty Peggy, as with the Chafyn Grove equivalent, I should have stood up against the bullies. But I didn't. I was still young, but not all that young. Should have known better.

Mention of radical politics and the street people brings forth a revealing memory, revealing of a sea-change in social mores. I was walking along Telegraph Avenue, axis of Berkeley's beads-incense-and-marijuana culture. A young man was walking ahead of me, dressed in the insignia of the flower-power generation. Every time a young woman passed him, walking in the opposite direction, he would reach out and tweak one of her breasts. Far from slapping him, or crying 'Harassment!', she would simply walk on by as if nothing had happened. And he would proceed to the next one. Today I find this almost impossible to believe, but it is a very secure memory. His demeanour did not appear especially lascivious, and his action was evidently not taken by the young women as the gesture of a male chauvinist pig. It seemed all of a piece with hippiedom, with the laid-back, peace-and-love atmosphere of sixties San Francisco. I am very glad to say that things have changed. Today's counterparts in age and class of that young man, and the young women he molested (as we should now say), would be among those most strongly outraged at behaviour which was then the norm for that age, class and political persuasion.

In spite of all the politics, I did an adequate job as a junior (indeed, exceptionally young) assistant professor. George Barlow and I shared the lectures on animal behaviour, and I included the 'selfish gene' lecture that I had initiated at Oxford. I like to think that the students at Oxford and Berkeley in the late sixties may have been the very first undergraduates in the world to hear of the new ideas that were to become fashionable, in the seventies and thereafter, as 'sociobiology' and 'selfish genery'.

Marian and I were made to feel very welcome at Berkeley, and we made good friends there. As well as George Barlow, these friends included David Bentley the neurophysiologist, Michael Land, now the world's leading authority on eyes throughout the animal kingdom, and Michael and Barbara MacRoberts, who later came to Oxford as spirited additions to the Bevington Road circle, as did the gently sardonic David Noakes, who was George Barlow's leading graduate student during my Berkeley years. George hosted a weekly ethology seminar for interested graduate students at his house in the Berkeley Hills, and those evening meetings recaptured for Marian and me something of the wonderful atmosphere of Niko's Friday evenings at Oxford.

I had never been to America before, and I did find some things bewildering. At my first meeting of the Zoology faculty, everyone spoke almost entirely in numbers. Who's doing 314? No, I'm doing 246. Nowadays the English-speaking world knows that Xology 101 means (sometimes patronizingly or even derisively) a freshman's introduction to Xology. But all that numerology was perplexing to me when I first arrived. And who, today, doesn't understand the verb 'to major'? But I recall reading an American campus novel and getting a little fed up with the twittering of sophomores and juniors and seniors when, like a breath of fresh air, 'An English major came into the room.' Aha, I thought, my mind immediately filled with visions of riding breeches and moustaches, a real character at last.

Marian and I both worked hard at our research. And we talked and talked and talked to each other about our shared scientific interests, on walks in Tilden Park up in the Berkeley Hills, on drives around the beautiful California countryside, at meals, on shopping expeditions over the Bay Bridge to the city of San Francisco, all the time. The atmosphere of our discussions was that of a mutual tutorial, each learning from the other, exploring arguments step by step, moving one step backwards, two steps forwards. The mutual tutorial is something I now strive to achieve in public discussions with colleagues, often filmed for my website or put out on DVDs. Those discussions with Marian were to be the basis of the joint experiments that we did later, after returning to Oxford.

My research at Berkeley was a continuation of my work on chick pecking. My doctoral research had been very Popperian. It made precise predictions about total quantities of choices made in a fixed time. But the model had always begged for more exact observational testing, using precise sequences of pecks as they happened, rather than total numbers of pecks per minute. In Berkeley I turned to the exact sequence, building a new apparatus which, unlike my Oxford one, was capable of recording exactly when each peck happened, rather than just counting pecks per minute. I also increased the peck rate by rewarding each peck with a blast of infra-red heat, which the chicks liked. They were rewarded equally, regardless of which key they pecked, but they still showed colour preferences and they still seemed to be choosing on the basis of the Drive Threshold Model. The pecks were recorded on magnetic tape using a sophisticated and expensive piece of equipment that had been built for George Barlow, known as the Data Acquision System – so called because of a typographical error in the word 'Acquisition' on the label.

One simple expectation of the Drive Threshold Model is that there should be long runs of pecks at the preferred colour (when the drive was above the threshold for that colour only), interspersed with runs of indifference (when the drive was above two thresholds). There should never be significant runs of pecks at the less preferred colour. Following the Attention Threshold Model, I expected that indifference to *colour* really meant preference for one *side*. Since the apparatus was programmed to present each colour to alternate sides, changing after every peck (with occasional random variations), I predicted sequences such as you see in the picture on this page, which represents real data from one particular experiment, and seems to confirm the prediction very nicely.

Of course, this picture is no more than an anecdote, one out of many experiments. I did statistical analyses to substantiate this prediction and several other predictions, using data from large numbers of experiments. The predictions of the Attention Threshold version of the Drive Threshold Model were upheld.

Some time during our second year at Berkeley, Marian and I were visited by Niko and Lies Tinbergen. Niko wanted to persuade us to return to Oxford, where he had obtained an attractive research grant to offer me, and where Marian could write up her doctoral research, which, as Niko could see, was going well at Berkeley. The Tinbergens returned to Oxford, leaving us to think about the offer. We decided to accept it, but meanwhile Niko had written of a new opportunity. Oxford had decided to appoint a new university

Successive columns represent sequences of pecks. Open circles represent preferred colour. Note how runs of pecks at the preferred colour (open circles) alternate with runs of pecks at one side or the other – perhaps the nearest side – amounting to runs of alternation between colours.

From M. and R. Dawkins, 'Some descriptive and explanatory stochastic models of decision-making', in D. J. McFarland, ed., *Motivational Control Systems Analysis* (London, Academic Press, 1974), pp. 119–68.

lecturer[1] in animal behaviour linked to a fellowship at New College, and Niko wanted me to apply. This teaching job would not preclude the research grant he had earlier promised me. I agreed to apply for the lectureship, and Oxford flew me over for the interview.

It was a magical trip, with what seemed like all before me. Music stamped the memory: Mendelssohn's Violin Concerto, which I listened to on the plane, spellbound by the Rocky Mountains below and by exciting prospects ahead. Oxford put on its very best performance, which is the Maytime blossoming of cherry and laburnum all along the Banbury Road and Woodstock Road. New College, too, played its golden fourteenth-century part and I was happy, my exuberance not dimmed when I was greeted on arrival by the news that Colin Beer, former member of the Oxford ABRG and now a professor at Rutgers University in New Jersey, had put in an unexpected late application for the lectureship. Even the fact that Niko had excitedly switched his allegiance from me to Colin didn't upset my optimistic mood. If Niko had decided that Colin was a better bet, that was good enough for me. I would still have the research position and, as I told the interviewing committee, if Colin were there in Oxford too, so much the better. They did indeed give the job to Colin, and I took up the research grant.

---

[1] The American equivalent would be 'assistant professor going on associate professor'.

COMPUTER FIX

This picture of me with my parents, taken at a family wedding (my sister Sarah was a bridesmaid, so not with us at this point), unfortunately does not show the bright red of the cap I wore as a Chafyn Grove schoolboy. In my first term at Oundle I don't think I was as happy as I looked for the camera. One of the best things about that school was Ioan Thomas, caught here encouraging an appetite for wonder at the living world.

Life at Over Norton: the battered Land Rover with which we crashed through rough country; Wessex Saddlebacks landscaping the equally rough country that was then the garden of our cottage, c.1951; my inventive father standing proudly by his patent pasteurizer; haymaking with the little grey Fergie.

In the summer holidays I earned my keep bale-sledging.
Below: Following in father's footsteps: moving some
family heirloom or other.

Above: Peter Medawar before the stroke that changed his life. *Top*: Niko in his element, at Ravenglass painting dummy eggs. *Right*: George Barlow, my Berkeley friend and guide, came to Oxford later on sabbatical, and we went punting together on the Cherwell. (That's not John Lennon, it's Tim Halliday the newt expert.)

Left, top: 'The deeply intelligent eyes, understanding what you meant even before the words came out . . . sceptical, quizzical tilt of the eyebrows, under the untidy hair.' Mike Cullen, sadly missed mentor to so many. *Left*, middle: What to peck? Chicks that had never seen overhead light. *Above*: Hunting the Surrey puma; intrepid explorer scours the landscape for wild beasts. *Below*: Wild beasts or frightened boys? California National Guard raggedly confront the Peace People in Berkeley.

Top left: Cricket commentary: Ted Burk and I recording behaviour with microphone and Dawkins Organ. *Middle left*: The Animal Behaviour Research Group after the move from Bevington Road. Marian is far left. I am slightly right of centre. *Bottom left*: A PDP-8 computer like the one that fed my addiction in 13 Bevington Road. *Bottom right*: Professor Pringle and (left to right) his colleagues, E. B. Ford, Niko Tinbergen, William Holmes, Peter Brunet, David Nichols. *Left*: Danny Lehrman (standing) and Niko Tinbergen (right) settling their differences. *Below*: Niko in his element again: will the ash fall from his cigarette before he finishes the shot?

Deep thought. *Above*, Bill Hamilton and Robert Trivers wrestling with a problem during Bill's visit to Harvard; *far left*, the endlessly invigorating John Maynard Smith in his beloved garden. *Left*, *The Selfish Gene* with the original Desmond Morris cover. *Below left*, with the tall, thoughtful, Lincolnesque George Williams. *Below right*, 'I must have that book!' Michael Rodgers, K-selected science publisher.

Marian and I left Berkeley in 1969 with mixed emotions, and it has remained for me a place of magic and pilgrimage: a dreamtime of lost youth, of clever and friendly colleagues, of clear, bright sunshine alternating with cooling mist over the Golden Gate, of the dawn-fresh scent of pine and eucalyptus, of flower children with decent and sincere, if naive, liberal values.

We crated and shipped our few belongings from our Berkeley apartment and drove right across the continent in our old, cream-coloured Ford Falcon station wagon, thickly encrusted with anti-war slogans and Eugene McCarthy election stickers, to New York. By prior arrangement, we sold the Ford on the quayside (amazingly, our pre-arranged customer, who had made his own laid-back Berkeley-style way to New York, showed up on time), boarded the liner *France* for Southampton, and prepared to resume our life at Oxford with many of our old friends still there and Colin Beer newly arrived. In the event, Colin preferred to spend his time in New College and was scarcely seen in the department, much to everyone's disappointment. He stayed only a year. Danny Lehrman – the same Daniel S. Lehrman whose theoretical critique had so influenced my doctoral thesis – had shrewdly kept Colin's position at Rutgers warm for him, and when it became clear that Oxford couldn't find a position in medieval French to match the professor-ship his wife held in America, Colin decided to return. Once again, the lectureship in animal behaviour was advertised, once again the long-suffering New College agreed to associate a fellowship with it, and once again Niko urged me to apply. Along with a shortlist of others I was again interviewed by two committees: a university committee chaired by Laughing John Pringle, and a college committee chaired by the genuinely laughing, almost preternaturally genial Warden, Sir William Hayter, former British Ambassador in Moscow.

This time I really wanted the job, and this time I got it. The news came as Marian and I were anxiously waiting with friends in an Indian restaurant in Oxford. Suddenly, we heard the sound of Mike Cullen's motor scooter as it pulled up outside. Mike burst into the restaurant, wordlessly pointed both index fingers at me, and vanished as quickly as he had arrived. I had got the job. With hindsight I don't think I should have got it at the time, given that the chief competition was the wildly brilliant Juan Delius, although I like to think that I grew into it and was worthy of it in the end. Juan was a dear friend and mentor, an immensely clever, knowledgeable and funny German-Argentinian. He once defined Argentinian humour for me: 'They enjoy slapstick, but if somebody slips on a banana skin it's only *really* funny if he breaks his leg.' The board at 13 Bevington Road was frequently adorned with wonderful notices in Juan's unique brand of English: 'What bastard has absconded my oles?' (Who has taken my stencil for drawing circles of different sizes?)

The life of a tutorial fellow of an Oxford college is in many ways a charmed one. I got a room in a glowing, oolitic limestone medieval building surrounded by famously beautiful gardens; a book allowance, a housing allowance, a research allowance; and free meals (though not free wine, contrary to envious rumours) in the stimulating and entertaining company of leading scholars of every subject except my own. The stimulating scholars of my own subject were to be found in the Zoology Department – where I spent the majority of my time.

I was introduced to the strange world of high-table conversation. After dinner there would sometimes be occasion to bring out the Senior Common Room Betting Book – either to record a new wager or to browse the old ones, all written in the same affected style as high-table conversation itself. Here's a brief sample, going back to the 1920s when the most assiduous betting man was the eccentrically brilliant G. H. Hardy, whose Lewis-Carroll-like mathematician's sense of humour seems to have infected his colleagues:

(7th Feb 1923) The Subwarden bets Prof Hardy his fortune till death to one halfpenny that the sun will rise tomorrow.

(6.8.27) Prof Hardy bets Mr Woodward 10,000 to 1 in halfpennies that he (Prof Hardy) will not be the next President of Magdalen, and Mr Woodward bets Prof Hardy 1 to 5,000 that he (Mr Woodward) will not be the next President of Magdalen.

(Feb 1927) Professor Hardy bets Mr Creed 2/6 to 1/6 that the New Prayer Book will go phut. Mr Smith, Mr Casson and Mr Woodward to adjudicate if necessary.

I'm amused that so obvious a value judgement could be subject to a bet. No wonder an odd number of adjudicators was needed.

Another bet even leaves the size of the stake up to later judgement:

(Dec 2 1923) Professor Turner bets the Steward of SCR a large sum that it would be a good thing to have a copy of the ABC (London) Railway Guide in SCR (Won by Prof Turner, A.H.S.)

(15 Feb 1927) Mr Cox bets Professor Hardy 10/- to 1/- that Rev. Canon Cox ('Fred') will not be the next Bishop of Nyasaland.

I love that parenthetic 'Fred'. Unfortunately the result of this bet was not recorded. I'd like to know whether 'Bishop Fred' presided over the see of my early home country. In failing to settle the question for me, Google did turn up the fact that a nineteenth-century Bishop of Nyasaland was Charles Alan Smythies – very probably related to the seven generations of my Smythies ancestors who were vicars.

(11th March 1927) Mr Yorke bets Mr Cox 2/6d that no verse occurs in the Gospel according to St Matthew the literal interpretation of which justifies or advocates self-castration. Won by Mr Cox.

(October 26th 1970) Professor Sir A Ayer bets Mr Christiansen that the Chaplain will be unable, if challenged without warning, to repeat twelve of the thirty nine articles to be found in the Book of Common Prayer. The stake to be one bottle of claret.

(24th Nov 1985) The Chaplain bets Dr Ridley a bottle of claret that Dr Bennett will be wearing a clerical collar at dinner on the occasion of the visit of the Bishop of London. (Chaplain won.)

(4th August 1993) Mr Dawkins bets Mr Raine £1 that Bertrand Russell married Lady Ottoline Morrell. Adjudicator Mlle Bruneau. (Dawkins lost and paid, 20 years late.)

Bets like the last one can't happen any more because it is so trivially easy for everyone to check such factual questions on their smartphones without rising from their Senior Common Room armchairs. Even then, it was scarcely necessary to appoint an adjudicator for a purely factual matter.

Back to 1970, when I was twenty-nine and newly returned to Oxford. The singing Elliott had gone the way of all silicon, but Moore's Law and the research grant that had lured me back to Oxford the previous year made it possible for me to have my 'own' computer, a PDP-8, which exceeded the Elliott in every respect except physical size and price. Also in accordance with Moore's Law (which was already going strong in those days), it was functionally much smaller yet physically larger than a modern laptop, and ludicrously it had a log book in which you were supposed to record every time you switched it on (of course I didn't). It was my pride and joy and a valued resource – together with me as sole programmer for everybody in 13 Bevington Road (which took its toll on my time). Now my addiction to computers could really take off, and I no longer had to indulge it nocturnally, as during my shameful affair with the Elliott 803.

Previously I had used only high-level compiler languages – human-friendly languages, which the computer translates into its own binary machine language. But now, in order to use the PDP-8 as a research tool, I had to master its 12-bit machine language, a task into which I threw myself with zest. My first machine-code project was the 'Dawkins Organ', a system for recording animal behaviour – equivalent to George Barlow's 'Data Acquision' apparatus but much much cheaper. The idea was to make a keyboard which an observer could use in the field, pressing buttons to indicate actions by an animal. Key-presses would be recorded on a tape recorder, which would later automatically tell the computer exactly when each action by the animal occurred.

My keyboard literally was a makeshift electronic organ, with each key playing a different note (inaudible except to the tape recorder). This part would be easy to make. The box would contain a simple two-transistor oscillator, the pitch of whose note was tuned by a resistance. Each key on the organ would connect a different resistor and hence play a different note. The observer was to take the organ into the field and watch an animal's behaviour like a work-study officer, pressing a specific key for each behaviour pattern. A tape recording of the sequence of notes would then constitute a timed record of the animal's behaviour. Theoretically, a person with a good ear listening to the tape could detect which key had been pressed, but this wouldn't be helpful. I needed to cast the computer in the role of person-with-good-ear. It could have been done electronically, with a series of tuned frequency-detectors, but that would have been an expensive hassle. Could the same feat – perfect pitch sensitivity in the computer – be achieved in software alone?

I was discussing the problem with my computer guru of the time, Roger Abbott, a clever engineer (and coincidentally organist) employed on the large research grant of Professor Pringle. Roger came up with an inspired suggestion. Every musical note has a characteristic wavelength which signifies its pitch. Computers are – and were, even in those days – so fast that the interval between wave crests within a musical note could be measured in hundreds of program cycles. Roger suggested that I should write a machine-code program to time the intervals between wave peaks: write, in other words, a little routine to act as a high-speed clock, counting how many jumping-back program loops it could walk through before being interrupted by the next wave crest (which, when averaged over lots of wave crests, tells it the pitch of the note). When a note ended (when more than a critical time elapsed since the last wave peak) the computer should make a note of the time and then wait for the next organ note. The computer's clocking loop, in other words, would be used not only to recognize the pitch of a musical note but also, on a hugely longer timescale, to measure the passage of time between notes.

Having got this central routine working, the rest was just a matter of slogging through the writing and debugging of a user-friendly program. This took rather a long time, but it ended successfully. The Dawkins Organ was a viable product. The user of the organ began each session by playing a scale on the tape – all the notes on the organ in ascending order of pitch. The taped scale would then be used to 'calibrate' the software – 'teach' the computer the repertoire of notes it would be asked to recognize. After the calibration scale was ended (by hitting the first note for the second time), all further notes on the tape would designate behavioural events. This calibration system had the advantage that the organ did not have to be carefully tuned. Any set of notes that were sufficiently distinct from each other would do, because the computer quickly learned which notes to listen out for.

So, when the tape was brought home and played into the computer, the computer knew exactly what the animal had done, and when. The nucleus of the program was the timing loop, but it was embedded in a substantial quantity of code to punch out, on paper tape, the names of all the behaviour patterns and the exact times when they occurred.

I published a paper on the Dawkins Organ,[1] and made the software available free of charge. Over the next few years Dawkins Organs were used by numerous members of the Oxford ABRG, and by some ethologists elsewhere in the world, for example in the University of British Columbia.

My addiction to machine code programming took me in a downward spiral. I even devised my own programming language, BEVPAL, with its own programming manual, a somewhat otiose exercise since the language was used by nobody except myself and, briefly, Mike Cullen. Douglas Adams amusingly satirized computer addiction of exactly the kind that hit me. The target of his satire was the programmer who had a particular problem X, which needed solving. He could have written a program in five minutes to solve X and then got on and used his solution. But instead of just doing that, he spent days and weeks writing a more general program that could be used by anybody at any time to solve all similar problems of the general *class* of X. The fascination lies in the generality and in the purveying of an aesthetically pleasing, user-friendly product for

[1] R. Dawkins, 'A cheap method of recording behavioural events for direct computer access', *Behaviour*, 40 (1971), pp. 162–73.

the benefit of a population of hypothetical and very probably non-existent users – not in actually finding the answer to the particular problem X. Another symptom of this kind of geekish addiction is that every time you solve a local problem and make the computer jump through yet another hoop, you want to rush out into the street and drag someone in to show them how elegant it is.

The productive camaraderie that a small building like 13 Bevington Road fosters came to an end around this time, and the animal behaviour group moved to the new zoology/psychology building, the huge, battleship-like horror on South Parks Road, then informally known as HMS Pringle after the ambitious Linacre Professor who persuaded the university authorities to build it – having failed to cajole them into building a pencil-thin skyscraper that would have disastrously overtopped Matthew Arnold's dreaming spires. I have mixed feelings about my part in later getting HMS Pringle officially named the Tinbergen Building, for it is widely deplored as the ugliest building in Oxford. It won an architectural award from the Concrete Society – enough said.

Around this time I published a short paper in *Nature*.[1] Every day hundreds of thousands of our brain cells die, and this was upsetting to me even at the age of twenty-nine. My Darwin-obsessed brain sought comfort in the idea that if the cell deaths were non-random, such apparently wholesale slaughter might be constructive, not purely destructive:

A sculptor changes a homogeneous lump of rock into a complex statue by subtraction, not addition, of material. An electronic data processing machine is most likely to be made by connecting components up in complex ways, and then enriching the connexions to make it even more complex. On the other hand, it could be con-

---

[1] R. Dawkins, 'Selective neurone death as a possible memory mechanism', *Nature*, 229 (1971), pp. 118–19.

structed by starting with extremely rich, even random intercon-
nexions, and then carving out a more meaningful organization by
selectively cutting wires.

. . .

The theory proposed here may seem fanciful at first. Further
reflexion shows, however, that its lack of verisimilitude is mainly a
consequence of the highly improbable postulate on which it rests;
namely, that brain cells are decreasing in numbers at a prodigious
rate daily. Because this postulate, however far-fetched, is an established fact, the present theory is not suggesting anything very
implausible in addition; rather the reverse, as it makes the process
seem less wasteful. All that is at issue is whether neurones die at
random, or selectively in such a way as to store information.

A curious little one-off, this paper is perhaps mildly interesting as an
early example of the kind of theory that later became fashionable
under the name – coined a year later, so I obviously didn't use it –
'apoptosis'.

Marian soon got her doctorate, and we began to collaborate on
research projects growing out of the many discussions – mutual
tutorials – from our Berkeley days. We planned a study that would
exemplify, and clarify, one of the fundamental concepts of the ethological school of animal behaviour studies, the Fixed Action Pattern.

Lorenz and Tinbergen and their school thought that much of
animal behaviour consisted of a sequence of little clockwork routines
– Fixed Action Patterns (FAP). Each FAP was thought to be like a piece
of anatomy, just as much a part of the animal's bodily equipment as,
say, the collar bone or the left kidney. The difference is that collar bones
and kidneys are made of solid material, whereas the FAP has a time
dimension: you can't pick it up and put it in a drawer, you have to
watch it play out in time. A familiar example of an FAP would be the
pushing movement that a dog makes with its snout when burying a
bone. These movements are identically replayed even when the bone

is on a carpet and there is no soil in which to bury it. The dog really does look like a (charming) clockwork toy, although the exact direction of the movement is influenced by the position of the bone.

Every animal has a repertoire of FAPs, like one of those dolls that you wind up by pulling a string and which then utters a saying randomly plucked from a finite repertoire. Once initiated, whichever saying is chosen goes through to completion. The doll doesn't switch messages halfway through. The decision of which of a dozen sayings to produce is unpredictable but, once taken, the consequences of the decision are followed through predictably as clockwork. That was the FAP doctrine in which Marian and I, as Tinbergenian ethologists, had been brought up; but was it a true reflection of reality? This was the question we wanted to answer – or, to be precise, the question we sought to re-express in terms that might make it answerable.

In theory, one could write down the continuous stream of animal behaviour as a sequence of muscular contractions. But if the FAP theory was right, the predictability of behaviour would render it a laborious waste of effort to write down every muscular contraction, even were it possible to do so. Instead, all we should need to do is write down the FAPs, and the sequence of FAPs would – on an extreme interpretation – be a complete description of that particular animal's behaviour.

But this would only work if FAPs really were equivalent to organs or bones – if it were true, in other words, that each pattern occurs as a whole, not breaking off part-way through or mixing with another pattern. Marian and I wanted to find a way to assess the extent to which this proposition was true. Both our doctoral theses had been concerned – in our two ways – with decision-making, and it was natural for us to translate the FAP problem into the language of decisions. In that language, the animal takes a *decision* to initiate an FAP; but, once initiated, the FAP goes on to its conclusion, with no further decisions until the end. At that point the animal's behaviour stream would enter a period of uncertainty, pending the next decision to initiate (and complete) an FAP.

We chose to study drinking in chicks as our example, and we hoped it would be representative.[1] Drinking in birds (other than pigeons and doves, which suck) is an elegant glissando of a movement, and it certainly gives a subjective impression of being initiated by a discrete decision, after which it always goes through to completion. But could we back up our subjective impression with hard data?

We filmed a side view of our chicks drinking, and then analysed the behaviour frame by frame to see if we could measure its 'decision structure'. We measured the position of the bird's head in successive frames of film, then fed the coordinates into the computer. The idea was to measure the predictability of the next frame, knowing the position of the head in previous frames.

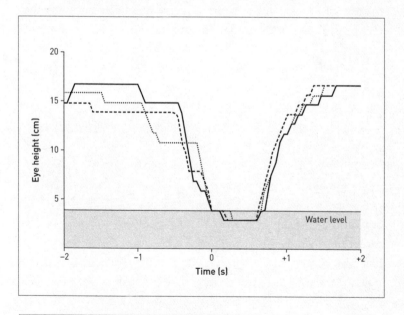

---

[1] R. and M. Dawkins, 'Decisions and the uncertainty of behaviour', *Behaviour*, 45 (1973), pp. 83–103.

The diagram shown on page 225 is a graph of eye height against time, for three drinks by the same chick, lined up (zero on the time axis) on the moment when the bill hit the water. You get a sense that from that moment on, indeed from just before it, the behaviour is stereotyped and predictable, but the early part of the downstroke is more variable and subject to decisions: decisions to pause and even (as we showed separately) to abort the drink.

But how should we measure predictability? The graph below shows one way. It represents a single drink in the same way as before. But each point on the graph of eye position has arrows attached to it. The length of the arrow signifies, for each frame of film, the likelihood (as totted up over all the drinks by all the chicks) that the eye height in the next frame will be lower, higher, or the same.

You can see that during the upstroke, when the bird is allowing the water to trickle down its throat, there is a high probability that the upstroke will continue its graceful curve in the upward

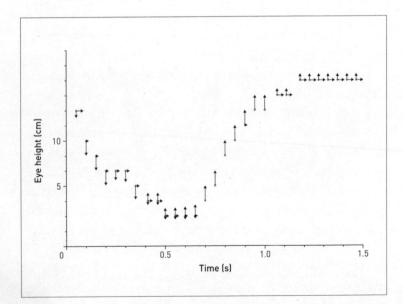

direction. A decision to perform an FAP is being carried out, with no further decisions during its course. But during the downstroke there is more unpredictability. For each frame of the downstroke, the height of the eye in the next frame is undecided between lower or the same, and there is even some likelihood that it might be higher – that is, that the drink will be aborted.

Could we use these arrows to compute an index of uncertainty or 'decisioniness'? The index we chose was based on information theory, devised in the 1940s by the inventive American engineer Claude Shannon. The information content of a message can be informally defined as its 'surprise value'. Surprise value is a convenient opposite of predictability. Classic examples are 'It is raining in England today' (low information content because no surprise) versus 'It is raining in the Sahara Desert' (high information content because surprising). For reasons of mathematical convenience, Shannon computed his index of information content in *bits* (short for 'binary digits'), by summing up the logarithm (base 2) of the prior probabilities that were open to doubt before the message was received. The information content of a penny toss is one bit, because the prior uncertainty is heads or tails – two equiprobable alternatives. The information content of a playing card's suit is two bits (there are four equiprobable alternatives and the base two logarithm of four is two, corresponding to the minimum number of yes/no questions you'd need to ask in order to establish the suit). Most real examples are not so simple, and the possible outcomes are usually not equiprobable, but the principle is the same and a version of the same mathematical formula conveniently does the trick. It was this mathematical convenience that led us to use the Shannon Information Index as our measure of predictability or uncertainty.

Once again we have a graph (overleaf) of eye height against time during a drink. The thin lines represent times of low predictability, or high probability of a decision intervening to change the future.

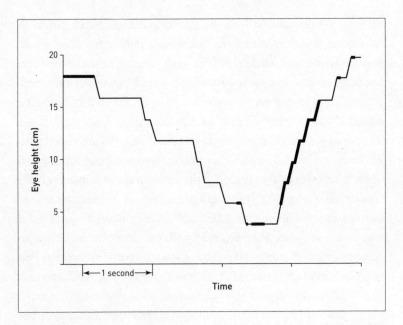

The thick black lines represent times of high predictability (information content less than an arbitrary threshold of 0.4 bits), during which a decision is being carried out and no new decision is expected. The upstroke is predictable once it starts, but the downstroke is not. The pause between drinks is predictable for the rather boring reason that the pause is most likely to continue into the next frame – it's hard to predict when the next drink will start.

As always, keep in mind that the particular behaviour, here drinking, is not of interest in itself. Drinking in chicks was a stand-in for behaviour generally, just as pecking was in my doctoral research. We were interested in the very idea of a decision and – in the case of drinking – whether we could identify moments of decision. We were trying to explore a way to demonstrate the very existence of a Fixed Action Pattern, rather than simply taking it for granted as ethologists were wont to do.

We adopted a different approach in our next project on decision-making, which was a study of self-grooming in flies. Ethologists often ask whether, if you know what an animal is doing now, you can predict what it will do next. Marian and I wanted to know whether, sometimes, you can predict what it will do in the more distant future *better* than you can predict what it will do in the more immediate future. This might be true, for instance, if behaviour is organized like human language. There are times when the beginning of a sentence predicts how the sentence will end better than it predicts the middle – which might contain any number of embedded adjectival or relative clauses, for instance. 'The girl hit the ball' is a sentence whose beginning demands something like the ending, whether or not there are embedded adjectives or adverbs or clauses in the middle: '**The girl** with red hair, who lives next door, vigorously **hit the ball**.'

We didn't find evidence of language-like grammatical structure in the grooming behaviour of flies (although see below). But we did find an interesting zigzag pattern in the way predictability decays over time: in other words, the immediate future may be less predictable than the (slightly) more distant future. I'll outline our research briefly here and not in detail, because it's a bit complicated.

Flies are not normally seen as beautiful, but the way they wash their faces and their feet is rather dear. Look next time a fly lands on you: you'll very probably see the behaviour. It may rub its front feet together, or wipe its great big eyes with them. It may rub the middle foot on one side against the hind foot on the same side, or clean its abdomen or wings with its hind feet. Somewhere inside that tiny head, decisions are spontaneously being generated, and a fair number of those decisions concern which bit of the body to clean next. The appeal of self-grooming behaviour for us was that the fly's choice of behaviour was unlikely to be externally stimulated. We presumed that external stimulation amounted to an ever-present need to keep clean – ever-present in the sense that, though important, it was unlikely to determine exactly when a particular grooming action would be chosen. Dirty wings would impair flight. Dirt would impair the highly sensitive tasting organs in the feet, which flies use to decide whether or not to stick out the tongue and eat. So cleaning is important. But presumably the decision about which bit to clean is not determined by the sudden arrival of a new piece of dirt. Rather, we suspected that these rapid, moment-to-moment decisions were internally generated by unseen fluctuations deep inside the nervous system.

We recognized eight distinct grooming acts, which we presumed would show up as FAPs if we had the time to do a frame-by-frame analysis like the one we did with the chicks drinking: FR (rub front feet together), TG (rub tongue between front feet), HD (wipe head with front feet), FM (rub one or other middle foot between front feet), BM (rub one or other middle foot between back feet), BF (rub back feet together), AB (wipe abdomen with back feet), WG (wipe wings with back feet). Using a Dawkins Organ, we recorded the sequences of these eight grooming acts, plus MV (move away) and NO (stand still, doing nothing).

The graph below shows the probability, given that the fly is now doing HD, that it will do FR next ('lag' = 1, probability very high), next but one (very low probability), next but two (high probability), next but three (low probability) etc. You can see that there is a

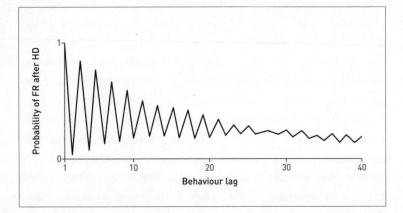

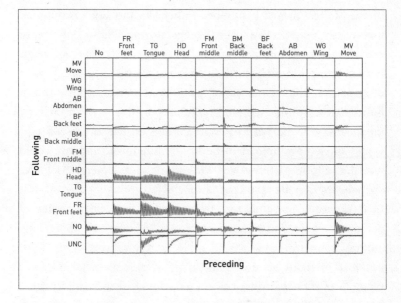

pronounced tendency to alternate, and also that there is a general die-away (as you'd expect) in predictability as we look to the more distant future – longer and longer 'lags'.

That picture was for the particular case of FR following HD. We plotted the same kind of graph for all the possible transitions, and put the graphs into a table (above).

You can see that many of the transitions follow the same zig-zag pattern, although some are exactly out of phase with each other. The bottom row (UNC) shows the uncertainty attached to predictions of the future, following each behaviour, calculated using the Shannon Information Index, in the same way as for the chick drinking study.

We also tried the experiment of using the human ear to identify patterns in animal behaviour. For this, we used a Dawkins Organ rendition of fly grooming behaviour, but eliminated the true intervals between musical notes. I told the computer to reduce all the intervals to a single standard short interval and then we simply listened to the 'music'. It sounded rather like 'modern' (as opposed to 'traditional') jazz. Also rather like the 'singing' Elliott computer of my juvenile insomniac dalliance – I suppose the comparison might be interesting. I thought the human ear might be a promising piece of apparatus to use in detecting patterns in animal behaviour, but didn't follow up the method; I merely report it here as an interesting curiosity. If the World Wide Web had existed in those days, I would surely have uploaded the fly washing music and you could now dance to it. As things are, I'm afraid those Dipteran Melodies are gone for ever, like the Lost Chord.

I cannot claim that our fly study, or the other studies of decision-making that preceded it, really tell us much about how animal brains work. I see them more as explorations of methods: not just methods of doing research on animal behaviour, but methods of *thinking*. Marian and I did a lot more work on the flies, but it's all published and I won't write any more about it here. It did, however, feed into my next big writing project: a long theoretical paper on 'Hierarchical organisation as a candidate principle for ethology'. This is the subject of a later section.

Meanwhile, in 1973, Niko Tinbergen won the Nobel Prize in Physiology or Medicine (jointly with Konrad Lorenz, his co-founder of ethology, and Karl von Frisch, the discoverer of the legendary bee dance). Just one year later, in 1974, Niko reached Oxford's mandatory retirement age of sixty-seven, and the university agreed to appoint a successor as Reader in Animal Behaviour. 'Reader' was a rather prestigious rank at Oxford, now, I think, fallen into desuetude in a move to bring the title of 'professor' into line with American custom by sprinkling it about more liberally – the rather unkindly dubbed 'Mickey Mouse professors'. I was very content where I was as lecturer, and had no ambition to apply for the job.

Most people thought of Mike Cullen as Niko's natural successor. Perhaps for that very reason, in order to make a clean break, the majority of the appointment committee went for David McFarland. As Hans Kruuk wrote in his biography of Tinbergen, 'one could hardly have found anyone more unlike Niko'. Though controversial in many quarters, David's appointment was in some ways an inspired one, at least if you take the view that a new appointment is an opportunity for a new departure. His science was highly theoretical, indeed mathematical. He brought to it the intuitions of a mathematician, and he surrounded himself with trained mathematicians and engineers who could do the algebra. The talk in the coffee room switched from gulls and sticklebacks in the field to feedback control systems and computer simulations.

Perhaps it was a microcosm of the way biology was changing. I was young and not yet set in my ways. 'If you can't beat them, join them' was my attitude. So I set to work to learn control theory from the engineers and mathematicians who now surrounded me. And what better way to learn it than hands-on? I again indulged my passion – or vice – for computer programming, and wrote a program for a digital computer ('my' PDP-8), enabling it to behave like an analogue computer. To this end, I invented yet another computer language, which I called SysGen.

Unlike the propositions in a conventional computer language like Fortran, which are executed sequentially, SysGen statements were executed 'simultaneously' – not *really* simultaneously, of course, because a digital computer does everything sequentially at bottom; but they could be written in any order. My task in writing the SysGen Interpreter program was to persuade the digital computer to behave as *if* the operations were simultaneous: a virtual analogue computer. As with an analogue computer, results were displayed as a set of graphs on an oscilloscope screen.

I'm not sure how useful SysGen was in practice, but inventing the language, and writing the Interpreter program for it, certainly helped me to understand not just control theory but also the integral calculus. It gave me a much better idea of what it means to integrate. I was mindful of my maternal grandfather's recommendation of *Calculus Made Easy*, by his old mentor Silvanus Thompson (who, as quoted earlier, was fond of saying, 'What one fool can do, another can'). Thompson introduces his explanation of integration with another phrase that has stuck in my brain: 'So we had best lose no time in learning how to integrate.' I had only half understood integration in Ernie Dow's lessons, and SysGen gave me the sort of hands-on experience that assists comprehension.

Similar in intention, but much easier and less time-consuming, was my attempt to understand Chomsky-style linguistics by the hands-on method. I wrote a computer program to generate random sentences, which might not have been very meaningful but were always scrupulously grammatical. This is easy – and that very fact is instructive – given that your programming language allows procedures (subroutines) to call themselves *recursively*. This was true of Algol-60, the programming language that I favoured at the time under the influence of Roger Abbott, who had brilliantly succeeded in writing an Algol compiler for the PDP-8. Algol subroutines could call themselves, unlike the contemporary version of that old workhorse of scientific programmers, IBM's Fortran language. Mention of Fortran reminds me of a nice in-joke told by Terry Winograd, pioneer of artificial intelligence. Some time in the 1970s I attended a fascinating conference in Cambridge on the state of the art of artificial intelligence programming, and Winograd was the star lecturer. At one point he gave vent to a wonderful piece of sarcasm: 'Now, you may be one of those who says, "Fortran was good enough for my *grandfather*, it's good enough for me."'

Given that your programming language allows procedures to call themselves recursively, writing programs to deliver correct grammar is remarkably – elegantly – easy. I wrote a program that had procedures with names like NounPhrase, AdjectivalPhrase,

PrepositionalClause, RelativeClause etc., all of which could call any other procedure, including themselves, and it generated random sentences like this one:

> *(The adjective noun (of the adjective noun (which adverbly adverbly verbed (in noun (of the noun (which verbed))))) adverbly verbed)*

Parse it carefully (as I have done here using brackets, although the computer didn't generate them but left them implicit) and you'll see that it is grammatically correct although not exactly dripping with information. It makes syntactic but not semantic sense. The computer could easily inject semantics (if not sense) by replacing 'noun', 'adjective' etc. with particular, randomly chosen instances of nouns and adjectives. Thus you could inject a vocabulary from a chosen domain, such as pornography or ornithology. Or you could inject the vocabulary of francophoney metatwaddle – as Andrew Bulhak was later to do when he wrote his hilarious 'Postmodernism Generator', which I quoted in *A Devil's Chaplain*:

> If one examines capitalist theory, one is faced with a choice: either reject neotextual materialism or conclude that society has objective value. If dialectic desituationism holds, we have to choose between Habermasian discourse and the subtextual paradigm of context. It could be said that the subject is contextualised into a textual nationalism that includes truth as a reality. In a sense, the premise of the subtextual paradigm of context states that reality comes from the collective unconscious.

This randomly generated garbage makes about as much sense as many a journal devoted to the metatwaddle of 'literary theory', and Bulhak's program is capable of generating a literally indefinite quantity of it.

Two more programming projects date from around this time in my life, both of which also, as it turned out, served to hone my skills for the future rather than deliver results of more immediately practical usefulness. The first of these was a program to translate from one computer language to another: specifically, from BASIC to Algol-60. It worked well for those two languages, and would have worked, with minor detailed changes, to translate from any computer language of that general algorithmic type to any other. My second project of this time was STRIDUL-8: a program to make the PDP-8 computer sing like a cricket.

I had been inspired to work on crickets by my Berkeley friend, the neurobiologist David Bentley; and my entomologically inclined graduate student Ted Burk (now a professor in Nebraska) was keen to do his doctoral thesis on them. David kindly sent me some eggs of the Pacific field cricket *Teleogryllus oceanicus*. They hatched in Oxford and soon we had a thriving colony, which Ted looked after, feeding them on lettuce. While Ted productively pursued his own research on the crickets' behaviour, I conceived a parallel project using computer-generated courtship song. That research project was never completed, but I did complete the writing of STRIDUL-8, and it worked pretty well.

My testing apparatus was a seesaw, made out of balsa wood so it was very light – as it had to be for crickets. It was really just a long balsa-wood passage, closed with netting at each end and over its top, and resting on a hinged fulcrum in the middle. Only one cricket was placed in the passage at a time, and it was free to walk from one end to the other as often as it liked. Whichever end it approached tipped down, as a seesaw should, and this fact was recorded by a micro-switch, which importantly also reversed the location of the sound. There were two little loudspeakers, one at each end of the seesaw. Cricket song was played through whichever of the two loudspeakers was at the opposite end of the seesaw from the cricket.

So, imagine you are a female cricket, sitting somewhere towards the west end of the corridor. Song is playing from the east. You like what you hear, so you start to walk east. When you near the east end, your weight tilts the seesaw to the east, tripping the micro-switch and thereby informing the computer, which now switches the song to the west end loudspeaker. So you turn and walk west, and the whole process happens in reverse. Preferred songs therefore generate a large number of seesaw reversals and these were automatically counted by the computer. Whether the cricket thought she was pursuing an ever-receding coy male, or whether she thought the male was jumping capriciously over her head, or whether she thought at all, is impossible to say. Unpreferred songs would generate only a small number of rockings of the seesaw. Indeed, if a song was positively aversive, the cricket would stay at the opposite end of the passage and not generate any seesaw tips at all.

That, then, was my apparatus for measuring how much crickets like songs of different types. Play Song *A* for five minutes of the alternating seesawing regime, then do the same with Song *B*, and so on for many trials, properly randomized etc. Count seesaw tippings as a measure of how much the cricket liked each song. The point about computer-generated songs, as opposed to real ones, was to try to dissect out, in classic Tinbergen fashion, what it is about their own species song that crickets like. The computer would vary how it sang, in systematic ways. The initial plan was to start with a simulation of the species' natural song and then change it – drop bits out, enhance other bits, vary the interval between chirps and so on. Later – it was a somewhat wild hope – I envisaged the computer being programmed instead to start with random song and 'learn' – or we could equivalently say 'evolve' – choosing 'mutations' step by step, until it progressively homed in on a synthetic preferred song. If the preferred song had turned out to be the natural song of *Teleogryllus oceanicus*, wouldn't that have been sensational? And then if I had done the same thing with *Teleogryllus commodus* and the computer had homed in on its rather different song. What bliss that would have been for the researcher!

In programming the computer to sing, I wanted to make it as versatile as possible. Versatility is what computers are good at. As with the analogue computer simulation, as with the language translation program, I wanted to program the general case. And this is where STRIDUL-8 came in: its language allowed you to specify any combination of pulses and intervals, and therefore any cricket song in the world. STRIDUL-8 had an intuitively reasonable bracket notation, which enabled the user to insert repeats, and repeats embedded within repeats, in a manner reminiscent of the grammar of language (see pages 229 and 235).

STRIDUL-8 worked well. Its simulations of cricket song sounded like real crickets to human ears, and it was easy to program the computer to sing like any cricket species in the world. However, when I demonstrated the system to Dr Henry Bennet-Clark, world authority on the acoustics of insect sounds, newly arrived from Edinburgh to take up a position in Oxford, he made a face and said 'Eeugh!' STRIDUL-8 could only specify the pattern in time of pulses of sound, each pulse corresponding to one stroke of the wings against each other. I had made no attempt to simulate the actual wave form produced by each wing stroke, and this is what Henry objected to. He was right. STRIDUL-8, as it stood, could not have done justice to the European tree crickets of whose song Henry once wrote that if moonlight could be heard that is how it would sound. Temporarily discouraged, I put my whole cricket song project on the back burner, while I attended to other pressing tasks, notably a challenging invitation from Cambridge. And unfortunately I never returned to it: my cricketing days were over. I've often regretted it. I think most scientists have sad loose ends, projects started, never finished. If I ever had vague intentions to return to the crickets, they were thwarted by Moore's Law: computers change so rapidly that if you leave a loose end of research untied for as long as I left mine, you find that the extant computers have all become newer, sexier models, and they have forgotten how to run your earlier programs. To find a computer that would run STRIDUL-8 today I'd have to go to a museum.

THE GRAMMAR OF
BEHAVIOUR

THE Oxford Animal Behaviour Research Group under Tinbergen had long maintained cordial relations with the corresponding sub-department at Cambridge, housed in the neighbouring village of Madingley. 'Madingley' was founded in 1950 by W. H. Thorpe – a distinguished scientist whose gently austere, almost ecclesiastical personality is best summed up by Mike Cullen's jest that it was entirely appropriate that when Thorpe needed a notation for recording birdsong, he transcribed it for the *organ*. Madingley celebrated its quarter-century in 1975 with a conference in Cambridge organized by Patrick Bateson and Robert Hinde, the leading figures of the Madingley group after Thorpe's retirement, both of whom later became heads of Cambridge colleges. Many of the speakers at the Madingley conference were past or present members of that group, but they invited some outsiders too, and David McFarland and I were honoured to be the Oxford contingent.

Nowadays, on the rare occasions when I agree to speak at such a conference, I confess that I usually find myself dusting off a previous talk and updating it. Younger and more vigorous in 1974, I took the risk of pushing the boat out and undertaking to write something entirely new for Madingley's jubilee conference and the book that came out of it. The topic I chose, 'hierarchical organization', had a track record in the history of ethology. It was the main theme of one of the boldest – and most criticized – chapters in Tinbergen's magnum opus *The Study of Instinct*, the chapter entitled 'An attempt at a synthesis'. I took a rather different approach – or, rather, several different approaches, and I too attempted a synthesis.

The essence of hierarchical organization, as I interpreted it, is the idea of 'nested embedment'. I can explain this by contrast with what it is *not*, and this is where I echo the discussion above about grammar. You might attempt to describe the stream of events – the stream of things an animal does, say – as a Markov Chain. What's that? I won't attempt a formal, mathematical definition such as was offered by the Russian mathematician Andrey Markov. An informal, verbal definition is this. A Markov Chain of animal behaviour is a series in which what an animal does now is determined by what it did previously, back a fixed number of steps but no further. In a first-order Markov Chain, what the animal does next can be predicted statistically from its immediately preceding action, and not from anything earlier. Looking at the last but one action (last but two action etc.) gives you no additional predictive power. In a second-order Markov Chain, you improve your ability to predict if you look at the previous two actions, but no further back than that. And so on.

Hierarchically organized behaviour would be very different. Markov Chain analysis, of any order, wouldn't work. The predictability of behaviour wouldn't decay smoothly as you look into the future, but would jump up and down in an interesting way – like with the blowfly grooming, but more interestingly than that. In an ideal case, behaviour would be organized in discrete chunks. And chunks within chunks. And chunks within chunks within chunks. That's what's meant by nested embedment. The clearest model for nested embedment is syntax, the grammar of human language. Think back to the program that I wrote to generate random grammatical sentences, and the example sentence that I quoted:

The adjective noun of the adjective noun which adverbly adverbly verbed in noun of the noun which verbed **adverbly verbed**.

The core sentence is in bold. You can read it and it is grammatically correct, without the various embedded relative clauses or prepositional clauses in the middle. We can build up the embedment as follows. The important point is that the build-up can occur *inside* the core sentence, or inside already embedded parts of the sentence. Read to yourself the emboldened parts of the following:

The adjective noun of the adjective noun which adverbly adverbly verbed in noun of the noun which verbed **adverbly verbed**.

The adjective noun of the adjective noun which adverbly adverbly verbed in noun of the noun which verbed **adverbly verbed**.

The adjective noun of the adjective noun which adverbly adverbly verbed in noun of the noun which verbed **adverbly verbed**.

The adjective noun of the adjective noun which adverbly adverbly verbed in noun of the noun which verbed **adverbly verbed**.

The adjective noun of the adjective noun which adverbly adverbly verbed in noun of the noun which verbed adverbly verbed.

In every member of the sequence above, you can read the emboldened part on its own, and discover that it is grammatically correct. You can delete the unbold, embedded bits and it might change the meaning but it doesn't stop the sentence from being grammatically correct.

If, on the contrary, you were to build up the sentence by adding the bits progressively from left to right, none of the series would be grammatical until you hit the end of the whole sentence.

The adjective noun [not a sentence]

The adjective noun of the adjective noun [not a sentence]

The adjective noun of the adjective noun which adverbly adverbly verbed [not a sentence]

The adjective noun of the adjective noun which adverbly adverbly verbed in noun [not a sentence]

The adjective noun of the adjective noun which adverbly adverbly verbed in noun of the noun which verbed adverbly verbed [finally, we have a sentence].

Only in the very last case does the sentence achieve closure and become grammatical. What I wanted to know was whether animal behaviour is organized as a Markov Chain, or whether it is organized in a nested embedded way, perhaps like syntax or perhaps in some other way hierarchically embedded. You can see that there were inklings of the idea lurking behind the research that Marian and I did on drinking in chicks and especially self-grooming in flies. Now, in my Madingley paper, I wanted to look more generally at the question of hierarchical organization, from a theoretical point of view as well as by looking at real studies of animal behaviour.

After defining various kinds of hierarchy in a convenient notation of mathematical logic, I considered possible evolutionary advantages of hierarchical organization. To illustrate what I called the 'evolutionary rate advantage' I borrowed from the Nobel Prize-winning economist Herbert Simon a parable of two watchmakers called Tempus and Hora. Their watches kept equally good time, but Tempus took much, much longer to complete a watch. Both kinds of watch had 1,000 components. Hora, the more efficient watchmaker, worked in a hierarchical, modular way. He put his components into 100 sub-assemblies of ten components each. These in turn were assembled into ten larger units, which were finally put together to complete the watch. Tempus, on the other hand, tried to put all 1,000 components together in a single large assembly operation. If he dropped one component, or if the telephone interrupted him, the whole caboodle fell to bits and he had to start again. He very rarely completed a watch, while Hora, with his hierarchical modular technique, was churning them out. The principle will be familiar to all computer programmers, and it surely applies to evolution and to the building of biological systems.

I also extolled another advantage of hierarchical organization, the 'local administration advantage'. If you are trying to control an empire from London, or in earlier times from Rome, you cannot micro-manage what happens in remote parts of the empire, because communication channels – in both directions – are too slow. Instead, you appoint local governors, give them broad policy directives, and leave them to take day-to-day decisions on their own. The same necessarily applies to a robot vehicle on Mars. Radio signals take several minutes to travel the distance. If the vehicle encounters a local difficulty, say a boulder, it sends the information back to Earth, and it takes four minutes to get there. 'Turn left to avoid boulder,' flashes back the urgent reply, and that again takes four minutes to reach Mars. Meanwhile, the wretched vehicle has long since ploughed into the boulder. Obviously the solution is to delegate local control to an on-board computer, and give the local computer only general policy instructions like: 'Explore the crater to the north-west, taking care to avoid boulders whenever you encounter them.' By the same token, if there are several vehicles exploring different parts of Mars, it makes sense for Earth to send general policy instructions to one senior computer on Mars, which sends more detailed instructions coordinating the activities of all its subordinate vehicles, each with its own on-board computer to take fine-grained local decisions. Armies and business corporations use similar hierarchical chains of command, and once again biological systems do the same.

Especially pleasing in this connection are the giant dinosaurs whose very long spinal cord imposed an inconvenient distance between the brain in the head and the seat of much of the action, the giant hind legs. Natural selection solved the problem with a second 'brain' (enlarged ganglion) in the pelvis:

> Behold the mighty dinosaur,
> Famous in prehistoric lore,
> Not only for his power and strength
> But for his intellectual length.
> You will observe by these remains
> The creature had two sets of brains–
> One in his head (the usual place),
> The other at his spinal base.
> Thus he could reason 'A Priori'
> As well as 'A Posteriori'.
> No problem bothered him a bit
> He made a head and tail of it.
> So wise was he, so wise and solemn,
> Each thought filled just one spinal column.
> If one brain found the pressure strong
> It passed a few ideas along.
> If something slipped his forward mind

'Twas rescued by the one behind
And if in error he was caught
He had a saving afterthought.
As he thought twice before he spoke
He had no judgment to revoke.
Thus he could think without congestion
Upon both sides of every question.
Oh, gaze upon this model beast,
Defunct ten million years at least.

Bert Leston Taylor (1866–1921)

'Thus he could reason "*A Priori*" / As well as "*A Posteriori*"' – I wish I'd written that. You'd have to look far before you found another poem with quite so many flashes of clever wit in almost every line.

Having established the advantages of hierarchical organization more generally, I moved on to see whether there was evidence of it in specific cases of animal behaviour. Beginning by re-analysing the data Marian and I had recorded from blowflies, I moved on to other data from the animal behaviour literature, which I ferreted out in the library. Among others, I included a large study on the behaviour of damsel fish, another on face-grooming behaviour by mice and another on the courtship behaviour of guppies.

I wanted to devise mathematical techniques for detecting hierarchical embedment, in an attempt at objectivity, unbiased by my own preconceptions. Here's just one of several computer-based methods I thought up. This one I dubbed Mutual Replaceability Cluster Analysis. My method started by counting frequencies of transitions between behaviour patterns, but then analysed the data in a special hierarchical way. I fed into the computer a table showing how many times each behaviour pattern in the animal's repertoire was followed by each other one. Then the computer systematically examined the data to see if it could find pairs of behaviour patterns

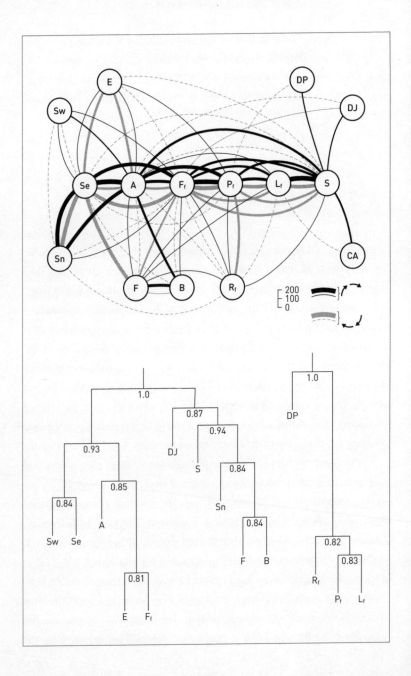

that were *mutually replaceable*. Mutually replaceable means that you could stick either of them in the place of the other and the overall pattern of transition frequencies would remain the same (or near enough the same, according to some previously defined criterion). Once a mutually replaceable pair had been identified, both members of the pair were renamed with a *joint* name, and the table of transitions contracted because it now had one fewer rows and one fewer columns. Then the contracted table was fed back into the cluster-analysis program, and the whole thing was repeated as many times as necessary to use up the whole list of behaviour patterns. As each pair of behaviour patterns was swallowed up in a cluster, or as each already swallowed cluster was swallowed up in a bigger cluster, the program moved up one node in a hierarchical tree. Opposite, for instance, is my Mutual Replaceability tree for the behaviour patterns of guppies, using data from a group of Dutch workers led by Professor G. P. Baerends (who, incidentally, had been Niko Tinbergen's first graduate student and later became one of the leading figures in European ethology).

The upper diagram shows the transition frequencies of guppy behaviour patterns, as measured by the Dutch scientists. Each circle is labelled with the code name of a behaviour pattern, and the thickness of the lines shows the frequency of transition from one to the other (black lines move from left to right, grey lines from right to left). The lower diagram shows the results of feeding the same data into my Mutual Replaceability Cluster Analysis program. The numbers represent the numerical index of mutual replaceability that I used to compare with the criterion for deciding to unite two entities (actually a rank correlation coefficient, if you happen to be interested). I got similar hierarchical trees for the damsel fish, the mice, Marian's and my blowflies etc.

Yet another way of thinking about hierarchy, which I used in my Madingley paper, is the hierarchy of *goals*. A goal is not necessarily a consciously held goal in the animal's brain (although it might be). I simply meant a condition which brings behaviour to an end. For example, complicated sequences of prey-catching behaviour in a cheetah would be brought to a close by the 'goal state' of a successful kill. But goals can be hierarchically embedded within each other, and that is a fruitful way to look at it. I made a distinction between 'action rules' and 'stopping rules'. An action rule tells the animal (or computer in the case of a computer simulation) exactly what to do and when to do it, including lots of conditional instructions (IF . . . THEN . . . ELSE etc). A stopping rule tells the animal (or computer simulation): 'Behave at random (or try out lots of possibilities) and don't stop until the following *goal state* is achieved' – say, a full stomach.

A pure action-rule program for a complicated task like hunting by a cheetah would become impossibly elaborate. Much better to use stopping rules. But not just one big stopping rule – behave at random until the goal state of full stomach is achieved. Any cheetah living by that rule would die of old age before achieving a square meal! Instead, the sensible way for natural selection to have programmed the behaviour would be with hierarchically embedded stopping rules. The global goal (continue until stomach is full) would 'call up' subsidiary goals such as 'walk around until gazelle sighted'. The goal state 'gazelle sighted' would terminate that particular stopping rule and initiate the next one: 'drop down and crawl slowly towards gazelle'. That would be terminated by the goal state 'gazelle now within striking distance'. And so on. Each of these subsidiary stopping rules would call up its own, internally embedded stopping rules, each with its own goal state. At much lower levels, even individual muscle contractions often conform to the design that engineers call 'servo-control'. The nervous system specifies a target state for a muscle, which contracts until the target state ('stopping rule') is achieved.

But I earlier introduced the idea of hierarchical embedment by using the analogy of human grammar. My Madingley paper finally returned to this fascinating topic, and asked whether there was any evidence that animal behaviour had something equivalent to grammatical structure. If it did, this would be extremely interesting, because it might give us some inklings of the evolutionary antecedents of human language. When true language, with true hierarchical syntax, finally evolved in humans, dare we speculate that it was able to build on a ready-made foundation of pre-existing neural structures that were put in place for different reasons, nothing to do with language, long ago?

The earliest attempt to look at this question was made by my Oxford colleague John Marshall, a linguist. He used courtship behaviour of male pigeons, taking data from the published etho-logical literature. There were seven 'words' of the pigeon lexicon: things like Bow (to the female), Copulate, etc. Marshall used his skills as a linguist to postulate a 'phrase structure grammar', just as Chomsky had before him for human language. For my Madingley paper, I translated Marshall's grammar into the (now largely obsolete) computer language that I favoured at the time, Algol-60. Readers familiar with computer programming will note that, once again, the program is heavily recursive – procedures call themselves, the very essence of hierarchical embedment as I have already explained. In the program, 'p' was replaced by 'If some probability condition, such as 0.3, is met . . .'

At the top of the diagram overleaf is Marshall's 'phrase structure grammar' for pigeon courtship behaviour. In the middle is my Algol-60 translation. And at the bottom are several sequences of 'behaviour' generated by my program.

Unfortunately, Marshall's analysis doesn't really allow us to draw any secure conclusion about pigeons. How do we know whether the grammar he proposed is 'correct'? In the case of human syntax, any native speaker of the language can immediately tell you if it is

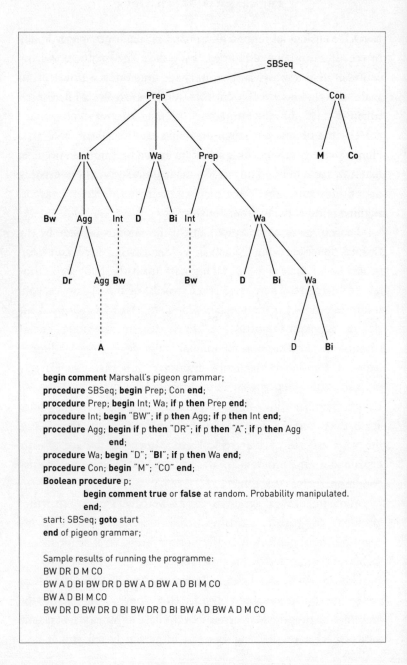

```
begin comment Marshall's pigeon grammar;
procedure SBSeq; begin Prep; Con end;
procedure Prep; begin Int; Wa; if p then Prep end;
procedure Int; begin "BW"; if p then Agg; if p then Int end;
procedure Agg; begin if p then "DR"; if p then "A"; if p then Agg
            end;
procedure Wa; begin "D"; "BI"; if p then Wa end;
procedure Con; begin "M"; "CO" end;
Boolean procedure p;
        begin comment true or false at random. Probability manipulated.
        end;
start: SBSeq; goto start
end of pigeon grammar;
```

Sample results of running the programme:
BW DR D M CO
BW A D BI BW DR D BW A D BW A D BI M CO
BW A D BI M CO
BW DR D BW DR D BI BW DR D BI BW A D BW A D M CO

correct. Marshall had no such checking mechanism. As with much of the research that I did during this period, my goal state was not so much to find something permanently true about particular animals as to find novel and exciting ways in which animal behaviour might be studied in the future.

The Madingley paper[1] represented a kind of closure for me, a climax to the first part of my scientific career, beginning in my early twenties and ending in my early thirties. At this point, I took off in an entirely new direction, never to return to those youthful mathematical pastures. That new direction, which was to define the rest of my career, and approximately the second half of my life, opened up with the publication of my first book, *The Selfish Gene*.

---

[1] R. Dawkins, 'Hierarchical organization: a candidate principle for ethology', in P. P. G. Bateson and R. A. Hinde, eds, *Growing Points in Ethology* (Cambridge, Cambridge University Press, 1976), pp. 7–54.

THE IMMORTAL GENE

IN 1973, strike action by the National Union of Mineworkers led to a crisis in which the Conservative government of Edward Heath imposed a so-called 'three-day week' in Britain. In order to conserve fuel stocks, electricity for non-essential purposes was rationed. We were limited to three days per week, and there were frequent power cuts. My cricket research depended on electricity but writing didn't – in those days I wrote with a portable typewriter on, of all outlandish surfaces, sheets of flat white stuff called paper. So I decided to call a temporary halt to my cricket research and begin work on my first book. This was the genesis of *The Selfish Gene*.

Selfishness and altruism and the whole idea of a 'social contract' were much in the air at the time. Those of us on the political left tried to balance our sympathy for the miners on the one hand with hostility to what some saw as their strong-arm tactics, holding society at large to ransom, on the other. Did evolutionary theory have any-thing to contribute to this important dilemma? The previous decade had seen a string of popular science books and television documen-taries, gallantly attempting to apply Darwinian theory to questions of altruism and selfishness, collective versus individual welfare, but actually getting the theory flat-out wrong. The error was always a version of what has been called 'evolutionary panglossism'.

As reported by my friend and mentor the late John Maynard Smith, his own mentor the formidable J. B. S. Haldane satirically coined three erroneous, or at best unreliable, 'theorems'. Aunt Jobiska's Theorem (from Edward Lear) was 'It's a fact the whole world knows . . .' The Bellman's Theorem (from Lewis Carroll) was 'What I tell you three times is true.' And Pangloss's Theorem (from Voltaire) was 'All is for the best in this best of all possible worlds.'

Evolutionary panglossians are vaguely aware that natural selection does a pretty effective job of making living things good at their business of living. Albatrosses seem beautifully designed for flying above the waves, penguins for flying beneath them (I happen to be writing this on a ship in Antarctic waters, marvelling through my binoculars at these prodigies of avian virtuosity). But panglossians forget, as it is so easy to forget, that this 'good at' applies to *individuals*, not to *species*. Good at flying, good at swimming, surviving, reproducing, yes, natural selection will tend to make individual animals good at those things. But there is no reason to expect natural selection to make *species* good at avoiding extinction, good at balancing their sex ratio, good at limiting their population in the interests of the common weal, good at husbanding their food supply and conserving their environment for the benefit of future generations. That would be panglossism. Group survival may emerge as a consequence of improved individual survival, but that is a fortunate by-product. Group survival is not what natural selection is about.

The panglossian error is tempting because we humans are blessed with foresight and can judge which actions are likely to benefit our species, or our town, or our nation, or the whole world, or any specified entity or interest group, in the future. We can foresee that to overfish the seas would be in the long run counterproductive for all fishermen. We can foresee a happier future if we limit our birth rate so that fewer individuals are born, to enjoy richer lives. We can decide that self-restraint now will pay dividends in the future. But natural selection has no foresight.

To be sure, a panglossian version of natural selection theory had been proposed, which, if only it worked, might achieve something like the 'all is for the best' utopia. But unfortunately it doesn't work. At any rate, it was one of my aims in *The Selfish Gene* to persuade my readers that it doesn't work. This was the theory called 'group selection'. This maddeningly seductive error – the Great Group Selection Fallacy or GGSF – ran all through Konrad Lorenz's popular 1964 book *On Aggression*. It also pervaded Robert Ardrey's best-sellers, *The Territorial Imperative* and *The Social Contract* – where I was especially affronted by the mismatch between Ardrey's erroneous message and the high quality of the English in which he expressed it.[1] I aspired to publish a book on the same theme as Ardrey's *Social Contract* (itself a sort of biological rewrite of Rousseau's famous treatise); but mine would be based on rigorous natural selection theory, not the GGSF. My ambition was to undo the damage done by Ardrey and Lorenz – and by many television documentaries of the time, whose promulgation of the error was so ubiquitous that in *The Selfish Gene* I even dubbed it 'the BBC Theorem'.

I was all too familiar with panglossism and the GGSF because I met them weekly in undergraduate essays. And indeed I, when an undergraduate myself, had laced many of my essays with the fallacious view that what really matters in natural selection is the survival of the species (my tutors never noticed). When I eventually came to write *The Selfish Gene*, my dream was that I would change all that. I was daunted by the knowledge that, in order to succeed, my book would need to be as well-written as Ardrey's and sell as prolifically

---

[1] Konrad Lorenz, *On Aggression*, translated by Marjorie Latzke (London, Methuen, 1964); first published in German as *Das sogenannte Böse* – 'the so-called evil' – in 1963. Robert Ardrey, *The Territorial Imperative: A Personal Inquiry into the Animal Origins of Property and Nations* (London: Collins, 1967), and *The Social Contract: A Personal Inquiry into the Evolutionary Sources or Order and Disorder* (London, Collins, 1970).

as Lorenz's. I jokingly spoke of it as 'my best-seller', never really believing that it would become one but giving self-consciously ironic voice to my wilder ambition.

Natural selection is a purely mechanical, automatic process. The world is constantly tending to become full of entities that are good at surviving, denuded of things that are not. Natural selection has no foresight, but brains have, and that is why panglossism is so appealing to us. Brains may agonize about the long-term future and project forward this century's self-indulgence into next century's catastrophe. Natural selection cannot do that. Natural selection can't agonize about anything. Natural selection can only blindly favour short-term gain, because every generation is automatically filled with the offspring of those individuals who did whatever it took, in the short term, to manufacture offspring more effectively than other individuals of their own generation.

And when you look carefully and hard at exactly what is going on as the generations flash by, your gaze is drawn irresistibly to the gene as the level at which natural selection really works. Natural selection automatically favours self-interest among entities that potentially can pass through the generational filter and survive into the distant future. As far as life on this planet is concerned, that means genes. Here's how I put it in *The Selfish Gene*, where I introduced the phrase 'survival machine' to describe the role of (mortal) individual organisms vis-à-vis their (potentially immortal) genes:

> The genes are the immortals . . . [they] have an expectation of life that must be measured not in decades but in thousands and millions of years.
>
> In sexually reproducing species, the individual is too large and too temporary a genetic unit to qualify as a significant unit of natural selection. The group of individuals is an even larger unit. Genetically speaking, individuals and groups are like clouds in the

sky or dust-storms in the desert. They are temporary aggregations or federations. They are not stable through evolutionary time. Populations may last a long while, but they are constantly blending with other populations and so losing their identity. They are also subject to evolutionary change from within. A population is not a discrete enough entity to be a unit of natural selection, not stable and unitary enough to be 'selected' in preference to another population.

An individual body seems discrete enough while it lasts, but alas, how long is that? Each individual is unique. You cannot get evolution by selecting between entities when there is only one copy of each entity! Sexual reproduction is not replication. Just as a population is contaminated by other populations, so an individual's posterity is contaminated by that of his sexual partner. Your children are only half you, your grandchildren only a quarter you. In a few generations the most you can hope for is a large number of descendants, each of whom bears only a tiny portion of you – a few genes – even if a few do bear your surname as well.

Individuals are not stable things, they are fleeting. Chromosomes too are shuffled into oblivion, like hands of cards soon after they are dealt. But the cards themselves survive the shuffling. The cards are the genes. The genes are not destroyed by crossing-over, they merely change partners and march on. Of course they march on. That is their business. They are the replicators and we are their survival machines. When we have served our purpose we are cast aside. But genes are denizens of geological time: genes are forever.

I had already convinced myself of this truth a decade earlier, in almost exactly the same words, when I gave the 1966 undergraduate lectures in Oxford that I have already described. I recalled on page 199 the rhetorical flourishes with which I tried to persuade the undergraduates of the centrality of the immortal gene in the logic of natural selection. Here are my 1966 words, and you can see how similar they are to the equivalent, more rhetorical paragraphs of *The Selfish Gene.*

> Genes are in a sense immortal. They pass through the generations, reshuffling themselves each time they pass from parent to off-spring. The body of an animal is but a temporary resting place for the genes; the further survival of the genes depends on the survival of that body at least until it reproduces, and the genes pass into another body . . . the genes build themselves a temporary house, mortal, but efficient for as long as it needs to be . . . To use the terms 'selfish' and 'altruistic' then, our basic expectation on the basis of the orthodox neo-Darwinian theory of evolution is that *Genes will be 'selfish'.*

When I recently found the text of that 1966 lecture (with its encouraging marginal note by Mike Cullen), I was surprised to realize that I had not then read George C. Williams's book, *Adaptation and Natural Selection*, published in the same year:[1]

> With Socrates' death, not only did his phenotype disappear, but also his genotype . . . The loss of Socrates' genotype is not assuaged by any consideration of how prolifically he may have reproduced. Socrates' genes may be with us yet, but not his genotype, because meiosis and recombination destroy genotypes as surely as death.
>
> It is only the meiotically dissociated fragments of the genotype that are transmitted in sexual reproduction, and these fragments are further fragmented by meoisis in the next generation. If there is an ultimately indivisible fragment it is, by definition, 'the gene' that is treated in the abstract discussions of population genetics.

[1] George C. Williams, *Adaptation and Natural Selection* (Princeton, NJ, Princeton University Press, 1966).

When I eventually read Williams's great book (some years later, I regret to say), his Socrates passage immediately resonated with me and I prominently acknowledged Williams's importance, as well as Hamilton's, for the theme of *The Selfish Gene* when I came to write it.

Williams and Hamilton were somewhat similar characters: quiet, withdrawn, self-effacing, deep-thinking. Williams had a dignity and mien which – perhaps enhanced by his high forehead and the cut of his beard – reminded many of Abraham Lincoln. Hamilton had more an air of A. A. Milne's Eeyore. But when I wrote *The Selfish Gene* I didn't know either man, just their published work and how central it was for our understanding of evolution.

Because genes are potentially immortal in the form of accurate copies, the difference between successful genes and unsuccessful genes really matters: it has long-term significance. The world becomes filled with genes that are good at being there, good at surviving through many generations. In practice that means good at cooperating with other genes in the business of building bodies that have what it takes to survive long enough to reproduce – for bodies are the temporary vehicles in which genes reside and which pass them on. Throughout *The Selfish Gene* I used that phrase 'survival machine' as my name for an organism. Organisms are the entities in life that actually do things – move, behave, search, hunt, swim, run, fly, feed their young. And the best way to explain everything that an organism does is to assume that it has been programmed, by the genes that ride inside it, to preserve those genes and pass them on before the organism itself dies.

I also used the word 'vehicle' as equivalent to 'survival machine'. It reminds me of an amusing occasion when a Japanese television crew came to interview me about *The Selfish Gene*. They all travelled to Oxford from London packed into a black cab, with tripods and floodlights and, as it seemed, arms and legs sticking out of every window. The director informed me, in halting English (the official interpreter couldn't make me understand him at all and had been sent off in disgrace), that he wanted to film me in the taxi as it drove around Oxford. This puzzled me, and I asked why. 'Hoh!' came the puzzled reply: 'Are you not author of Taxicab Theory of Evolution?' I afterwards guessed that the Japanese translators of my writings must have rendered 'vehicle' as 'taxicab'.

The interview itself was quite amusing. I rode in the taxi alone except for the cameraman and sound man. In the absence of the official interpreter, there was no interviewer, and I was bidden simply to talk *ad lib* about *The Selfish Gene* while we did a scenic tour of Oxford. The taxi driver doubtless had the streets of London intricately mapped in his enlarged hippocampus, but he didn't know Oxford. It therefore fell to me to guide him, and my otherwise measured discourse on selfish genes was punctuated by frantic shouts of 'Turn left here!' or 'Turn right at the traffic lights and then get in the right-hand lane!' I hope they managed to find the unfortunate interpreter before returning to London.

In *The Selfish Gene*, I criticized the panglossian idea that animals have foresight and work out what would be good for the long-term future of their species or group. What is wrong with this is not the idea that animals 'work out what would be good'. There is in any case no suggestion that the 'working out' is conscious. No: what is wrong is the idea of the species or group as the entity whose benefit is maximized. Biologists often legitimately use the language of 'working out what would be good' as a shorthand route to sound Darwinian reasoning. The trick is to identify the correct level in the hierarchy of life at which the shorthand metaphor of conscious reasoning is applied. It is quite OK to put yourself in the position of an individual animal and ask: 'What would I do if I were trying to achieve the goal of propagating my genes?'

The Selfish Gene is filled with imagined soliloquies in which a hypothetical animal 'reasons' to itself: 'Should I do X or Y?' The meaning of 'should' is: 'Would X or Y be better for my genes?' This is legitimate, but only because it can be translated into the question: 'Would a gene for making individuals do X (in this situation) become more frequent in the gene pool?' The subjective soliloquy is justified by the fact that it can be translated into the language of gene survival.

One might be tempted to interpret 'Should I do X or Y?' as meaning 'Would X or Y be more likely to prolong my own life?' But if long life is bought at the expense of not reproducing – that is to say, if we pit individual longevity against gene survival – natural selection will not favour it. Reproduction can be a dangerous business. Male pheasants that are gorgeously coloured to attract females also attract predators. A drab, inconspicuous male would probably live longer than a brightly coloured, attractive male. But he'd be more likely to die unmated, and the genes for safety-first drabness are less likely to be passed on. Gene survival is what really matters in natural selection.

The following is a legitimate shorthand, put into the mouth of a male pheasant: 'If I grow drab feathers I shall probably live a long time, but I won't get a mate. If I grow bright feathers I'll probably die young, but I'll pass on lots of genes before I die, including the genes for making bright feathers. Therefore I should take the "decision" to grow bright feathers.' Needless to say, 'decision' doesn't mean what a human would ordinarily mean by it. Conscious thought is not involved. Organism-level shorthand can be confusing, but it works so long as you remember always to keep open the pathway to a translation back into gene language. No pheasant actually takes a 'decision' to grow bright feathers or drab. Instead, genes for growing bright feathers or drab have different probabilities of surviving through the generations.

It really can be helpful, when trying to understand from a

Darwinian point of view what animals do, to see them as robot machines, 'thinking' about what steps to take in order to pass on their genes to future generations. Such steps may involve behaving in certain ways, or growing organs of a particular shape or character. It can also be helpful to think metaphorically of *genes* as 'thinking' about what steps to take in order to pass themselves on to future generations. Such steps will usually involve manipulating individual organisms via the processes of embryonic development.

But it is never even *metaphorically* legitimate to treat animals as thinking about what steps to take in order to preserve their species, or their group. Differential group or species survival is not what happens in natural selection. What happens is differential gene survival. Therefore, legitimate shorthands are of the form 'If I were a gene, what would I do to preserve myself?' Or – and ideally this should be exactly equivalent – 'If I were an organism, what would I do to preserve my genes?' But 'If I were an organism, what would I do to preserve my species?' is an illegitimate shorthand. So is – this time for a different reason – 'If I were a species, what would I do to preserve myself?' The latter metaphor is illegitimate because a species, unlike an individual organism, is not the kind of entity that even metaphorically behaves as an agent, doing things, acting upon decisions. Species don't have brains and muscles, they are just collections of individual organisms that do. Species and groups are not 'vehicles'. Individual organisms are.

I should point out that neither in my lectures of the 1960s nor in *The Selfish Gene* did I see as very novel the idea of the gene as the fundamental unit of natural selection. I thought of it – and clearly said so – as implicit in the orthodox neo-Darwinian theory of evolution: that is to say, the theory first clearly formalized in the 1930s by Fisher, Haldane, Wright and the other founding fathers of the so-called Modern Synthesis such as Ernst Mayr, Theodosius Dobzhansky, George Gaylord Simpson and Julian Huxley. It was only after *The Selfish Gene* was published that both critics and admirers came to see the idea as revolutionary. That was not how I thought of it at the time.

Having said that, however, I should add that not all the founding fathers of the Modern Synthesis were clear about this important implication of the theory that they collectively put together. To the end of his centenarian life, the authoritative German-American taxonomist Ernst Mayr expressed hostility to the idea of gene selectionism, in terms that suggested to me that he misunderstood it. And Julian Huxley, the founding father who actually coined the phrase 'Modern Synthesis', was an out-and-out group selectionist, without clearly realizing it. The first time I met the great Peter Medawar, he startled my student self with a deliciously sacrilegious remark, delivered with his characteristically patrician, yet impish style. 'The trouble with Julian is that he really doesn't *understand* evolution.' Fancy saying that – of a Huxley! I could hardly believe my ears and, as you see, I have never forgotten it. I later heard another Nobel Prize-winner, the French molecular biologist Jacques Monod, say something a bit similar, though not about Huxley: 'The trouble with natural selection is that everybody *thinks* he understands it.'

I mentioned that I began *The Selfish Gene* when power cuts interrupted my cricket research. I had completed only the first chapter of the book when I happened to meet an editor from the publishers Allen & Unwin. He was paying a routine visit to the Department of Zoology in search of possible books, and I told him about my embryonic project. He sat down and read that first chapter on the spot, liked it and encouraged me to continue. But then – unfortunately from one narrow point of view, fortunately from others – the industrial unrest came to an end and the lights came back on. I shoved my chapter in a drawer and forgot about it as I resumed my research on crickets.

From time to time during the next two years I contemplated

returning to the book. The impetus was especially strong when I read and lectured about new publications that were beginning to appear in the early 1970s and proved beautifully compatible with the thesis of my gestating book. Most notable among these were papers by the young American biologist Robert Trivers, and others by the veteran British professor John Maynard Smith. Both these authors made use of the intuitive shortcut I mentioned (the philosopher Daniel Dennett would now call it an intuition pump):[1] the shortcut of imagining that an individual organism behaves 'as if' consciously calculating the best policy for preserving and propagating its genes.

Trivers treated a parent animal *as if* it were a rational agent calculating what economists call the 'opportunity cost' of an action. A parent has to pay the costs of rearing each offspring. Among these costs might be food, including time and effort spent gathering it, time spent protecting the child from predators, and the risks incurred by the parent in doing so. Trivers wrapped them all up in one metric which he called Parental Investment or PI. Trivers' key insight was that that PI must be an *opportunity cost*: the investment in any one child is measured as *lost opportunities to invest in other children*. Trivers used the notion to develop a penetrating theory of 'parent–offspring conflict'. The decision on the best time to wean a child, for example, is subject to a 'disagreement' between the child and its mother, both behaving as rational economists whose 'utility function' is the long-term survival of their own genes. The mother 'wants' to terminate suckling earlier than the child does, because she places greater 'value' than he does on her future offspring, who will benefit from early weaning of the present child. The present child also 'values' his future siblings, but only half as highly as his mother does because of the way Hamilton's Rule pans out. Therefore there

---

[1] Daniel C. Dennett, *Intuition Pumps and Other Tools for Thinking* (New York, Norton, 2013).

is a period of 'weaning conflict', an uneasy phase of transition between the early time when both parties 'agree' that suckling should continue and the later time when both parties 'agree' that it should end. During this phase, when the mother 'wants' weaning but the child doesn't, observers of animal behaviour should see the symptoms of a subtle battle between mother and child. In passing I should add that, long after *The Selfish Gene* was published, the Australian biologist David Haig cleverly showed how many of the ailments of pregnancy can be explained in terms of the same Triversian conflict going on inside the womb – not about weaning in this case, obviously, but about other aspects of the allocation of necessarily scarce resources.

Parent–offspring conflict was obviously a subject tailor-made for my book, and Trivers' brilliant paper on the subject was one of the spurs that encouraged me to take my first chapter out of the drawer where it had languished since the end of the power strike. It became the inspiration for chapter 7 of *The Selfish Gene*, 'Battle of the generations'. Chapter 8, 'Battle of the sexes', also made use of Trivers' ideas, this time showing how males and females might compute their opportunity costs differently. When might a male, for example, desert his mate, leaving her 'holding the baby' and 'in a cruel bind' while he seeks a new one? Trivers also influenced chapter 10, 'You scratch my back, I'll ride on yours'. His paper in this case was an earlier one, on reciprocal altruism, which showed that kin selection is not the only evolutionary pressure towards altruism. Reciprocation – the repaying of favours – can also be very important, and it works across species, not just within them as kin selection does. So Trivers' name was added to those of Hamilton and Williams among the four authors who had greatest influence on *The Selfish Gene*. I also asked him to write the foreword – which he graciously did, although at that point we had never met.

The fourth was John Maynard Smith, who later became a

beloved mentor. As a boy I had met the book that he would refer to as 'my little Penguin', and I was much taken by the smiling author photograph: shock of nutty professor hair askew like the pipe in his mouth, thick round glasses in need of a clean – the sort of man I was immediately drawn to. I also liked the biographical note, which explained that he had been an engineer designing aircraft, but gave it up and went back to university to study biology because he noticed that 'aircraft were noisy and old fashioned'. Many years later, a new edition of that book, *The Theory of Evolution*, was published by Cambridge University Press and I was honoured to be invited to write the foreword.[1] I included the following tribute to this genial hero:

> Readers of 'campus novels' know that a conference is where you can catch academics at their worst. The conference bar, in particular, is the academy in microcosm. Professors huddle together in exclusive, conspiratorial corners, talking not about science or scholarship but about 'tenure-track hiring' (their word for jobs) and 'funding' (their word for money). If they do talk shop, too often it will be to make an impression rather than to enlighten. John Maynard Smith is a splendid, triumphant, lovable exception. He values creative ideas above money, plain language above jargon. He is always the centre of a lively, laughing crowd of students and young research workers of both sexes. Never mind the lectures or the 'workshops'; be blowed to the motor coach excursions to local beauty spots; forget your fancy visual aids and radio microphones; the only thing that really matters at a conference is that John Maynard Smith must be in residence and there must be a spacious, convivial bar. If he can't manage the dates you have in mind, you must just reschedule the conference. He doesn't have to give a formal talk (although he is a riveting speaker) and he doesn't have to chair a formal session (although

---

[1] John Maynard Smith, *The Theory of Evolution* (Cambridge, Cambridge University Press, 1993; first published London, Penguin, 1958).

he is a wise, sympathetic and witty chairman). He has only to turn up and your conference will succeed. He will charm and amuse the young research workers, listen to their stories, inspire them, rekindle enthusiasms that might be flagging, and send them back to their laboratories or their muddy fields, enlivened and invigorated, eager to try out the new ideas he has generously shared with them.

My relationship with John didn't get off to an outstandingly good start, however. I first met him in 1966 when, as Dean of Biological Sciences, he interviewed me for a job at the University of Sussex. I was already pretty much committed to going to Berkeley. However, there was a job going at Sussex and Richard Andrew, their resident expert on our shared subject of animal behaviour, pressed me with flattering urgency to apply. I told Richard about my near-commitment to go to Berkeley, but he said there was no harm in doing the Sussex interview anyway, so I thought: What the hell, why not? I'm afraid my 'what the hell' attitude didn't endear me to Maynard Smith in the interview. I said I wouldn't lecture about animal taxonomy. He said it was part of the job. I rather arrogantly said: Well, I've got a job offer at Berkeley and I'm not quite sure why I am doing this interview anyway. He was nice about it when he and Dr Andrew took me to lunch, but, as I said, it was not a good start to what was later to prove a delightful friendship.

In the early 1970s, Maynard Smith began the long series of papers in which, together with colleagues such as Geoffrey Parker and the late George Price, he deployed a version of the mathematical theory of games to solve a number of problems in evolution. These ideas were immensely congenial to the idea of the selfish gene, and Maynard Smith's papers constituted the other major stimulus that led me to dust off my old chapter 1 and write the whole book.

Maynard Smith's particular contribution was the notion of the evolutionarily stable strategy or ESS. 'Strategy' in this sense can be construed as 'preprogrammed rule'. Maynard Smith set up mathematical models in which preprogrammed rules with names like (for the particular case of animal combat) Hawk, Dove, Retaliator, Bully, are let loose in an imagined (or simulated) world to interact with each other. Once again, it is important to understand that the animals implementing the rules are not assumed to be consciously aware of what they are doing, or why. Each preprogrammed rule has a *frequency* in the population (like genes in a gene pool, although the link with DNA doesn't have to be made explicit in the models). The frequencies change in accordance with 'payoffs'. In the social and economic sciences where game theory originated, payoffs can be thought of as equivalent to money. In evolutionary game theory, payoffs have the special meaning of reproductive success: high payoffs for a strategy lead to increased representation in the population.

The key point is that a successful strategy is not necessarily one that wins its particular contests against other strategies. A successful strategy is one that numerically dominates the population. And since a numerically dominant strategy is by definition likely to encounter copies of *itself*, it will stay numerically dominant only if it flourishes in the presence of copies of itself. This is the meaning of 'evolutionarily stable' in Maynard Smith's 'ESS'. We expect to see ESSs in nature, because if a strategy is evolutionarily *un*stable, it will tend to disappear from the population as rival strategies outbreed it.

I won't expound evolutionary game theory any further here because I did that in *The Selfish Gene*, and the same applies to Trivers' ideas on parental investment. Here it is sufficient to say that the publications of Trivers and Maynard Smith in the early 1970s rekindled my interest in the ideas of Hamilton that had inspired me in the 1960s, and moved me to return to the book whose first chapter had slumbered in a drawer since the end of the power strikes. Maynard Smith's game-theoretic ideas dominated the chapter on aggression, and inspired my treatment of many topics in later chapters.

So, finally, in 1975, having finished my 'hierarchical organiz-ation' paper, I took the sabbatical leave to which I was entitled, stayed at home every morning, and devoted myself to my typewriter and *The Selfish Gene*. Indeed, so devoted was I to the task that I didn't attend the crucial meeting at which New College was electing a new Warden. A colleague slipped out of the meeting and tele-phoned urgently to tell me the vote was extremely close and to beg me to come quickly. I now think that, although my sabbatical leave entitled me to do so, my absenting myself from such a crucial vote was an act of self-indulgent irresponsibility. The meeting would have taken only a few hours of my time, and the repercussions of my lost vote potentially might have been felt for many years. Fortunately, the man for whom I would have voted got in anyway (and became an excellent Warden), so I don't have to endure a burden of guilt for changing the course of college history. Actually, his rival would have been very good too, and college meetings would certainly have been amusing as he was justly reputed to be the wit-tiest man in Oxford.

I wrote *The Selfish Gene* in a frenzy of creative energy. I'd com-pleted three or four chapters when I spoke to my friend Desmond Morris about publication. A legendarily successful author himself, Desmond arranged a meeting with Tom Maschler, doyen of London publishers. I met Mr Maschler in his high-ceilinged, book-lined room at Jonathan Cape in London. He'd read my chapters and liked them, but urged me to change the title. 'Selfish', he explained to me, is a 'down word'. Why not *The Immortal Gene*? With hindsight, he was very probably right. I can't now remember why I didn't follow his advice. I think I should have done.

I didn't, in any case, pursue him as a publisher because matters were rather forcefully taken out of my hands. At lunch one day in New College Roger (now Sir Roger) Elliott, Oxford's Professor of Theoretical Physics, said he had heard I was writing a book, and asked me about it. I told him a little of what I was trying to do, and he seemed interested. As it happened, he was a member of the Board of Delegates of Oxford University Press, and he passed the word to Michael Rodgers, the appropriate editor at that ancient publishing house. Michael wrote and asked to see my chapters. I sent them to him.

And then the whirlwind hit – beginning with his characteristically loud voice, over the telephone: 'I've read your chapters. Haven't been able to sleep since. *I MUST HAVE THAT BOOK!*' Well, some people might resist that kind of persuasion, but not me. Michael was clearly my kind of publisher. I signed the contract and set to work with redoubled urgency to complete the book.

I now find it quite hard to comprehend how we all used to tolerate the burden of writing in the age before computer word processors. Pretty much every sentence I write is revised, fiddled with, re-ordered, crossed out and reworked. I reread my work obsessively, subjecting the text to a kind of Darwinian sieving which, I hope and believe, improves it with every pass. Even as I type a sentence for the first time, at least half the words are deleted and changed before the sentence ends. I have always worked like this. But while a computer is naturally congenial to this way of working, and the text itself remains clean with every revision, on a typewriter the result was a mess. Scissors and sticky tape were tools of the trade as important as the typewriter itself. The growing typescript of *The Selfish Gene* was covered with xxxxxxx deletions, handwritten insertions, words ringed and moved with arrows to other places, strips of paper inelegantly taped to the margin or the bottom of the page. One would think it a necessary part of composition that one should be able to read one's text fluently. This would seem to be impossible when working on paper. Yet, mysteriously, writing style does not seem to have shown any general improvement since the introduction of computer word processors. Why not?

The Selfish Gene went through two fair copies typed up by Pat Searle, the motherly secretary of the Animal Behaviour Research Group. Each one went to Michael Rodgers and came back with his helpful, handwritten annotations. In particular, he excised some purple passages that my romantically youthful enthusiasm had pushed way over the top. In Peter Medawar's metaphor of the writer as organist, 'a scientist's fingers, unlike a historian's, must never stray toward the diapason'. The end of chapter 2 of *The Selfish Gene* is about as purple as science prose should get, and I blush to recall (and am glad I haven't preserved) the paragraph that followed it. Here's the paragraph of paler purple that survived Michael's moderating pen. It's the end of the chapter on the origin of life and the spontaneous arising in the primeval soup of 'replicators', which later moved into the world of 'vehicles' – living organisms.

Was there to be any end to the gradual improvement in the techniques and artifices used by the replicators to ensure their own continuation in the world? There would be plenty of time for improvement. What weird engines of self-preservation would the millennia bring forth? Four thousand million years on, what was to be the fate of the ancient replicators? They did not die out, for they are past masters of the survival arts. But do not look for them floating loose in the sea; they gave up that cavalier freedom long ago. Now they swarm in huge colonies, safe inside gigantic lumbering robots, sealed off from the outside world, communicating with it by tortuous indirect routes, manipulating it by remote control. They are in you and in me; they created us, body and mind; and their preservation is the ultimate rationale for our existence. They have come a long way, those replicators. Now they go by the name of genes, and we are their survival machines.

That paragraph encapsulates the central metaphor of the book, and also its science-fictiony feel. Indeed, I began my preface with the words:

> This book should be read almost as though it were science fiction. It is designed to appeal to the imagination. But it is not science fiction: it is science. Cliché or not, 'stranger than fiction' expresses exactly how I feel about the truth. We are survival machines – robot vehicles blindly programmed to preserve the selfish molecules known as genes. This is a truth which still fills me with astonishment. Though I have known about it for years, I never seem to get fully used to it. One of my hopes is that I may have some success in astonishing others.

And the opening lines of chapter 1 continued the science fiction mood:

> Intelligent life on a planet comes of age when it first works out the reason for its own existence. If superior creatures from space ever visit earth, the first question they will ask, in order to assess the level of our civilization, is: 'Have they discovered evolution yet?' Living organisms had existed on earth, without ever knowing why, for over three thousand million years before the truth finally dawned on one of them. His name was Charles Darwin.

Niko Tinbergen hated that opening, when the book was published and he read it. He didn't like anything that suggested that humanity is an intelligent species, and he felt deeply wounded by the terrible effects we have had on the world. But that really wasn't the kind of point I was making.

I should say something about the last chapter: 'Memes: the new replicators'. Given that the rest of the book thrust the gene to centre stage as the starring replicator in the evolution of life, it was important to dispel the impression that the replicator has to be DNA. In keeping with the science fiction mood of the opening, I pointed out that on other planets the evolution of life could be fostered by a completely different system of self-replication – but that, whatever it was, it would have to have certain qualities, such as high fidelity of copying.

Casting around for an example, I could have used computer viruses if they had been invented in 1975. Instead, I lit upon human culture as a new 'primeval soup':

> But do we have to go to distant worlds to find other kinds of replicator and other, consequent kinds of evolution? I think that a new kind of replicator has recently emerged on this very planet. It is staring us in the face. It is still in its infancy, still drifting clumsily about in its primeval soup, but already it is achieving evolutionary change at a rate which leaves the old gene panting far behind.
>
> The new soup is the soup of human culture. We need a name for the new replicator, a noun which conveys the idea of a unit of cultural transmission, or a unit of *imitation*. 'Mimeme' comes from a suitable Greek root, but I want a monosyllable that sounds a bit like 'gene'. I hope my classicist friends will forgive me if I abbreviate mimeme to *meme*. If it is any consolation, it could alternatively be thought of as being related to 'memory', or to the French word *même*. It should be pronounced to rhyme with 'cream'.
>
> Examples of memes are tunes, ideas, catch-phrases, clothes fashions, ways of making pots or of building arches. Just as genes propagate themselves in the gene pool by leaping from body to body via sperms or eggs, so memes propagate themselves in the meme pool by leaping from brain to brain, via a process which, in the broad sense, can be called *imitation*.

I went on to discuss various ways in which the idea of memes might be applied, for example to the spread and inheritance of religion. My primary intention, however, was not to make a contribution to the theory of human culture, but to downplay the gene as the only conceivable replicator that might lie at the root of a Darwinian process. I was trying to push 'Universal Darwinism' (the title of a later paper, based on my lecture to the 1982 conference commemorating Darwin's death). Nevertheless I am delighted that the philosopher Daniel Dennett, the psychologist Susan Blackmore and others have run, so productively, with the meme ball. More than thirty books have now been published with the word 'meme' in their title, and the word has made it into the *Oxford English Dictionary* (whose criterion is that it must be used, without attribution or definition, in a significant number of published places).

Publication of one's first book is a heady time for a young author. I made frequent trips to the stately neo-classical OUP building in Walton Street, and sometimes the London office in Ely House, to meet the various people involved in the complex business of production, design, marketing and so on. When it came to jacket design, the science fiction mood of the book led me again to the elegantly porticoed North Oxford door of Desmond Morris. As well as being a biologist, television personality, anthropological collector, (implausible) raconteur[1] and best-selling author, Desmond is an accomplished surrealist painter. His paintings have an unmistakably biological feel. He has created a dreamscape in which other-worldly creatures live and move and have their evolution – for they do evolve, from canvas to canvas: just what was needed for *The Selfish Gene*. He was delighted by the idea of providing a jacket design, and Michael Rodgers and I went to look at the paintings on his walls

---

[1] I suspect him of being the original source of a widely circulated anecdote about the film star Diana Dors. She and he came from the same Wiltshire town and were childhood friends. Her real surname was not Dors but Fluck. She was invited back to open some fête or other, and the vicar, thinking to introduce her by the name the locals would have known, genially asked them to welcome the lovely 'Diana . . . Clunt'.

and in his studio. *The Expectant Valley* stood out, not just for its bold colours and air of brooding fecundity, but also more mundanely in that it provided a convenient space to accommodate the title. We chose it with pleasure and I believe it enhanced the sales of the book.

As it happened, Desmond had an exhibition around this time in a small gallery in Walton Street near the OUP building, and *The Expectant Valley* was one of the paintings on sale. Its price, £750, happened to be exactly equal to the advance the publishers had paid me for my book. The coincidence was too much to resist and, after repeated visits to the gallery during which I became fond of many of the paintings, I bought *The Expectant Valley*. I think Desmond was a bit embarrassed, and he kindly threw in another, slightly similar painting, *The Titillator*. The two go rather well together.

The Selfish Gene was published in the autumn of 1976; I was thirty-five. It was reviewed widely, surprisingly so for a first work by an unknown author, and I still don't really know why it received the attention it did. There was no launch party and no obvious fanfare organized by the publishers. Some months after publication it came to the notice of Peter Jones, one of the producers on the BBC's 'flagship' science series *Horizon*. Peter asked me if I would like to present a documentary on the subject, but I was much too shy at that time to dare appear on television, and I recommended John Maynard Smith instead. He did a good job – he had a wonderfully warm and engaging manner – and the documentary, which had the same name, *The Selfish Gene*, must have given a good boost to the book's sales, at least in Britain. But the broadcast came too late to account for the wide review coverage the book received.

I don't do it any more, but for that first book I kept a scrap-book of reviews, and I have just been glancing at them again. There were more than 100, and a rereading doesn't generally bear out the common perception of the book as controversial. Almost all the reviews were favourable. Among early reviewers were the psychiatrist Anthony Storr, the anthropologists Lionel Tiger and Francis Huxley (son of Julian), the naturalist Bruce Campbell and the philosopher Bernard Williams, whom I came to know much later as one of those entertaining conversationalists whose wit had the capacity to 'raise the game' of any companion. There were hostile reviews from two biologists identified with the political left, Steven Rose and Richard Lewontin, and – more subtly barbed – from Cyril Darlington on the opposite side of the political spectrum. But these were rare. Most of the reviewers got the message, expounded it fairly and were very nice about the book. Especially warming for me were the highly favourable reviews by Peter Medawar and W. D. Hamilton. Hamilton even hit the particular nail I had originally targeted in my quest to answer Lorenz, Ardrey and the panglossians of the 1960s and the 'BBC Theorem':

This book should be read, can be read, by almost everybody. It describes with great skill a new face of the theory of evolution. With much of the light, unencumbered style that has lately sold new and sometimes erroneous biology to the public, it is, in my opinion, a more serious achievement. It succeeds in the seemingly impossible task of using simple, untechnical English to present some rather recondite and quasi-mathematical themes of recent evolutionary thought. Seen through this book in their broad perspective at last, these will surprise and refresh even many research biologists who might have supposed themselves already in the know. At least, so they surprised this reviewer. Yet, to repeat, the book remains easily readable by anyone with the least grounding in science.

There was nobody in the world whom I would rather have surprised in such a way than 'this reviewer'. I was also touched by the way Bill Hamilton ended his beautifully written review with poems, one by Wordsworth, and the other by Housman, whose Shropshire

Lad I often found myself identifying with Bill's complex personality:

> From far, from eve and morning
> And yon twelve-winded sky,
> The stuff of life to knit me
> Blew hither: here am I
>
> . . .
>
> Speak now, and I will answer;
> How shall I help you, say;
> Ere to the wind's twelve quarters
> I take my endless way.

Not a bad epitaph for an evolutionary scientist, and Bill Hamilton was probably the greatest evolutionary scientist of the second half of the twentieth century. While this volume of autobiography was in its final stages, I found a treasure among a bundle of old papers, with Bill's handwriting at the top: it was a copy of the last page of his lecture notes, containing a rewriting of another Housman poem, 'The Immortal Part', to incorporate the idea of the 'immortal' gene. I have no memory of the lecture he is referring to, or when he gave it, and the paper is undated. I have reproduced it in the web appendix.

Long after *The Selfish Gene* was published, Bill became my close colleague at Oxford and I saw him almost daily at lunch in New College. I am humbly proud of the role played by my book in bringing his brilliant ideas to a wider audience. But I like to hope, too, that there are other ways in which the book changed the way my professional colleagues think about their subject. I like to think it is no accident that, if you visit a biological field station in the Serengeti, or Antarctica, or the Amazon or the Kalahari, and listen to the active researchers talking shop over their beers in the evenings, what you hear will be laced with talk of genes. It's not that they're talking about the molecular antics of DNA – although that is interesting too – but the underlying assumption of these conversations is that the behaviour of the animals and plants under study is aimed at preserving genes and propagating them through succeeding generations.

LOOKING BACK
DOWN THE PATH

Publication of *The Selfish Gene* marks the end of the first half of my life and a convenient point at which to pause and look back. I'm often asked whether my African childhood led me to become a biologist. I'd like to answer yes, but I'm not confident. How can we know whether the course of a life would have been changed by some particular alteration in its early history? I had a trained botanist for a father and a mother who knew the name of every wildflower you could normally expect to see – and both of them were always eager to satisfy a child's curiosity about the real world. Was that important to my life? Yes, it surely was.

My family moved to England when I was eight. What if they hadn't? At the eleventh hour I was sent to Oundle rather than Marlborough. Did that arbitrary change seal my future? Both were boys-only schools. A psychologist might suggest that I'd have turned out a socially better-adjusted person if I'd been sent to mixed schools. I scraped into Oxford. What if I'd failed, as probably I nearly did? What if I had never had tutorials with Niko Tinbergen, and therefore followed my earlier plan to do biochemical research for my DPhil rather than animal behaviour? Surely my whole life would then have been different? Probably I would never have written any books.

But perhaps life has a tendency to converge on a pathway, something like a magnetic pull that draws it back despite temporary deviations. As a biochemist, might I have eventually returned to the path that led to *The Selfish Gene*, even if I had then given it a more molecular slant? Perhaps the pull of the pathway would have led me to write (again biochemically slanted) versions of every one of my dozen books. I doubt it, but this whole 'returning to the path' idea is not uninteresting and I shall . . . er . . . return to it.

The hypotheticals that I posed are relatively large. Take something utterly trivial yet, I shall argue, momentous. I've already speculated that we mammals owe our existence to a particular sneeze by a particular dinosaur. What if Alois Schicklgruber had happened to sneeze at a particular moment – rather than some other particular moment – during any year before mid-1888 when his son Adolf Hitler was conceived? Obviously I have not the faintest idea of the exact sequence of events involved, and there are surely no historical records of Herr Schicklgruber's sternutations, but I am confident that a change as trivial as a sneeze in, say, 1858 would have been more than enough to alter the course of history. The evil-omened sperm that engendered Adolf Hitler was one of countless billions produced during his father's life, and the same goes for his two grandfathers, and four great-grandfathers, and so on back. It is not only plausible but I think certain that a sneeze many years before Hitler's conception would have had knock-on effects sufficient to derail the trivial circumstance that one particular sperm met one particular egg, thereby changing the entire course of the twentieth century including my existence. Of course, I'm not denying that something like the Second World War might well have happened even without Hitler; nor am I saying that Hitler's evil madness was inevitably ordained by his genes. With a different upbringing Hitler might have turned out good, or at least uninfluential. But certainly his very existence, and the war as it turned out, depended upon the fortunate – well, unfortunate – happenstance of a particular sperm's luck.

> A million million spermatozoa,
> All of them alive:
> Out of their cataclysm but one poor Noah
> Dare hope to survive.
>
> And among that billion minus one
> Might have chanced to be
> Shakespeare, another Newton, a new Donne–
> But the One was Me.
>
> Shame to have ousted your betters thus,
> Taking ark while the others remained outside!
> Better for all of us, froward Homunculus,
> If you'd quietly died!
>
> Aldous Huxley

If his father had sneezed at a particular hypothetical moment, Adolf Hitler would not have been born. Nor would I, for I owe my improbable conception to the Second World War – as well as to much less momentous things that happened. And of course all of us can take the argument back through countless previous generations, as I did with my hypothetical dinosaur and the destiny of the mammals.

Taking on board the contingent frailty of the event chain that led to our existence, we can still go on to ask – as I did a moment ago – whether the course of a named individual's life is sucked back, magnetically, into predictable pathways, despite the Brownian buffetings of sneezes and other trivial, or not so trivial, happenings. What if my mother's joking speculation were really true, if the Eskotene Nursing Home really had muddled me up with Cuthbert's son and I had been brought up as a changeling in a missionary household? Would I now be an ordained missionary myself? I think geneticists know enough to say no, probably no.

If my family had stayed on in Africa and I had persisted at Eagle rather than moving to Chafyn Grove, and then been sent to Marlborough rather than Oundle, would I still have got into Oxford and met Niko Tinbergen? It is not unlikely, for my father would have been hell-bent on my following him and half a dozen earlier Dawkinses into Balliol. Despite taking other forks in the road, pathways can converge again. The likelihood that they will do so depends on genuinely investigable matters such as the relative contribution of genes and education to adult abilities and proclivities.

We can leave rarefied speculations about hypothetical sneezes and converging pathways, and return to familiar territory. As a man looks back on his life so far, how much of what he has achieved, or failed to achieve, could have been predicted from his childhood? How much can be attributed to measurable qualities? To the interests and pastimes of his parents? To his genes? To his happening to meet a particularly influential teacher, or happening to go to summer camp? Can he list his abilities and shortcomings, his pluses and minuses, and use them to understand his successes and failures? This is the familiar territory I meant, and it is that trodden, for example, by Darwin at the end of his autobiography.

Charles Darwin is my greatest scientific hero. Philosophers are fond of saying that all philosophy is a series of footnotes to Plato. I sincerely hope that is not the case, because it doesn't say much for philosophy. A far better case could be made that all of modern biology is a series of footnotes to Darwin. And that would be a genuine compliment to the science of biology. Every biologist treads in Darwin's footsteps and, in all humility, none of us could do better than to follow his example. In the closing pages of his autobiography he essayed a retrospective itemization of personal faculties lacked or possessed. Again in humility I shall do the same, taking his method of self-assessment as a model to be followed.

> ... I have no great quickness of apprehension or wit which is so remarkable in some clever men, for instance Huxley.

Here, at least, I can claim mental kinship with Darwin, although in his case the modesty was exaggerated.

> My power to follow a long and purely abstract train of thought is very limited; I should, moreover, never have succeeded with metaphysics or mathematics.

Again, same for me, despite the ludicrously ill-founded reputation for mathematical ability that I briefly enjoyed – or endured – in Bevington Road days. John Maynard Smith, as a mathematical biologist himself, engagingly expressed wonderment at how it is possible to 'think in prose'. He said it in the *London Review of Books* in 1982, at the end of a joint review of *The Selfish Gene* and its sequel (aimed at professional biologists), *The Extended Phenotype*:

> I have left till last what is to me the strangest feature of both books, because I suspect it will not seem strange to many others. It is that neither book contains a single line of mathematics, and yet I have no difficulty in following them, and as far as I can detect they contain no logical errors. Further, Dawkins has not first worked out his ideas mathematically and then converted them into prose: he apparently thinks in prose, although it may be significant that, while writing *The Selfish Gene*, he was recovering from a severe addiction to computer programming, an activity which obliges one to think clearly and to say exactly what one means. It is unfortunate that most people who write about the relation between genetics and evolution without the intellectual

prop of mathematics are either incomprehensible or wrong, and not infrequently both. Dawkins is a happy exception to this rule.

Back to Darwin's autobiographical soliloquy:

So poor in one sense is my memory, that I have never been able to remember for more than a few days a single date or a line of poetry.

That might well have been really true of Darwin, and it doesn't seem to have held him back. My ability to remember poetry word for word hasn't helped my science much, although it has enriched my life and I would not ever wish to lose it. Perhaps, too, a feeling for poetic cadence has some influence on writing style.

My habits are methodical, and this has been of not a little use for my particular line of work. Lastly, I have had ample leisure from not having to earn my own bread. Even ill-health, though it has annihilated several years of my life, has saved me from the distractions of society or amusement.

My habits are anything but methodical, and that – not ill health in my case – has surely annihilated what might have added up to years of more productive life. The same accusation could be levelled at the distractions of society or amusement (and playing with computers in my case), but life is for living as well as producing. I have had to earn my own bread. But – while happy to ignore the attacks I have (yes, really) received for being white, male and adequately educated – I cannot deny a measure of unearned privilege when I compare my childhood, boyhood and youth to others less fortunate. I do not apologize for that privilege any more than a man should apologize for his genes or his face, but I am very conscious of it. And I am grateful to my parents for giving me what will strike some as a favoured childhood. Others might consider it less than a blessing to have been sent away to the spartan regime of boarding school aged seven, but even there I have reason to be grateful to my parents, for whom this style of education was a great expense, necessitating sacrifices from them.

Darwin had earlier let his modesty guard drop a little when he considered his – by any standards formidable – powers of reasoning:

> Some of my critics have said, 'Oh, he is a good observer, but has no power of reasoning.' I do not think this can be true, for the *Origin of Species* is one long argument from the beginning to the end, and it has convinced not a few able men. No one could have written it without having some power of reasoning.

Mr Darwin (never Sir Charles, and what an amazing indictment of our honours system that is), that last sentence should win a prize for world-class understatement. Mr Darwin, you are one of the great reasoners and one of the great persuaders of all time.

I am not a good observer. I'm not proud of it and I try eagerly, but I am not the naturalist my father and his father would have wished. I lack patience and have no great knowledge of any particular animal or – despite one privilege of my upbringing – plant group. I know the songs of only half a dozen common British songbirds, and can recognize only about the same number of con-stellations in our night sky or families of our wildflowers. I am much better at the phyla, classes and orders of the animal kingdom – and so I should be, having studied zoology at Oxford: for no other university placed such an emphasis on that classical approach to the subject.

The evidence suggests that I am a reasonably effective persuader. Needless to say, the subjects about which I persuade are small beer compared to Darwin's – except in the sense that, amazingly, the job of persuading people of Darwin's own truth is still not over, and I am one of the labourers in Darwin's vineyard today. But that story belongs in the second half of my life, during which the majority of my books were written: it belongs in the companion volume that should follow in two years' time – if I am not carried off by the unpredictable equivalent of a sneeze.

ACKNOWLEDGEMENTS

For advice, help and support of various kinds, I would like to thank Lalla Ward Dawkins, Jean Dawkins, Sarah and Michael Kettlewell, Marian Stamp Dawkins, John Smythies, Sally Gaminara, Hilary Redmon, Sheila Lee, Gillian Somerscales, Nicholas Jones, John Brockman, David Glynn, Ross and Christine Hildebrand, Bill Newton Dunn, R. Elisabeth Cornwell, Richard Rumary, Alan Heesom, Ian McAlpine, Michael Ottway, Howard Stringer, Anna Sander, Paula Kirby, Stephen Freer, Bart Voorzanger, Jennifer Jacquet, Lucy Wainwright, Bjorn Melander, Christer Sturmark, Greg Stikeleather, Ann-Kathrin Ehlers, Jan and Richard Gendall, Rand Russell.

TEXT ACKNOWLEDGEMENTS

Every effort has been made to trace the copyright holders, but any who have been overlooked are invited to get in touch with the publishers.

'To the Balliol Men Still in Africa' by Hilaire Belloc reprinted by permission of Peters Fraser & Dunlop (www.petersfraserdunlop.com) on behalf of the Estate of Hilaire Belloc.

Extract from *Iris Murdoch: A Life* by Peter J. Conradi © Peter J. Conradi, 2001, reprinted by permission of A. M. Heath & Co Ltd and W. W. Norton.

Extract from *The Autobiography of Bertrand Russell* by Bertrand Russell © 2009 The Bertrand Russell Peace Foundation, reprinted by permission of Taylor & Francis Books UK and The Bertrand Russell Peace Foundation Ltd.

Lyrics from 'A Song of Reproduction' reprinted by permission of the Estates of Michael Flanders & Donald Swann 2013. Any use of Flanders & Swann material, large or small, should be referred to the Estates at leonberger@donaldswann.co.uk.

Extract from 'Summoned by Bells' from *Collected Poems* by John Betjeman © 1955, 1958, 1962, 1964, 1968, 1970, 1979, 1981, 1982, 2001 reprinted by permission of John Murray (Publishers) and The Estate of John Betjeman.

Extract from 'A Hike on the Downs' from *Collected Poems* by John Betjeman © 1955, 1958, 1962, 1964, 1968, 1970, 1979, 1981, 1982, 2001 reprinted by permission of John Murray (Publishers) and The Estate of John Betjeman.

Extract from *The Loom of Years* by Alfred Noyes © 1902 reprinted by permission of The Society of Authors as the Literary Representative of the Estate of Alfred Noyes.

'Blue Suede Shoes' by Carl Lee Perkins © 1955, 1956 Hi Lo Music, Inc. © Renewed 1983, 1984 Carl Perkins Music, Inc. Administered by Wren Music Co., Division of MPL Music Publishing, Inc. All rights reserved. International copyright secured. Used by permission of Music Sales Limited.

Extract from *The Silent Traveller in Oxford* by Chiang Yee © 1944 Signal Books Ltd.

PICTURE ACKNOWLEDGEMENTS

Section two

Emperor Swallowtail (*Papilio ophidicephalus*): © Ingo Arendt/Minden Pictures/Corbis.

Section three

The Great Hall, Oundle School, Northamptonshire: © Graham Oliver/ Alamy; Ioan Thomas, 1968: Oundle School Archive.

Niko Tinbergen painting hens' eggs to resemble gulls' eggs, *c.*1964: Time & Life Pictures/Getty Images; Mike Cullen, 1979: Monash University Archives, photo Hervé Alleaume; the Surrey Puma hunt: photo courtesy Virginia Hopkinson; People's Park demonstrators and the National Guard, Berkeley, 19 May 1969: © Bettmann/Corbis; punting in Oxford: photo courtesy Lary Shaffer; Peter Medawar at University College, 26 November 1960: Getty Images.

RD and Ted Burk, November 1976: Time & Life Pictures/Getty Images; Danny Lehrman and Niko Tinbergen: photo courtesy Professor Colin Beer; Niko Tinbergen filming: courtesy Lary Shaffer.

William D. Hamilton and Robert Trivers, Harvard, 1978: photo courtesy Sarah B. Hrdy; Michael Rodgers: photo courtesy Nigel Parry; RD and George C. Williams: photo by Rae Silver courtesy John Brockman; John Maynard Smith: Corbin O'Grady Studio/Science Photo Library; *The Selfish Gene*: courtesy Keith Cullen.

INDEX

The Greatest Show on Earth
The Evidence for Evolution
Richard Dawkins

'This is a magnificent book of wonderstanding: Richard Dawkins
combines an artist's wonder at the virtuosity of nature with a
scientist's understanding of how it comes to be'
MATT RIDLEY, AUTHOR OF *NATURE VIA NURTURE*

Charles Darwin, whose 1859 masterpiece *On the Origin of Species* shook
society to its core, would surely have raised an incredulous eyebrow at the
controversy over evolution still raging 150 years later.

The Greatest Show on Earth is a stunning counter-attack on creationists,
followers of 'Intelligent Design' and all those who still question evolution
as scientific fact. In this brilliant *tour de force* Richard Dawkins pulls
together the incontrovertible evidence that underpins it: from living
examples of natural selection to clues in the fossil record; from plate
tectonics to molecular genetics.

The Greatest Show on Earth comes at a critical time as systematic
opposition to the fact of evolution flourishes as never before in many
schools worldwide. Dawkins wields a devastating argument against this
ignorance whilst sharing with us his palpable love of science and the
natural world. Written with elegance, wit and passion, it is hard-hitting,
absorbing and totally convincing.

'A voice of reason in irrational times, Richard Dawkins
is both theorist and explainer of one of the greatest
discoveries of the human mind'
THE TIMES

ANDREW WILSON

DEATH IN A DESERT LAND

Death in a Desert Land is not authorised
by Agatha Christie Ltd

SIMON &
SCHUSTER

London · New York · Sydney · Toronto · New Delhi

A CBS COMPANY

First published in Great Britain by Simon & Schuster UK Ltd, 2019
A CBS COMPANY

This paperback edition published 2020

1 3 5 7 9 10 8 6 4 2

Simon & Schuster UK Ltd
1st Floor
222 Gray's Inn Road
London WC1X 8HB

Simon & Schuster Australia, Sydney
Simon & Schuster India, New Delhi

www.simonandschuster.co.uk
www.simonandschuster.com.au
www.simonandschuster.co.in

A CIP catalogue record for this book
is available from the British Library

Paperback ISBN: 978-1-4711-7350-9
eBook ISBN: 978-1-4711-7349-3

Typeset in the UK by M Rules
Printed and bound by CPI Group (UK) Ltd, Croydon, CR0 4YY

MIX
Paper from
responsible sources
FSC
www.fsc.org   FSC® C020471

To M. F.

DEATH
IN A
DESERT
LAND

Prologue

Dearest Father,

I can hardly believe the words that I am about to set down. It sounds ridiculous, quite preposterous, but I am in fear of my life.

There have been many occasions where I have been forced to confront the prospect of dying. The terrifying ascent and descent of the Finsteraarhorn, when I was convinced I was going to slip into a glacial crevice or, in the midst of a ferocious snow storm, be swept off the edge of a precipice and fall thousands of feet to my death.

I remember, too, the time when I travelled across the desert sands to Hayyil, a perilous journey that many have not survived. When I finally reached the feared city I was taken prisoner. There I heard it said that in that place murder was considered so normal it was likened to the spilling of milk. There are those, many of whom were born in Arabia, who seemed

1

offended by my spirit of adventure, as if I was an affront to the female sex. Indeed, in Hayyil I was told that a woman should only leave her house on three occasions: to marry her bridegroom, on the death of her parents, and in the event of her own death.

And then there was that no man's land of the soul that I inhabited after Dick's death. I have never been so low as then, when it felt as though I had nothing left to live for, when I had been tempted to put an end to it all. The shock of the news of his shooting at Gallipoli almost stilled my heart. I was at a lunch party in London when one of our number casually mentioned how sad it was that Dick Doughty-Wylie had been killed in action. How were they to know of my attachment to him? Afraid that I was going to be ill, I excused myself from the lunch, my head reeling. I learnt later that Dick had left his pistol behind on his boat and witnesses told me that he strode into village houses, which conceivably could have been packed to the rafters with Turkish soldiers, holding nothing more than his cane. As he reached the top of the hill, a swell of Hampshires, Dublins and Munsters behind him, he was shot in the head and buried where he fell. What a terribly sad end to a glorious life.

The only conclusion I can come to as to why Dick would advance unarmed is that he could not face the situation back at home: a wife who threatened to commit suicide if he were to leave her, a woman – me – whom he loved but knew he could never marry.

I must resist the urge to get swept back into the

past. But there is a reason why I cite these occasions, moments when I have faced the possibility of non-existence.

I believe, and have done so since I was a girl, that there is nothing that lies beyond the here and now. When I am gone I know that I will know nothing, be nothing. I don't expect to meet dear Dick, or my darling brother Hugh, in some paradise of an afterlife.

So you see it is not death that frightens me. It is the thought – quite natural and justifiable – that someone may want to wrest my life away from me before I am quite ready. I have lived a good life, a great deal fuller and richer than many women on this earth, but am I ready to die? Lately, I have become weak, I have suffered from illness; but not to see that rare bloom of a daffodil in my garden in Baghdad? Not to walk through the date palms on a spring day or take a swim in the waters of the Tigris on a hot summer's evening? Not to sit under the shade of a tamarind tree and eat a ripe fig? Never again see the delights of my little museum that houses the treasures of the past? And this is nothing compared to the important work still to be done with King Faisal and the continuing improvements in an independent Iraq. No, I am not ready to go just yet.

It could be my fancy, but I have become convinced that someone wishes to do me harm. An uneasiness of spirit has come over me. I feel as though I am being watched, studied, but when I look up there is no one in the room. In the early hours, when I have been in bed reading, I have noticed how Tundra suddenly stirs, the dog's ears pricked, her bright eyes turning on some invisible enemy, a low growl beginning to form in the back of her throat. I have gone to the window, looked out into the purple night, but I have been unable to see or hear anything beyond the stirring of the palms in the breeze. I have taken to sleeping with my gun under my pillow. I need to be ready, prepared. As someone once told me, 'Every Arab in the desert fears the other.'

    Ever your affectionate daughter,
    Gertrude

*

Dearest Father,

    I cannot write very much because my hand is trembling so; I apologise if you cannot read my words.

    Enclosed is a drawing that I received this morning. As you can see, it shows a grave at Ur, which has recently come to light during the dig overseen by Mr Woolley and his team. Next to a stick figure you will see a set of initials. They are my own: G.L.B.

    If the missive's purpose was to unnerve me then it has worked. I feel shaken to the core. My hope is that this was the only function of the letter, that it was designed to unsteady me and nothing more.

    In my previous letter, I spoke of my irrational fears that someone wants me dead; now I am afraid that *this is indeed the case. You know that I do not have a melodramatic streak. I do not strive to create drama where there is none. I have always borne my miseries with fortitude. I am of a practical bent and not prone to fancies. I wish I did not have to set the words down; to see them written before me gives them a certain reality that makes me tremble. But set them down I must.*

If I were to be found dead – and if there was an indication that my death was not due to the onset of some terminal illness – then it is safe to assume that I was murdered. And my murderer? I suggest you look no further than Ur.

Ever your affectionate daughter,
Gertrude

Chapter One

'So what do you make of it?' asked Davison, as he took the two letters from me.

I did not answer immediately. Too many questions were crowding my head. I took a sip of iced soda water and gazed across the terrace of the hotel to the Tigris below. A lonely boatman was singing a queer, discordant song that brought back to me the ghosts of the past.

'And the handwriting is definitely that of Gertrude Bell?'

'It seems so,' said Davison, peering at the scrawl again. 'Someone who knows about these kinds of things has compared the letters to others she wrote to her father and stepmother and although it's difficult to say for certain, there are particular elements of style – such as the distinctive way she formed her "d"s for example, with a curious backwards slope – that suggest they were indeed written by Miss Bell.'

'I've always understood that she died of an illness, pneumonia or something bronchial,' I said, remembering the obituaries I had read of the famous adventurer and Arabist

7

when she died in July 1926. At that time I had been in the midst of my own troubles, drowning in a sea of grief after the death of my mother, valiantly trying to hold together a marriage that was falling apart at the seams, battling a creative block that was driving me to the edge of reason.

'Yes, that was the story put out by the family,' said Davison. 'But according to the doctor who examined her, Miss Bell died from barbiturate poisoning. A bottle of Dial tablets was found by her bedside at her house here in Baghdad. Of course, no one wanted to draw attention to the fact that her death, or so it was thought, was a suicide.'

'And tell me again how these letters came to light.'

'It had been thought that her family had taken possession of the bulk of Miss Bell's archive, diaries, numerous photographs, documents relating to the archaeological museum she founded, letters and so on. Indeed, as I'm sure you know, her stepmother published two volumes of Gertrude's letters only last year. But then, just last month, these unsent two letters were discovered in a tin box that served as a place to store seeds. It was only when one of Gertrude's former servants, Ali, a gardener whom the family continued to employ, started to look for a particular type of seed that he came across them. Of course, he couldn't read the contents – he's a local and Miss Bell was always proud of the fact that she communicated with her servants in Arabic – but Ali knew that they had been written in his mistress's hand. And he knew enough English to realise that the initials G.L.B. were those of Miss Bell. He did the right thing and took the letters to our man in Baghdad. Apparently, the drawing distressed him a great deal. He thought it represented some kind of curse.'

'I can imagine it would have that effect,' I said. Although I had missed the British Museum's Treasures of Ur exhibition earlier in the year, I had seen a similar drawing of dozens of stick figures reproduced in the *Illustrated London News*. Reading about the discovery of the skeletons – which were thought to be victims of human sacrifice that dated to 2,500 years before the birth of Christ – had sent a chill through me. 'Do you know if Miss Bell had any enemies?'

As Davison smiled, his intelligent grey eyes sparkled mischievously. 'Plenty, I would have thought. She was hardly the easiest of women to get along with. Headstrong – independent if one wants to be polite, bloody infuriating if one is speaking plainly. Sorry, I—'

'Davison, you know I always prefer plain speaking. Did you know her well?'

'We only met a few times, once out here in Baghdad, another time in Egypt, and then, of course, in London.'

'And is there anything I should know about her background? Her work for you at the Secret Intelligence Service or for any of the other covert government departments, for instance?'

Davison looked away from me, his gaze settling on a cluster of black rocks on the other side of the riverbank. 'Now, is that a sacred ibis down there?' He started to raise himself out of the wicker chair to take a closer look. 'I do believe it is. You know what, I've never seen one of those. Fascinating, of course, especially if you're interested in

Egyptian mythology. Venerated and mummified by the ancient Egyptians, you know, a representation of Thoth.'

I could feel my cheeks begin to colour with frustration.

'But, on closer inspection,' he said as he squinted down at the river, 'it could be a northern bald ibis, said to be one of the first birds that Noah released from his Ark, that bird being a representation of fertility. Anyway, whatever it was, it's gone now.'

As he turned his head to me, Davison assumed a pose of the utmost seriousness. He managed to freeze his features into a mask of implacability before the skin on his cheeks started to turn pink, his eyes sparkled once more and he burst into a loud fit of laughter.

'I'm sorry, Agatha,' he said, taking up a starched linen napkin to wipe the beads of sweat from above his upper lip. 'It was too good an opportunity to miss to tease you. I know it's not really a laughing matter, but you should have seen your face! You looked like you wanted to slap me – or at the very least walk out of the hotel and take the first Orient Express back to London.'

'You can laugh as much you like,' I said, fighting the urge to smile, 'but there was a time, not too far in the past, when you didn't trust me enough to provide me with all the information I needed to help you. Remember?'

'But Tenerife was different,' he said, lowering his voice to a near whisper. 'You know the reason why I was so reluctant to share certain details of my life with you.'

'That may be so,' I said. Although it would have been easy to do so, I decided not to embarrass Davison, and instead turned the conversation back to the current case.

'Now that you've had a jolly laugh at my expense, why don't you tell me what you know?'

'Very well,' he said, as he crossed his legs. 'Yes, you're right. Miss Bell did work in secret intelligence during the war.'

'In what capacity?'

'She was stationed in Cairo, where it was her mission to provide us with evidence about the links between the Germans and the Turkish Empire, particularly in eastern and northern Arabia. Because she had done all this travelling, trekking across the desert, gossiping with sheikhs over strong coffee, she had an unparalleled insight into certain alliances which would otherwise have remained obscure. She wrote reports for the *Arab Bulletin*, which I'm sure you know provided the British government with a stream of very helpful secret information.'

I thought back to my own time in Cairo, where I had lived with my mother for three months during the winter of 1907. What a stupid girl I had been. At seventeen years old, I had only been interested in romance – endless flirtations with dashing men in the three or four regiments stationed out there – and my appearance.

'Miss Bell sounds like she was an exceptional woman,' I said, feeling distinctly unworthy in comparison. 'Am I right in remembering that she took a degree from Oxford?' My education could be described as patchy at best – for great swathes of my childhood I did not even go to school – and, the more I heard from Davison, the more I was beginning to feel envious of Miss Bell's extraordinary achievements.

'Yes, the first woman to take a First – and a brilliant one

at that – in Modern History. And in two years, instead of the usual three. She always seemed the most intelligent person in the room. That had its benefits, particularly for the department, but of course she was not the most subtle of individuals. I remember once, at some grand dinner, sitting opposite her and hearing her describe one of the diplomat's wives, in a dismissive voice, as a "nice little woman". That was always her insult of choice for women she deemed her inferior, which was the majority.'

In that instant, I felt a certain relief that Miss Bell was no longer with us – I doubted she would have liked me – and then, almost immediately, I felt ashamed for thinking ill of the dead.

'But why do you think she believed someone at Ur wanted to kill her? Did she know anyone there?'

Davison took out two photographs from the inside pocket of his jacket and passed them over to me. 'This is Leonard Woolley, of whom you've no doubt heard, the man in charge of the dig down at Ur.' I studied the image of a man dressed in shorts and a jacket, a man with a puckish face, who was sitting cross-legged on the ground peering intently at a clay slab in his hands. 'Woolley and Miss Bell knew each other during the war when he was head of intelligence at Port Said and she was stationed in Cairo. From all accounts, they seemed to get on well. The only thing we've managed to dig up is a possible suggestion that the two did not see eye to eye in regards to the dividing up of the treasures at Ur.'

'What do you mean?'

'Well, in Miss Bell's role as head of antiquities in Iraq

it was her duty to decide which objects she should set aside for her museum here in Baghdad, and which ones she allowed the team at Ur to transport back to Britain and America. Apparently, Woolley was upset that Miss Bell insisted on keeping for the museum an ancient plaque showing a milking scene. I've been told that Woolley valued it at around ten thousand pounds. She also managed to secure a gold scarab, which experts believe is worth one hundred thousand pounds.'

I could not disguise my astonishment. 'Really? As much as that?'

'Yes, and she won it on the toss of a rupee.'

'I can imagine Woolley would be annoyed. But surely nobody is suggesting that's the reason why he might want her dead?'

'We both know that murder has been done for an awful lot less.'

'Indeed we do,' I said, taking a moment to pause to look at the river, with its traffic of *gufas* and other vessels. 'So what do you have in mind? You told me something of your plan before we left London, but I'm assuming there is something more specific you want me to do?'

'Yes, there is,' said Davison, all traces of his former joviality now erased from his face. 'We need to know for certain whether there is any truth in Miss Bell's suspicion that she was going to be murdered. For that, I'm asking you to travel down to Ur. I've already discerned that you would be welcome there. There is a Mrs Woolley, you see, and she normally dislikes other women on site. She is the queen of the camp and likes to be treated as such. She cannot endure the prospect of competition from other members of the female sex, but I am told that for you she would make an exception. The reason why you are most suitable for this assignment, the reason why your name was mentioned to me by the head of the division, Hartford, was because of Mrs Woolley's enthusiasm for *The Murder of Roger Ackroyd*. Something of a literary snob by all accounts, but her passion for that book is—'

'I see,' I said, feeling uncomfortable with the prospect of further praise of my work. 'Can you tell me any more about Mrs Woolley?'

Davison did not say anything for a few seconds. As he began to form his thoughts I noticed a pair of horizontal lines crease their way across his forehead, making him look a good deal older.

'Perhaps it's better if you forge your own opinion of her,' he said, draining the last of his brandy. He raised his hand to call over the waiter. 'But know this: Miss Bell told various acquaintances that she believed Mrs Woolley to be a dangerous woman.'

'And is this her, in this photograph here?' I asked, gesturing towards an image of a woman who, although middle-aged, still possessed a certain striking beauty. The photograph showed her sitting on the desert floor examining a shard of an old pot.

'Yes, that's her, all right,' said Davison. 'From what I've heard of her, Katharine Woolley is a Jekyll and Hyde character, charming one moment, cold and cruel the next. There is also some mystery surrounding the death of her first husband, Lieutenant Colonel Bertram Keeling, whom she married at St Martin in the Fields in March 1919. Six months later, Keeling, who was only thirty-nine, shot himself at the foot of the Great Pyramid.'

'And did Keeling work in intelligence too?'

'As a matter of fact he did – during the war, in Cairo. But at the time of his death he was Director-General of the Survey of Egypt and President of the Cotton Research Board.'

'A good cover for espionage if ever I heard one. Do you know if he had any dealings with Miss Bell? Had they a history I should know about?'

'Not as far as we know. But the very nature of these things means that a great deal of what occurred during the war remains a secret.'

'But doesn't it seem odd to you that Keeling and Miss Bell, both of whom worked in Cairo in intelligence during the war, went on to die as supposed suicides?' I asked. 'What if someone wanted them dead and made the murders *look* like suicides?'

'It's a possibility, of course. But we've never thought about connecting the two cases because—'

'Because the suggestion is that your own government, or an agency acting on its behalf, may have something to do with their deaths?'

'I wouldn't put it quite in those terms,' Davison said dismissively.

'I'd rather know the whole truth, if you have access to it,' I said.

'Yes, of course, but I promise on this occasion there is nothing else I can tell you. I'll put out some feelers, see what I can come up with, but at the moment there really is nothing to link their deaths.'

'That's not quite true,' I said, taking up the photographs of the archaeologist and his wife. 'There is something that links them together: the Woolleys.' I tried to picture a sequence of possible events, the scenes flashing through my mind like a series of imagined tableaux. 'Why would a man kill himself six months after getting married? That doesn't seem right to me, as he would surely still be in the first flush of romance. Of course, he may have realised that he had made a terrible mistake or he could have faced the prospect of ruin. Perhaps he had saddled himself with debt or embroiled himself in an impending scandal in his personal life. Those need to be ruled out. With suicide, there are so many factors one needs to take into account, but there's something about that case that strikes me as odd. And then, seven years later, Miss Gertrude Bell, at the peak of her achievements, takes her own life by an overdose of barbiturates. That too doesn't ring true. These letters written by her to her father just before her death, there is something very queer about them. Why weren't they sent? How did they end up in that seed tin? Why have they just turned up now?'

Davison was looking at me with a mix of admiration and bafflement. 'I'm at a loss to know what to say,' he said. 'I'm afraid I don't have any answers.'

'The "suicide" of Colonel Keeling, Katharine Woolley's first husband, in 1919,' I continued. 'The "suicide" of Miss Bell in 1926, whom we know had dealings with Leonard Woolley and who described Katharine as dangerous.

Then the recent discovery of these letters, letters written by Gertrude Bell in which she directs us to Ur to look for her killer. Could the murderer be either Leonard or Katharine Woolley?'

'But what could be their motive?'

'Something that is hidden out of sight, at least for the moment,' I said. 'It could be connected with their intelligence work – we know that Colonel Keeling, Miss Bell and Leonard Woolley all served in secret operations during the war. Perhaps that's something you can look into?'

Davison nodded and scribbled in his notebook. Although we were sitting in the shade of the terrace the breeze had dropped and the heat was becoming unbearable. I shifted in my seat and took another sip of my soda water, which was now lukewarm. 'Of course, there is another possibility.'

'There is?'

'Oh yes,' I said, pausing for a moment. 'The Woolleys, the husband and the wife, could have been responsible for both murders.'

'What do you mean? As if they had some kind of pact?'

'Perhaps,' I said. 'People have done stranger things for love – or some warped version of it.' I thought back to the case in Tenerife and the mess Davison had got himself into over his feelings for a young man, whose partly mummified body had been found in a cave. And I thought of my former husband Archie and the scandal surrounding my disappearance for ten days at the end of 1926.

'Yes, indeed, but best not to dwell on that,' said Davison, as he noticed the cloud of melancholy that had started to steal over me. 'So, what do you think? If I stay here in

Baghdad are you happy to travel down to Ur? See what you can dig up? As I said, I'm certain you're the perfect person for this.'

As Davison continued to talk – about what an extraordinary job I had done in Tenerife, how I had brilliantly applied my skills as a novelist to the business of solving crimes – I thought of my old life as a conventional wife and mother. Archie's affair with Miss Neele, followed by the nasty rash of newspaper headlines that followed my disappearance, the ridiculous rumours that Archie was somehow responsible, the allegations that I had staged the whole thing as some cheap publicity stunt, had taken their inevitable toll. And then there was the interview I had been persuaded to give to the *Daily Mail* earlier in the year which was designed, in that dreadful phrase, to 'put the record straight'. Little did anyone know how much I had drawn on my skills as a novelist during that meeting.

I often wondered, when I woke in the middle of the night and was unable to get back to sleep, whether I could have saved my marriage. If I had been more attentive to Archie . . . if I had been a better wife . . . if I had never taken up writing and had simply devoted myself to him and his concerns, and laughed at the inane jokes of his golfing friends and never had a complicated thought in my head. Would that have made any difference? Of course, it was all too late now. The divorce had gone through. We were no longer man and wife. But if I was no longer a wife, who was I? A mother, of course: yes, always. An author? After that awful period of writer's block following my mother's death, I had produced a couple of books of which I was not proud. But I hoped I was back on track now. After all, I had no option: writing was the way I earned my living. But what else?

'Agatha – are you all right?' It was Davison. 'Did you hear what I was saying?'

'Sorry, it's this heat,' I said, feeling a little dizzy. 'I'm not sure Baghdad entirely agrees with me.'

'Yes, you do look a little pale. I say, why don't I walk you back to your room?'

'That would be very kind, thank you,' I said, as Davison took my arm. 'I really do think it's best if I lie down.'

But I had no intention of taking a rest.

Chapter Two

After Davison had safely seen me into my room – and arranged to meet me for breakfast at the hotel the following day in order to say goodbye – I took a quick bath and changed into a silk blouse. I walked out onto the balcony and watched the sunset turn the Tigris red. In the distance, I could hear the call to prayer echo around the ancient city.

I had been in Baghdad for only a day and everything about it seemed so exotic, so wonderfully oriental. Even though I had travelled the world, or a good deal of it, during my 1922 tour with Archie, I had never visited anywhere like this. I was desperate to explore on my own and so, armed with my guidebook, I ventured downstairs, through the cool of the lobby with its marble floor, and out into the street.

The contrast between the hotel and the world outside could not have been greater and for a moment I had to steady myself by a palm tree as I took it all in. The main thoroughfare, Rashid Street, that ran along parallel to the Tigris, was full of a pulsating mass of people, all touting goods for sale: a cacophony of voices proclaimed fish squirming in barrels; partridges squashed together in undersized cages; pots and pans of every description; goods made from reeds (brushes, baskets, mats, even shoes); cheese and yoghurt from the milk of buffaloes; spices of every description, most of which I had never seen or even heard of; tables piled high with strange vegetables and odd-looking fruits; tailored clothes for the discerning Arab gentleman; cheap-looking garments clearly more suited to the servant classes; items of jewellery that shimmered with delicate mother-of-pearl and startling lapis lazuli; a stall that sold a range of oddly shaped musical instruments that I wasn't sure could be blown, plucked or rattled; and tools for the land and garden, some of which looked like they could inflict a nasty head injury.

As I walked, my sense of smell was assaulted by aromas sweet and foul and, at times, I had to take my handkerchief from my handbag to cover my mouth and nose. I had been warned about this before leaving England. It had been one in a litany of objections raised by my family, and some friends, when they heard that I intended to travel to Baghdad. Sewers in the streets, pestilence and the threat of disease, the danger from the natives, the unstable political situation, the awful climate, the insects, the length of the journey ... all had been used in an effort to dissuade me from travelling to the Near East.

Of course, I could tell no one about the real reason for my visit: my work for Davison and the Secret Intelligence Service. My sister, Madge, and Carlo, my secretary and friend, had all been in favour of me taking a holiday – after all, my daughter Rosalind was away at boarding school in Bexhill – and had approved when I had popped into Cook's, the travel agents, and booked a passage to the West Indies. Then, just a few days before I was due to leave, I had received a message from Davison. The letters surrounding the mysterious death of Gertrude Bell had come to light and he needed someone to investigate further. Although I was initially reluctant, as soon as he had mentioned passage on the Orient Express, and then the chance of visiting Ur, famous for the exquisite treasures that I had seen featured in the press, I told him that I would cancel my holiday. The West Indies could wait.

'You're going where?' screeched my sister when I told her of my change of plan.

'To Baghdad, via Damascus, travelling on the Orient Express as far as Stamboul and then the—'

'You must be out of your mind,' Madge continued, before launching into a tirade of a dozen reasons why such a journey was not only unwise but dangerous. 'You must be unhinged, Agatha. Please tell me you're not having another one of your queer episodes.'

'I'm perfectly well, thank you,' I said. 'You know what the doctors told me.'

'I know, but I'm just worried for you, that's all,' she said. 'What if anything were to happen to you out there? What about Rosalind? Don't you think you owe it to her not to do anything foolhardy? Of course, James and I would look after her,' she said, referring to her husband, 'but even so, it would be beastly for her to lose you, after all that's happened. And what's wrong with the West Indies? They would be so much more restful.' She paused. 'And why have you gone and altered your plans? I just don't understand what's behind all of this.'

'Don't worry, I'll be extremely safe,' I said. I had already prepared a story to explain why I had changed my mind. I told her – and Carlo – of how I had met a naval officer and his wife at a dinner party and how they had enthused about the delights of Baghdad. It was perfectly civilised, they had said; in fact Mesopotamia was the birthplace of civilisation, the area where the cuneiform script had been invented, the oldest form of writing in the world. That had seemed to silence Madge and she retreated in a sulk.

'All by yourself?' said my faithful secretary Carlo, when I told her of the trip. 'But you don't know anyone there, or anything about it.'

'You're right, but it's about time I did something a little more exciting,' I replied. 'After all, Rosalind is happy at school, you can go and take that trip to stay with your sister. I'll only be away for a matter of weeks. I'll be back for Christmas.'

As I gazed up at the turquoise and gold dome of the Haydar Khana mosque and heard snatches of Arabic all around me – a language I could not understand – the thought of an English Christmas seemed like an impossibility.

I continued wandering along Rashid Street until I came to the Souk al-Safafeer, full of men hammering copper into pitches and pots, but did not stay long as the deafening noise was bringing on a headache. The dizziness I had first felt on the terrace of the hotel had also returned.

I felt breathless and in desperate need of a glass of water. I made my way out of the first exit I saw, into a side street, but instead of walking in the direction of Rashid, I must have taken a wrong turning because I found myself in a darkened alleyway that ran between the backs of rundown houses. I looked around. There was no one to ask for directions to the main street. I tried to take a deep breath to calm myself, but I could feel my heart beating, a fast rhythm that only made me more afraid. The stench from a nearby trough was overwhelming and I tasted bile in my mouth. In that instant, even though I could see no one, I was sure that I felt someone's eyes on me.

As I walked down the alley, back towards what I thought was the direction of the copper souk, a door in one of the houses opened and a figure appeared. He was a boy on the verge of manhood. He exclaimed something in a guttural language I could not understand, but within a matter of seconds it was clear what he wanted as his long, bony figures started jabbing at my handbag. I told myself that it would be unlikely he would grab my bag; I knew there were extremely severe penalties for even minor crimes. I felt a sweaty hand encircle my wrist soon followed by a sharp pull on the bag.

'No, I'm sorry, I'm not going to give it you,' I said loudly, in the grandest voice I could muster.

The boy replied with an obscene-sounding flow of invective.

'I'm not carrying any money and there is nothing of value inside,' I said, relieved that I had left my poisons safely locked up inside one of my cases in my hotel room. It was clear that he couldn't understand a word of English, but I persisted. 'The handbag only contains items of sentimental value,' I added, pleading with my eyes. 'Photographs of my daughter, back in England. My dog, Peter.'

The boy was standing close enough now for me to feel his breath on my skin and see the light line of down across his upper lip. It was obvious he was not yet old enough to shave. But there was something in his eyes – fear, hesitancy, kindness even – that told me he didn't like what he was doing.

'Your parents wouldn't be very proud of you, would they?' I said. 'What would your mother think?' Perhaps he did understand something of this, as his eyes widened at the mention of the word 'mother'. But instead of loosening his grip as I had hoped, his hands tightened around my bag.

'Help!' I shouted. 'I'm being attacked. Help!'

Even before the words were out, I heard the sound of footsteps behind me. The boy let go of my bag and ran down the alley, turned a corner and disappeared. I fell back against the wall and clutched my handbag to my chest.

'Are you hurt?' The voice was that of an American.

I turned to see a tall, slim man dressed in a light linen suit. He was about my age and impossibly handsome, with dark, slicked-back hair and a moustache.

'That swine didn't hit you, did he?' he said, as he looked for any signs of injury.

'No, I'm all right, just a little shaken,' I said.

'You've got to be careful around here in the evenings. There are some unpleasant types walking the streets.' He paused. 'What are you doing back here?'

'I was wandering around the copper souk and took an exit into an alleyway and the next thing I knew I was lost,' I said, knowing that I sounded rather foolish.

'Well, there's no need to worry now,' he said, kindly. 'You're safe and that's all that matters.' He looked down the alley. 'I've half a mind to go after that kid. Show him what it means to feel scared.' He took another look at me and decided against it. 'Let's get you back to your hotel – you are staying in a hotel?'

'Oh yes, the Carlton,' I said.

'I'm sorry, I should have introduced myself,' he said, as we started to walk down the alley. 'My name is Harry Miller.'

'Miller?' I said. 'Why, that's my maiden name. My father was Frederick Alvah Miller from New York. You're not part of the same family?'

'No, we're from Philadelphia,' he replied, as we stepped into the copper souk. 'And your married name?'

'Mrs Agatha Christie,' I said, my words getting lost amid a thousand hammer blows.

'Sorry, this damnable noise! I can't hear a thing. Follow me.'

He led me out through a maze of stalls, tiny compartments filled with a variety of pots and pans and metal-beating equipment, staffed by men who looked exactly alike to me.

'How do you know your way around?' I asked.

'I've been coming here on and off for a while,' he said, as we reached the relative quiet of Rashid Street. 'Now, first things first. I wouldn't advise it – knowing it wouldn't get you very far – but I must ask: do you want to report the incident to the police?'

'Oh no, the boy didn't take anything, and it was silly of me to wander off like that.'

'Well, what do you say we go and have a drink? It will help steady your nerves.'

I thought his approach rather forward. 'I'd much sooner just return to my hotel, the Carlton, if you don't mind, Mr Miller.'

'Very well, but let me at least walk you back, Mrs . . . ? I didn't catch your name.'

'I'm Mrs Agatha Christie.'

'Well, how do you do, Mrs Christie,' he said with an amused glint in his dark eyes.

'I'm feeling much better now. I must thank you for rescuing me back there. I really don't know what would have happened if you had not come along.'

'Yes, it could have turned nasty,' said Mr Miller. 'And what brings you to the "God-given" city?'

'I'm here on holiday,' I said, as we continued to walk in the direction of the hotel. 'I wanted to see something of the world, something out of the ordinary.'

'You'll certainly get that here. Have you done much travelling before?'

'A little. South Africa, New Zealand, Australia, Hawaii and Canada. But that was back in 1922.'

'Sounds wonderful. And you're travelling alone?'

Mr Miller obviously wasn't wasting much time. 'Yes, yes, I am,' I said. 'I did make one acquaintance on the train journey here, a Mrs Clemence, who lives in Alwiyah. She keeps asking me to go and stay with her, but I am rather keen to get down to Ur.'

'You're going to Ur? That's where I'm heading too. I've been picking up some supplies in Baghdad, but I work down there, on the archaeological site, as the photographer.'

'How extraordinary,' I said. 'Actually, would you like to come into the hotel for a drink after all?' There were a few questions I wanted to put to Mr Miller.

'I'd be delighted,' he said.

A few minutes later we settled ourselves in the comfort of the library, Mr Miller with his Scotch and me with a lime and soda water.

'I really must thank you once again for what you did earlier, Mr Miller,' I said.

'It was nothing – only what any passing gentleman would do. And call me Harry, please. But I hope you won't let this incident put you off Baghdad. It's a wonderful city in many respects. Still has something of the mystique of the *Arabian Nights* about it, even after all the recent troubles.'

'Yes, you're right. I won't let it warp my view.'

'By the way, I'm intrigued,' he said. 'What brings you to Ur?'

'After reading about the site in the newspapers I was so keen to see it for myself. All those beautiful objects dug up from the desert sands.'

He leant a little closer to me. 'How did you manage to secure an invitation to Ur? Mrs Woolley usually turns down most requests for visits from other women.'

'With the help of a mutual friend – and also the fact that I believe she is a fan of one of my books.'

'You're a writer?'

'Yes.'

'What kind of books?'

'Detective fiction. Light thrillers. Short stories. Nothing serious, I'm afraid.'

'Would I have heard of any?'

I reeled off the titles, but Mr Miller looked none the wiser.

'Sorry, but I'm not a great fan of having my nose in a book,' he said, colouring slightly. 'Never have, even as a boy. I'd always much rather look at the world through a camera lens. You probably think me very stupid.'

'Not at all. If we were all the same then the world would be a very dull place indeed,' I said, regretting the words as soon as they were out of my mouth. I was sounding like one of the old maids who used to congregate at my grandmother's house. 'Tell me, Mr Miller,' – I couldn't bring myself to call him Harry, not yet at least – 'it must be fascinating to work as a photographer with the team at Ur. How long have you worked there?'

'This will be my third season,' he said. 'I keep telling myself that this year will be my last but something keeps drawing me back.'

'I can imagine, it must be quite magical to see those objects appear out of the sands, after thousands of years buried underground.' I recalled some of the photographs I had seen in the *Illustrated London News*.

'Oh yes, it is. And it's just as extraordinary – I'm talking about the darkroom now – to see the images emerge out of nowhere and form themselves on photographic paper.

Sometimes, in that darkroom, in the house in the middle of the desert, I feel I've had the equivalent of a religious experience. I know that may sound silly, but . . .'

'It doesn't sound silly in the least,' I said.

'I said the same thing once to Father Burrows – he's a Jesuit priest and the man in charge at Ur of reading and translating the cuneiform tablets – and he looked at me with the strangest expression on his face. Like he was hearing a phrase in a foreign language he couldn't understand.'

'Who else is there? At Ur?'

'There's quite a crowd at the moment, which is unusual,' said Miller. 'Of course, there are the people who work on the dig, Mr and Mrs Woolley. There is the epigraphist Father Burrows, the secretary Cynthia Jones, then there's the architect, Lawrence McRae, and his nephew, a young boy, Cecil. One has to pity him really, he's not quite right in the head, if you get my meaning. Lost his parents in an accident, and so McRae took him under his wing. He'd be in some institution if it weren't for his uncle's kindness. And then there are the current visitors – the Archers: Hubert, a railway millionaire from the mid-west, and his wife, Ruth, and daughter Sarah. The only reason Mrs Woolley tolerates their presence, especially that of the wife and daughter, is the fact that they promise to invest in the dig to the tune of what could be thousands of dollars. Archer is thick with the directors of the Penn Museum in Philly and, in addition to the sponsorship of the dig, promised them even greater riches in the future. The guy's one of those hangovers from the last century who believe in the word – the literal word – of the Bible.'

'I see – so it's the lure of the Old Testament that brought him to Ur?'

'Yes, and Mr Woolley is convinced that he will soon find evidence that Ur was indeed the birthplace of Abraham.'

'It sounds like a fascinating group of people,' I said. I did not, of course, mention the fact that one of the party could be the murderer of Gertrude Bell. 'And I hear Miss Bell did a great deal to help the preservation of the treasures here in Baghdad. I believe she drafted the antiquities law in Iraq and held an important position to that effect in this country?'

Miller had fallen silent.

'King Faisal must have thought very highly of her,' I continued. 'And then there's the archaeological museum here in Baghdad, which she set up and which I believe houses some objects discovered at Ur. It was so sad to hear the news of her death. I hear that she had been ill, but still such a tragic loss.'

Tears had formed in Miller's dark eyes and as he bit his top lip he reached out for his glass and drained it of the whisky.

'I'm sorry, I didn't realise,' I said. 'You must miss her a great deal.'

He nodded and then coughed, choking back the emotion. 'Yes, she was very dear to me,' he said, blinking. 'Excuse me, you must think me a fool. I'm fine most of the time – I lose myself in my work, have a few drinks in the evening – but then it steals up on you when you least expect it and floors you. Grief, I mean.'

'I know exactly what you mean,' I said, remembering

my chaotic state of mind after the death of my mother and then the way I went to pieces after I had discovered the truth of Archie's love affair. Losing him had been a kind of bereavement. At times, I wished that he had died instead of finding happiness with that other woman, a thought which then made me feel wicked and full of shame.

'I think it's admirable you can talk about it,' I said. 'So many of my fellow countrymen wouldn't dream of talking about their feelings in this way, and it does them no good whatsoever.' I wasn't totally in favour of men – or women when it came to it – unburdening themselves of their emotions, but it was clear that Mr Miller needed to hear something reassuring. 'Was she a very close friend?'

'I just don't understand it,' he continued. 'Her death, I mean. She had so much to live for. Not only her work here in Iraq, seeing the fruits of her labours, enjoying the friendship she had built up with the king, her delight at witnessing the museum come to life. But also, she had her personal happiness too, something she richly deserved.' He looked completely lost and broken. 'Why would she throw that away? No, it doesn't make sense.'

'What are you saying, Mr Miller?'

Anger fired up his eyes now. 'What I'm saying is that there is no way she would have committed . . .' He didn't want to say the words. 'There's no way she could have died . . . in that way, like the doctor said.'

'I thought she died of natural causes?'

'That was just a story put out by her father and step-mother to stop unseemly speculation in the press,' he said. 'The doctor who attended her, Dunlop, discovered a bottle of pills by her bed. Sedatives. I know she had problems sleeping – all of us do, sometimes – but there's no way she would have taken an overdose. It just wasn't in her nature.' He stared into his empty glass and fell silent for a few moments. 'Anyway, there's nothing that can be done now, it's all in the past.' He caught the eye of the waiter and signalled for him to pour him an extra-large measure of Scotch.

'Could there be another explanation?' I asked.

'Yes, I suppose that must be it – an accidental overdose. She had lost a good deal of weight, I believe, in the last few years. Her system was weakened – she'd suffered an attack of bronchitis, and then pleurisy. Perhaps she took a couple of pills to sleep and forgot she'd taken them and so took a couple more. And that was too much for her . . .'

It was obvious he had not thought of the possibility of murder – or, if he had, he was keeping that to himself. Had he heard about the existence of the two unsent letters, and the sinister drawing, that had been discovered in Miss Bell's house in Baghdad? Davison had told me that the servant who had come across them had been sworn to secrecy on the issue – the man would be severely punished and his family would suffer if he so much as breathed a word of what he had unearthed – but often these things leaked out. I decided it would be wise to keep the contents of the letters to myself, at least for the time being.

'And when are you travelling down to Ur, Mr Miller?' I said, changing the subject.

'In a couple of days' time,' he said, grateful that we had left the death of Miss Bell behind, for now at least. 'I'm just waiting for some chemicals to arrive and some supplies for the team and then it's back to work. From what Woolley tells me we should be in for an interesting season.'

He looked at me as if seeing me for the first time, his eyes assessing my hair, my skin, my clothes, and, as his gaze finally came to rest on my figure, I felt myself blushing. 'And it sure will be fascinating to see what Mrs Woolley makes of you.' He nodded his head, as if to confirm a thought he chose not to articulate, before he added, 'Yep, there could be fireworks there. Mark my words. Fireworks.'

Chapter Three

This was certainly no Orient Express, I said to myself as I sat in the stifling heat of a shabby compartment in the rackety train that would take me from Baghdad to Ur Junction. I fanned myself with a newspaper, but I still felt as though I couldn't breathe. The train hauled its way out of the station, passed through the ugly outskirts of Baghdad, and began its slow journey across the desert. I had a compartment to myself – perhaps I was the only person stupid enough to travel like this? – so fortunately no one had to see my reddened neck and face and the sheen of perspiration that was fast spreading across my skin.

In order to try and cool myself I closed my eyes and pictured the grandeur and elegance of the train that I had boarded at Calais. Saying the name 'the Simplon-Orient Express' silently to myself was enough to transport me back. The glass and silverware, sparkling so brightly it hurt one's eyes. The soft touch of the mahogany in the dining car. The soothing voices of the uniformed men, staff so attentive they seemed blessed with second sight. I recalled the deep sense of pleasure I enjoyed, sitting in the *wagon-lit*, opening up my brochure from Cook's (a document I had gazed on more than a dozen times already) and seeing the planned route ahead: Calais, Paris, Lausanne, Simplon, Milan, Venice, Belgrade, Sofia, Stamboul. I was sure nothing in life would give me more pleasure than the anticipation I felt when I first stepped onto the train. I didn't even mind the fact that I was due to travel second class; Davison's office, which had arranged the tickets, had informed me that all seats in first had been taken. Yet my heart did sink when I entered my compartment to find that I had to share with a talkative and somewhat over-familiar woman.

'Hello,' she had said, standing up and stretching out a plump hand. 'I'm Elizabeth Clemence, but you can call me Betty. Everyone does.'

As soon as I introduced myself she seized on the fact that I was an author. Was I that writer who had caused a stink a couple of years back by disappearing? I had to admit that yes, that had been me. Had it been true that my husband had been suspected of my murder? Nothing more than the speculation of a sensational press, I told her. So why had I disappeared? Could I really not remember anything of those ten or eleven days? My doctors had warned me not to dwell on the issue, I said – to do so could only bring about another episode, and so I brought that line of enquiry to a close.

After a few minutes' silence, Betty Clemence started up again by asking me where I was travelling to – was I going to Italy? No, I replied, a little further than that. But

where exactly? Well, I told her, I was going to Baghdad.
At this, she simply exploded. What a coincidence! She
couldn't believe it! Oh, she couldn't wait to tell her husband! Why, she lived there, and I had to stay with her. It
would be simply criminal if I were to pay for a hotel. But
what was I doing in Baghdad? I could not, of course, tell
her the real reason – that I was travelling to Iraq at the
request of the British Secret Intelligence Service to investigate the suspicious death of Miss Gertrude Bell – and so
I related the story about meeting a naval officer who had
recommended Baghdad to me. And, she asked, what was
his name? I could not remember, I said. Was it Rogers?
Fletcher? Aylesbury-Eyreton? The name had escaped me,
I said. Surely, if he has passed through Baghdad – and he
sounded like a person of some importance – then he would
have made himself known to her or her husband, Geoffrey.
And what were my plans once in Iraq? I reeled off a list
of things I had read about in my guidebook, before Betty
Clemence started to inform me of the delights of the city:
the tennis and lunch parties, the gardens, the wonderful
people (all of whom were English, of course, many of
them having retired from the higher ranks of the military
or diplomatic service). I got the distinct impression that,
to her, Baghdad was nothing more than a slightly more
exotic Bournemouth.

'And did you ever meet Miss Gertrude Bell?' I asked
when she finally paused for breath.

'Oh yes, an awful woman,' she said, puffing out her
cheeks. 'So rude. Some of the things that came out of her
mouth. Do you know she once had the audacity to suggest

that my life out in Baghdad was superficial? That I was wasting my time flitting from lunch party to tennis party without a thought in my head. I had half a mind to tell her what I thought of her – some of the circles she would mix in, you would not believe, my dear – but propriety and good manners prevented me. She was obviously unbalanced – unhinged, yes, that's the word. Sad, of course, what happened at the end, but I can't say I was surprised.'

'You weren't?'

'No, she had no husband, no children. What had she made of her life? What had she to show? Nothing but a few dusty artefacts clustered together in a museum.'

'But what of Iraq? Surely she did something there to—'

'Better if she'd never bothered, that's what my Geoffrey says. She should have left well alone. And I have to say I agree. Who knows what all this meddling will do? It's like opening a can of worms. Better to let the British take charge, just as we do in India. After all, these people have no idea how to govern themselves. Don't you agree?'

Before I could answer, she had launched into yet another line of enquiry. She wanted me to tell her my exact route to Baghdad. Was I not getting off the train at Trieste and then taking a boat to Beirut? That was her preferred itinerary; it was, of course, the best method, tried and tested. When I told her I was taking the Orient Express as far as Stamboul from where I would get another train via Damascus she informed me that that would be a disaster. Surely I could change my ticket and accompany her on her journey? When I told her that I could not she proceeded to outline the exact and innumerable ways in which my journey could and would go wrong.

'I expect it will be awful, but never mind, it's too late now,' I said. 'But I am looking forward to my travels, particularly Ur. Have you been there?'

Her eyes, fat currants set within her generously proportioned face, lit up at the mention of the name and she proceeded to tell me all about it. How she believed Mr Woolley's discoveries would change the world of archaeology forever. How there was every possibility that Leonard would find evidence that Ur had indeed been the birthplace of Abraham. And how thrilling – and also quite disturbing – it must have been to uncover those bodies. 'Just think how lovely it must have been for the king and queen to believe that their servants were going to accompany them after death. Their every need catered for, even in the afterlife. Woolley told me that there were no signs of struggle, suggestive of the idea that the servants went to their deaths willingly. So wonderfully romantic.'

There was nothing romantic about death, I wanted to tell her, but held my tongue.

'And what of Mr Woolley himself?' I asked.

She outlined some facts I already knew: how Leonard Woolley had been born in London, the son of a vicar; how he had worked with T. E. Lawrence at Carchemish, the archaeological site that lay about sixty or so miles north of Aleppo; and about how he had been imprisoned by the Turks during the war. She had heard him give a couple of lectures in Baghdad and yes, she had been greatly impressed by the vitality with which he talked, his ability to conjure up the past as if it were something real and concrete.

'What is his wife like?' I asked.

Betty raised her eyes in unspoken disapproval. 'The less said about Katharine Woolley the better,' she sniffed.

'What do you mean?'

'She's an odd woman. Strange.'

'In what way – or ways?'

'You'll see for yourself when you arrive at Ur,' said Betty. 'Like Geoffrey always says, no good ever comes from being a so-called independent woman.' I could have taken offence at this; after all, now that Archie had abandoned me and I was forced to support myself, this was the category I fell into. 'Look at what happened with poor Miss Bell.'

'Did they know each other? Miss Bell and Mrs Woolley?'

'They were not the best of friends.'

'I wonder why not? You would think they would have so much in common.'

'I wouldn't be surprised if there's something wrong with her.'

'With Mrs Woolley? What do you mean?'

'She's not all there.'

'You mean she's mad?' I asked.

'Let's just say it's Len I feel sorry for.' Mrs Clemence took up a book with an air of finality, a gesture which served as an indication she had nothing more to add on the subject. An uncomfortable silence – something I had not experienced since first stepping onto the train at Calais – settled over our carriage and I turned to my guidebook.

At that moment a knock at the compartment door

roused me and my thoughts returned to my current journey through the desert. The bearer informed me that we were about to stop for something to eat and, sure enough, within a matter of minutes the banshee cry of the train's brakes sounded all around us. I was ushered out onto a deserted platform and into an isolated restaurant, where I was presented with a series of dishes drowning in fat. As I ate, I again thought back to the conversation I had had with Mrs Clemence on the Simplon-Orient Express. What she had chosen to keep from me was just as interesting – no, it was much more interesting – than the information she felt free to part with. I opened my notebook and jotted down a series of key points about what I knew of the case so far.

July 1926: Gertrude Bell (GB) found dead in her bed in her house in Baghdad. Dial poisoning. Overdose? Family kept suicide quiet, made out she died of natural causes.

October 1928: Discovery of two letters by GB, and one drawing sent to her, in her house in Baghdad. Letters point to her fear that someone may be about to murder her. Why didn't she send the letters to her father? Drawing is an illustrated plan of one of the graves at Ur. Initials G.L.B. added next to one of the bodies. At the time of her death GB believed her future murderer would be found at Ur. What if the murderer is no longer there? After all, it is now two years since her death.

Possible factors that could have a bearing on the case:

- Suicide of Lieutenant Colonel Bertram Keeling, killed himself at the base of the Great Pyramid of Cheops, Cairo, September 1919. Did he know GB? Was the death really a suicide – could he have been murdered?
- Intelligence connection? Keeling, Gertrude Bell and Leonard Woolley all worked in British Intelligence during the war.
- The personality of Mrs Katharine Woolley. Clash between KW and GB?
- The treasures of Ur and their division between London/Philadelphia and Iraq. Clash between LW and GB? Relationship between GB and photographer Harry Miller?
- Enemy of GB, identity unknown.

By the time I stepped back onto the train I had a clearer idea of what I needed to find out. I would watch, listen, and ask the occasional question. Just as Mr and Mrs Woolley and their team excavated the past, so I would do my own spot of gentle digging. In the course of their work they uncovered treasures of exquisite beauty, but they also unearthed the unmistakable signs of human sacrifice. Darkness was falling across the desert and, in the distance, I could hear the scream of what sounded like a wounded animal being hunted, most probably to its death.

Chapter Four

I made the final part of the long journey across the desert by car. As soon as I stepped out I was assaulted by a great torrent of enthusiasm. 'Welcome to Ur!' I immediately recognised the impish face that greeted me as that of Leonard Woolley. His blue eyes sparkled with life, and, dressed in shorts, long socks, jacket and an open shirt, he looked more like an adolescent than a man of middle age. He gestured to an Arab boy to take my bags.

'It's such a pleasure to have you here, Mrs Christie,' said Woolley, shaking my hand. 'I can't tell you how much my wife is looking forward to meeting you. She has a hundred questions to ask you about your writing, particularly that book about the doctor. Very clever, yes, very clever indeed. But she's indisposed at the moment, having one of her headaches.' His face seemed to freeze for a moment, before he began again. 'But never mind. I'll give you a tour around the site, then I really must get back to work and leave you to settle in your room. Terribly basic, I'm afraid, but better than nothing. Then later, just before sunset, I'll take you up to the ziggurat,' he said, pointing to an enormous baked-brick structure that dominated the skyline. 'And then we can have dinner. Meals are usually quite a simple affair here, but we try our best.'

Woolley led the way down a dusty path towards a single-storey brick house with a veranda built for shade at the front.

'How was your journey?' he asked as we walked.

'Long, and hot, but—'

'You survived the food?'

'Yes, just about,' I said, smiling. Then the smile froze on my lips as we passed through a gate surrounded by a barbed wire fence that encircled the enclosure. I felt like I was stepping inside a prisoner-of-war camp.

'It's a shame about the wire fence, but it can't be helped,' said Woolley. 'The desert is full of marauders and thieves. As you no doubt know we are digging up a great deal of precious and semi-precious stones, as well as beautiful objects crafted from gold and lapis lazuli. Some of the finds have been valued in the thousands of pounds, if not the hundreds of thousands. Out here life is cheap, I'm afraid.'

I thought about the death of Gertrude Bell. Was that why she had been killed? Because of some dispute over a piece of ancient jewellery or precious metal?

'But don't worry, you will be perfectly safe,' said Woolley, misinterpreting the expression on my face as concern over my own safety. 'There's a jolly group we have with us at the moment. You'll meet them all at some point I'm sure, but look – over there is Father Burrows and our secretary, Miss Jones. Let's go over and I'll introduce you.'

I trailed behind Woolley as he continued to talk – about the climate, the ferocious heat of the summer months (which meant that, in effect, the season of the dig ran only from October to March), the terrible rainstorms in the autumn, and the extraordinary power of the sands. One spring he had returned to find a whole wing of the house buried up to the roof in sand, something that took three days to shift. 'But what we've uncovered makes up for the slight discomforts we experience, wouldn't you agree, Father?'

'Yes, indeed,' said a tall, spindly man wearing round, wire-framed glasses and a white clerical collar.

'Father Burrows, this is Mrs Christie, the famous author I'm sure you've heard so much about.'

'I rather think not,' I said, feeling myself blushing.

'And this, Mrs Christie, is the person whom we all rely on – the one who makes the operation run like clockwork. The indispensable Miss Cynthia Jones.'

'Mr Woolley, how you flatter so,' said Miss Jones, a kindly looking spinsterish type with lank mousey hair and large brown owl-like eyes. She turned to me and smiled sweetly. 'How do you do?'

'I'm a little tired, and dusty, after the journey, but thrilled to be here,' I said.

'I know,' Woolley interjected. 'Miss Jones, why don't you show Mrs Christie to her room, and we can have a tour of the site once she's settled in.'

'I'd be delighted,' said Miss Jones.

'I'll go and check on Katharine,' said Woolley. 'Poor thing, she's been forced to lie low again. These damnable migraines are the bane of her life. And then I must get back to the dig. Please excuse me, Mrs Christie.'

Woolley and Father Burrows retreated among talk of cuneiform tablets, royal cemeteries and seams of clay.

'You must forgive us if we appear rather caught up in our work, Mrs Christie,' said Miss Jones. 'Because the dig is confined to such a short season, in effect less than six months, it means that there is always so much to do. And with the richness of the finds, we never seem to have enough time.'

'I completely understand,' I said. 'It must be absolutely fascinating. I don't know much about archaeology – although last year I was on the island of Tenerife, where I saw at second hand something of the work of Professor Wilbor. Have you met him?'

Miss Jones said she had not, but she had read something of his work on the Guanche culture of the Canary Islands. A memory flashed into my mind of a man's crumpled body at the bottom of a dry river bed, a nasty smear of blood on the rocks, a bird-of-paradise flower spiked through his eye.

'That journey from Baghdad to Ur Junction is exhausting,' said Miss Jones. 'I don't know why the train has to take so long – and all through the night. Come with me. I'll show you to your quarters.'

As we stepped under the shade of the veranda I noticed a well-fed ginger cat curled up in a round basket fashioned from reeds. He had that look of unknowability so peculiar to cats, a quality that I found frustrating; it was one of the reasons why I preferred dogs.

'Look, there's Tom,' said Miss Jones.

I bent down to stroke him, but before I could do so I felt a light slap on the back of my hand.

'Best not to touch him,' said Miss Jones. 'The only person he seems to like is Mrs Woolley. Anybody else who tries to get close to him is rewarded with a vicious scratch or a nasty bite. We've tried to get rid of him several times, but Mrs Woolley will not have it. And her word is the law around here.'

'Oh, I'm sorry,' I said.

'You weren't to know,' she said, as we stepped into an open courtyard. 'On the left here is the room of Mr McRae, the architect, and his nephew, Cecil. I'll tell you about them later. And on the right is the antiquities room, where the treasures are studied and stored. That's always locked, of course, as one cannot be too careful. Mr Woolley has a key for that and I have a spare. There is a darkroom for Mr Miller, the photographer, who is in Baghdad for another day or so.' I didn't tell Miss Jones that I had already met him. 'And ranged around the courtyard are the other bedrooms.'

A number of Arab servants busied themselves with household tasks, carrying containers full of water, clearing up plates and dishes from breakfast, and sweeping up what seemed like an insurmountable amount of dust and sand. We passed into a large living room with apricot-coloured walls and a floor comprised of burnt bricks partly covered in rush matting. In the corner there was a makeshift library, mostly stocked with books on ancient history and archaeology, and in the centre of the room stood a long, rectangular wooden table and several chairs.

'All this won't be what you're used to, but it might provide you with some colourful material if you were ever to write a book set in the Near East,' said Miss Jones.

I did not reply; instead, I asked, 'How long have you been here?'

'A few years now,' she replied.

'And what do you do?'

'You mean, the reason why Mr Woolley finds me so – what was it – indispensable?' She smiled as she said this. 'I'm just a glorified dogsbody really. I don't do any of the transcribing of the tablets, but I type up Father Burrows's notes. I make sure every new find is logged and described, together with the appropriate reference numbers. I deal with the correspondence between the staff here and the museums in London and Philadelphia. And also I'm rather good at making tea.'

'That's a relief, I was worrying that all I would find out here is that peculiar brand of coffee everyone seems to drink,' I said.

'The kind that keeps you up all night – it's like tar, isn't it?' she said. 'Or something you'd find at the bottom of a rather dirty bucket.'

The comment made me laugh, but at this she shot me a look of warning. 'Shh – best keep your voice down,' she said in a whisper. 'We're about to pass by Mrs Woolley's room. She can't be disturbed when she is having one of her attacks.'

'I'm sorry,' I said again. I thought about what Mrs Clemence had said to me on the Orient Express. 'What exactly is wrong with her?'

Miss Jones did not respond. Either she had not heard me or she had deliberately chosen to ignore the question. 'Here's your room,' she said, opening the door. 'Very basic, but you should be comfortable enough.'

As I stepped into the darkness my heart immediately sank; it was nothing like the splendour of the Carlton. The walls and the floor, made from baked bricks, were bare of decoration apart from an old print of Baghdad and, by the side of the single bed, a threadbare Persian rug. In the corner there was a desk and chair, together with a bowl and jug for washing, and by the window, which looked out into the courtyard, was an old basket chair that had clearly seen better days.

'The whole house was built from reused bricks, found on the site, some of which are twenty-five centuries old, or so I'm told,' said Miss Jones, her eyes taking on a misty, dreamlike quality. 'So for all intents and purposes we live just like the people who settled in Ur in ancient times. At night, when the house is quiet, you can almost feel like you're stepping back in time. I'm not a superstitious person, Mrs Christie, but even I can feel the ghosts of the past here.' Her owl-like eyes blinked, a gesture which seemed to bring her back to the present. 'I'm just across the courtyard, so if you need anything, please just ask,' she added. 'Now, where are your bags? That stupid boy's no doubt put them in the wrong room. Excuse me, I'll go and see where he is.'

I walked over to the open window and looked out at the dusty courtyard. The sound of voices drifted over the hot air and then two figures – those of a handsome middle-aged man and an awkward-looking adolescent – came into view.

'I just don't understand,' I could hear the boy say. He seemed on the verge of tears.

'Women – one of life's greatest mysteries,' said the man in a lilting Scottish accent, his pale face and auburn hair shaded by a hat.

'It's not funny, Uncle,' the boy replied. 'How would you feel if your heart was always being broken?'

'I'm sorry, Cecil, but as I've said before, women can be extremely cruel. Best to avoid them, if you can.'

'Sarah says she could never love me – do you think that's true?' The boy's face was covered in spots, his dark hair had been plastered down on his head with grease, and his elongated frame reminded me of a large, ungainly bird. 'She said I was ugly. That I was stupid. That I should never talk to her again.'

'I think it's best if you put Sarah out of your mind. You know she won't always be here, don't you? Her mother and father will move on soon and then no doubt return to America. Of course, you could always write to her, but—'

'I don't want to write to her, I want to . . .' he began, his face colouring.

'Now, now, let's not get upset. Try to see—'

'If I can't have her, then no one will. I can make sure of that.'

'Cecil, you're talking nonsense now. Let's go and have a cup of tea.'

The older man, whom I took to be Lawrence McRae, the architect, placed a hand on his nephew's shoulder and led him across the courtyard and into the house. The voices quietened and a moment later Miss Jones returned, followed by the Arab boy and my bags.

'Sorry about this, but for some unfathomable reason Sahid left your bags outside Mrs Archer's room. She's staying here with her husband, Hubert, and daughter Sarah. I'm sure you'll meet them later when they've returned from their excursion to Eridu. Have you heard of it?'

'No, I'm afraid I haven't.'

'It's supposed to be the oldest city in the world. One of the five cities built before the great deluge.'

'Are they the visiting Americans?'

'That's right,' she said. 'He's very interested in sponsoring the expedition. Made his money from the railroads. He's here because he wants to see evidence that Ur was the birthplace of Abraham.'

'And has Mr Woolley uncovered such evidence?'

'Not yet, but he's determined that he will.'

'And what of Archer's wife and daughter?' I asked. 'What are they like?'

'Ruth is a lovely woman, hasn't a bad bone in her body. Her daughter, Sarah, meanwhile, is another matter.'

'What do you mean?'

'She is beautiful – blonde hair, blue eyes, alabaster skin, alluring figure. A vision of perfection. I suppose she can't help the way she looks, but when she's out of sight of her father you should see the way she teases some of the men on the site. It's not fair. A man has – well, he has certain appetites and urges. I've told her she should be more careful, the way she leads the men on. And then there is the upset she causes poor Mrs Woolley. It's obvious to everyone they don't like each other – I think Katharine is jealous of the girl's youth and beauty and Sarah can't abide the way that she tries to steal male attention away from her. I'm sure Katharine would have got rid of the girl had it not been for the fact that she knows they need Archer's mountain of money. So, if you notice an odd atmosphere at dinner tonight, you'll know what it's all about.'

'I see,' I said, slightly at a loss for words.

'I suppose we should be grateful for the fact of Mr Miller's absence – he's the photographer I mentioned earlier.'

'In what way?' I asked as I sat down on the hard mattress.

'He plays them off, one against the other. One minute he is fetching and carrying for Mrs Woolley, the next flirting with the young girl. I wouldn't be surprised if he tries a spot of lovemaking with you when he returns.'

I tried to stop myself from blushing at the memory of our encounter in Baghdad.

'So life in the camp is rather the opposite to what I was imagining,' I said. 'I brought a number of novels with me in case I might get bored. But it looks as though there will be little chance of that.'

'We'll keep you entertained, that's for certain,' said Miss Jones with a sardonic smile. 'What with the treasures from the site, the history, the personnel and our visitors, I doubt you'll have a dull moment.'

I seized the opportunity. 'Talking of visitors, you must have had a fascinating mix of people coming to see the site?'

'Oh yes, plenty. But most of them come just for the day.'

'Did you ever meet Miss Bell?'

Cynthia Jones blinked at the mention of her name.

'Gertrude. Yes, of course. We were great friends.' Tears came into her eyes and she steadied herself against the wall. 'Such a sad loss.'

'I'm sorry I brought up her name,' I said. 'If you'd rather not talk about her, I totally understand. It's just that she sounded like such a fascinating, unusual woman.'

'I never met anyone like her and doubt I ever will again,' she said. 'Of course, she wasn't everyone's cup of tea. She couldn't bear empty-headed women – she simply loathed the type of woman who used her husband's position to do nothing more than take tea or read novels or gossip. She had a desire to really live, it was quite frightening sometimes, almost as if she knew that she didn't have long in this world.'

'Yes, I had read about her illness in the newspapers,' I said, feigning ignorance of what I knew of the manner in which she had died.

'Oh, that business about passing away in her sleep was just a piece of nonsense.'

'You mean that . . . that she took her own life?'

'Suicide?' she said. 'I don't believe Gertrude would ever do that. She had too much left to live for.'

'If not suicide – then what?'

Miss Jones paused, walked over to the window and stared out into the courtyard before turning back to me. 'I'm not sure I should say anything.'

'What are you suggesting?'

She looked nervously around her, as if there were people – invisible to us – who might be listening to our conversation.

'Miss Jones—'

'Please, call me Cynthia.'

'Cynthia ... if you feel you want to say something in confidence to me, then please unburden yourself. It's clear that something is worrying you.'

'You must promise that you won't say anything,' she said, making sure the bedroom door was closed. 'This must be strictly between us.'

'Yes, of course. I promise.'

Cynthia bit her lip, took hold of my hand and led me towards the bed. 'It's funny we've only just met, but it's as if I've known you for years,' she said, smiling as we sat down. 'I feel I could tell you anything.'

'That's good,' I said, not entirely comfortable in my role as Miss Jones's new-found confidante. 'Please, feel free to tell me anything you wish.'

'Thank you,' she said, a troubled look in her eyes. She took a deep breath and clasped my hand. 'I always thought it was odd the way that Gertrude died. I can understand, of course, the way her death was hushed up. That's entirely natural, as the family didn't want any scandal. But what I never understood was that the doctor found some pills by her bed; sedatives, I think they were. I don't know. To me, it seemed a little contrived, almost as if it were staged.' She looked at me for encouragement.

'Yes, I understand. Go on.'

'When I heard the news I was terribly upset. It was such a big loss. I realised how much I would miss her. I used to go to Baghdad to stay with her in her house. She would come down here to visit and argue over the finds. We had so much to talk about – our families back in England, our love of travel, of history, of the old Mesopotamia and the new Iraq. And I think I was one of the few women whom she could tolerate.'

'But what of her death, Cynthia?'

'That's just what I was coming to. You see, a few months before she died I was in Baghdad, ready to return to Britain. The heat was already too much for me then. I don't know how Gertrude could stand the summers out here. Anyway, while I was there we were talking about the season at Ur. I noticed that she had turned pale and had started to tremble. I made her sit down and tell me what was bothering her.'

Miss Jones lowered her voice to a whisper. 'She told me that she was afraid that she might be killed. And if she were to die she was sure that her murderer was one of the team. Here on this dig. I couldn't believe it when she told me. I thought she must be talking nonsense, but she was adamant. There was someone here who wanted her dead.'

'Why did she think that?'

'She said she had received some threatening letters, together with a drawing, I'm not sure of what, which made her fear for her life.'

'Did she give you a name?'

She looked around her as if she were half expecting to see someone standing in the corner of the room.

'Take another deep breath,' I said, squeezing her hand. 'Cynthia – did she give you a name?'

She nodded, her eyes haunted by the memories and secrets of the past.

'Who was it? Who did Miss Bell believe might murder her?'

'You promise not to say anything? I could get into a great deal of trouble. I don't want to lose my position here.'

'Yes, I've already given you my word.'

'Very well. It was—'

Chapter Five

'Tom! Can someone please bring Tom to me?' The voice sounded high-pitched and strained and was coming from a room nearby. It had interrupted Miss Jones just as she was about to give me a name.

'Oh, that's her now,' said Cynthia, jumping up with a look of terror in her eyes. 'Mrs Woolley. I must go to her.'

'What do you mean? Was it Mrs Woolley that Miss Bell was afraid of?'

'She must have woken up,' she said, turning away from me. 'I have to see to her.'

I accompanied Cynthia Jones out of the room, following her like a shadow until we stood outside Katharine Woolley's bedroom.

'Oh, where is that damned cat?' asked Cynthia, turning to an Arab boy to whom she asked something in Arabic, presumably the same question.

'This headache is simply killing me,' said Mrs Woolley from inside the room. 'The least someone can do is to bring me Tom. Len! Where are you?'

'Could you just stay there while I go and try and fetch the cat?' Cynthia asked me. 'Hopefully he'll still be in his basket.' She tapped on the door. 'Katharine, it's Cynthia. I'll go and find him. I won't be a minute.'

'Do please hurry up,' she said from behind the door. 'You know he's the only thing that can bring me comfort when I'm feeling like this. And don't scare him like you did last time. You know what happened then – you got a nasty bite.'

A moment or so later the door opened to reveal a pale-faced Katharine Woolley, dressed in a white nightgown. Before I could say anything she took one look at me – her dark eyes seemed tortured, haunted somehow – and then closed the door. Through the partition I could hear the sound of strange mutterings, snatches of conversation she appeared to be having with herself. Perhaps the woman really was as unstable as people had suggested.

'Mrs Woolley,' I said, gently knocking on the door. 'There's no need to distress yourself. If you let me in I can help. It's Mrs Christie. I think you are expecting me.'

There was no response. 'I believe you suffer from ter-rible headaches. I might be able to provide you with some relief. I was a nurse during the war, you see. I came across a number of cases such as yours.' Still no answer. 'Mrs Woolley? Can you hear me?'

The door opened again and Katharine Woolley gestured for me to step inside. The shutters on the windows had been closed and the room was cool and dark. In contrast to the stark interior of my room, Mrs Woolley's quarters seemed feminine, tastefully decorated and quite beautiful.

In addition to the bed, the room was furnished with an old-fashioned desk made from dark wood, its surface covered with sheets of paper. There was a wardrobe with clothes spilling out, a table with a gramophone and a messy dressing table complete with various pots of lotion and jars of cosmetics. The walls had been decorated with a number of exquisitely worked sketches, some in pencil, some in charcoal, many of which hung at awkward angles as if they had been disturbed.

'Yes, all my own work,' she said, as she caught me looking at the drawings.

'They are very good,' I said. 'Not that I know much about art.'

'You have your own metier,' she said. 'Sorry, it's so rude of me not to properly welcome you here. It's these beastly headaches, they are driving me insane.' Katharine raised her hand to her temples and sat back down on the bed. 'You must excuse me.'

'I completely understand,' I said. 'And I'd rather not have a special welcome. But I must say I am very excited to be here.'

'You've met Len, at least?'

'Mr Woolley, oh yes,' I said. 'He was full of enthusiasm for your work here. I can't wait to see more of the dig and, of course, the ziggurat. He said he would show me that later, when he's returned.'

'And who else have you met? Father Burrows, I expect.'

'Yes, he seems an interesting character.'

'That's one way of putting it,' she said. As she smiled to herself she winced. 'Oh dear, it seems to be getting worse. I thought a nap would help, but I'm afraid it's only made the pain more intense.' Her long, bony fingers gripped the sides of her head and her mouth formed itself into an unpleasant grimace. In that moment she looked nothing like the photograph I had seen of her; features which I knew to be elegant and refined had been transformed into something careworn, even a little ugly. 'Did you say you knew of a way to relieve the pain?' she asked as she reached out to grip a bed sheet. 'Anything you could do, I'd be most grateful.'

'Yes, of course,' I said.

'And where's that girl got to? Why isn't she back with Tom yet?'

'Don't think about that for the moment,' I said. 'If you can take some deep breaths. Would you mind if I placed my hands on your neck and shoulders?'

'No, please do,' she said, closing her eyes.

'Could I wash my hands first? They are still a little dusty from the journey.'

'Of course – there's a bowl of water over there,' she said, pointing towards the dressing table. 'I'd show you but I can feel my vision blurring.'

I walked over to the dressing table, briefly stopping by the writing desk as I did so. I cast a quick look towards Mrs Woolley to check she wasn't watching me before I examined the sheets of paper that I had noticed on her desk.

'It's a work in progress, nothing more,' said Mrs Woolley from the bed. 'Don't be embarrassed. I know I would probably have done the same.'

As I poured some water into a bowl I could feel myself blushing in the dark.

'It's just a silly story I'm working on,' she continued. 'Notes towards a novel. I'd love to talk to you about the craft of writing at some point.'

'I'm not sure if I'm the right person.'

'Don't be so modest,' she said. 'But we've got plenty of time for that.'

'Now, when did the headaches start?' I asked, after drying my hands and returning to the bed.

'A few years ago,' she replied. 'But they seem to be getting worse.'

'Do you have any other symptoms? Tiredness? Nervousness?'

'Yes, both of those,' she said. 'Sometimes I feel as though I'm going quite mad. I think the headaches make me say certain things . . . And I see people doing such odd . . .'

Her voice died away as I laid my hands on her head, drew her dark hair back from her face and began to massage her temples and then the muscles in her shoulders. I could feel the stubborn knots of tension under her skin resisting the pressure exerted by my fingers and so pressed harder and deeper into the tissue.

'Does that hurt?' I asked.

'Yes, but please continue. I think I can just bear it.'

After working on her neck and shoulders I placed my hands over her eyes and started to gently manipulate the muscles around the sockets. Mrs Woolley fell silent and, as I worked and the tension started to melt away from her, I took the opportunity to study her in more detail. Her face was perfectly proportioned and refined, with high cheekbones and a clearly defined jawline. I estimated her to be more or less the same age as I was, in her late thirties, or forty at the very most. She had a kind of dark, almost exotic beauty that struck me as quite unusual, not the sort one normally found among the English. Had she some foreign blood in her? Could she have some Jewish ancestry?

'Does that feel better,' I said as I ended the procedure by gently stroking the skin on her temples.

'That feels marvellous,' she said, opening her eyes. 'Whatever you did, it's cleared the headache completely.'

'I'm pleased it worked.' I moved around to stand in front of Mrs Woolley.

'Len tries to help, he really does, as do some of the others, but his touch is nothing like yours.' She smiled. 'Now, what do you say to a cup of tea? It's the least I can do after welcoming you in such a rude and beastly manner.'

'That would be very nice indeed,' I said.

Mrs Woolley walked to the door and clapped her hands. A moment later the boy I knew as Sahid appeared and was issued with the appropriate orders. As she moved to open the shutters I noticed a dark sparkle had returned to her eyes.

'I'm sure the last thing you expected to do when you arrived was give a massage to a strange creature like me.'

So Katharine Woolley had a sense of humour; yet murderers, I had to remind myself, could be as capable as the next man – or woman – of making self-deprecating comments. 'I was a little surprised, I must admit, Mrs Woolley,' I said, 'but it was clear you were in some pain.'

'Please call me Katharine,' she said. 'Pull that chair up and come and sit down. You must be exhausted after your journey.'

'Thank you,' I replied, taking hold of a wicker chair and drawing it a little nearer towards the bed. 'I'm grateful that you've allowed me to visit. I know that you've got more than your fair share of visitors at the moment.'

'You mean those ghastly Americans? Oh, please don't look shocked. You'll see what I mean later. The man, this millionaire and his wife and daughter – don't get me started on that thing – are only here because of the depth of their pockets. Len needs them to provide the necessary funds for the continuation of the dig. It takes everything in my power to keep me from telling them what I really think of them.'

I was rather at a loss to know how I should reply and so I said nothing. There was a knock at the door and the boy appeared with the tea, which he set down on a low table between us.

'You, on the other hand, are in a different category altogether,' Katharine said, dismissing the boy with a wave of the hand. 'When I heard that there was a prospect of you coming to visit here I could not believe it. You must realise, Mrs Christie—'

'Agatha, please,' I interrupted.

'You must realise – Agatha – that what you have achieved is something quite out of the ordinary. I'm talking about *The Murder of Roger Ackroyd*, of course.'

Despite my initial reluctance, we continued to talk about that book and the subject of detective novels in general before it was my turn to ask her a few questions. I told her that I had been lured to the Near East not only by the fascinating series of articles I had read about Ur, but also by the enigmatic figure of Miss Gertrude Bell. On mention of the name, Katharine lowered her head so I could not see the expression in her eyes.

'Did you know her?'

'It was impossible not to know Gertrude,' she replied.

'Such a tragic end,' I said, deliberately vaguely in the hope of provoking a response. 'But her achievements were so varied. I suppose one must remember her for the part she played in helping make this country, her dealings with the king, and, of course, her role as head of antiquities. She must have been a great deal of help.'

'Help? No, rather the opposite if you must know.'

'In what way?' I asked.

'Let's just say that Gertrude and I didn't always see eye to eye.'

'You weren't friends?'

'No, we were decidedly not friends. That's not to say I was not sad when I heard of her death.' She said the words without any trace of emotion, almost as if she were saying them by rote.

'I've heard that she took her own life. Is that what you think happened?'

'I suppose it must have been the case. I didn't realise she had so many tragedies in her life. I think she'd lost a man she was close to and of course there was the death of her brother a few months before and then various illnesses – pleurisy, bronchitis. I think she was very weak at the end. I believe she'd found it difficult to sleep. Perhaps she took too many pills without realising it.'

'You don't believe she could have been murdered?'

'Murdered?' Katharine's voice jumped at the word. 'Whatever gave you that idea?'

I decided I would keep the existence of the threatening letters to myself; neither did I say anything about the drawing. 'I don't know,' I replied. 'There are just a few details about her death that don't quite add up.'

'Such as?'

I hesitated a moment before I began to tell her something of what I had learnt. 'Did you know, for instance, of the relationship between Miss Bell and Mr Miller who I believe works as the photographer here at Ur? It seems unlikely for her to have committed suicide if she had someone like Mr Miller in her life.'

'Whatever he saw in her I just don't understand,' she hissed, before she could stop herself.

So that relationship had been a source of rancour, after all. Had Mrs Woolley formed some kind of attachment with the handsome Mr Miller too? Was it jealousy – that age-old motive for murder – that lay behind all of this?

'Do you know how old she was?' she continued. 'She was nearly sixty! She was old enough to be his mother, if not his grandmother. It was almost indecent. And the way she treated poor Harry was completely unacceptable. Oh, I can't think about it any more – I feel that headache returning.' Her hands flew up to her temples once more and she began to pace the room like a cat confined in a space against its will. She suddenly stopped and looked at me with suspicion. 'And how do you know about that? About what happened between Gertrude and Harry?'

'I met Mr Miller in Baghdad,' I said, deciding to tell the

truth. 'He told me a little of his friendship with Miss Bell and his sadness on her passing. He couldn't believe that she would take her life.'

'And did he think that she had been murdered?'

'No, I don't think so,' I answered. 'I believe he thinks her death was a tragic accident.'

Katharine Woolley walked over to the dressing table, sat down and looked at herself in the mirror. 'I look an absolute fright – have you seen my complexion? So tired and drawn and grey. Can you see the lines?' she asked as she squinted into the glass. 'Look – here and here,' she pointed at her forehead and the corner of her mouth. 'And as for these shadows under my eyes!'

Mrs Woolley reached out for one of the jars on the crowded surface and began to massage some white cream deep into her skin.

'It's all this talk of Miss Bell. It's making me age by the minute. After all, I don't want to end up looking like she did.'

I thought the comment unnecessarily cruel, but I let it pass. 'Let's not dwell on it any more,' I said. 'Why don't you take some deep breaths. And here, drink this.' I poured out a tumbler of water from a carafe for her and watched as she drank it. 'Let's talk about something else. What about your writing? Can you tell me about that?'

After she had calmed down she went on to tell me a little of her novel, which was about a young woman who disguises herself as a man in order to have various adventurers in Iraq. She then went on to reveal how she had first met Leonard Woolley – she had arrived at Ur in the spring of 1924, keen to do something different with her life – but she made no mention of the first husband who had shot himself (or so it was said) at the foot of the Great Pyramid of Cheops in 1919.

The war continued to cast a shadow over the shared past of our generation; it was a perennial subject that often needed to be addressed when strangers first met. I told her about my time as a VAD nurse, working at the Town Hall Hospital in Torquay, and she, in turn, confided in me a little of her service in a prison camp on the Russian-German border. Some of the horrors she had seen, when she had tried to care for the army of Ukrainians who had been imprisoned there, could never be adequately described. I knew exactly what she meant. It was time to change the subject and so I asked her about her background. She had been born in Kings Norton – her father had been a merchant in the Midlands – and she had spent two years studying history at Somerville College, Oxford, before she had been forced to leave due to ill health.

'What a curse it is to have this body,' she said, her face darkening with a sudden melancholy. I felt saddened that a line of questioning which had been designed to brighten her state of mind had the opposite effect on her. 'Actually, would you mind if I had some time to myself? I'm feeling so terribly drained after that massage.' She did not look tired, however; rather, her eyes flashed with a manic quality that frightened me. It was clear that something that had been said during the course of our conversation was still distressing her.

'Of course,' I said, feeling more than a little embarrassed.

I stood up and moved towards the door. 'It's probably a good idea for you to get some more rest.'

'Look at me – I'm a wreck!' she gasped. 'I can't let the others see me like this. And still in my nightdress at this hour of the day! You must think me quite insane.'

I did not answer.

'Now where is that stupid girl with the cat? Why is it taking her so long?' She opened the door and shouted, 'Cynthia! What are you doing?'

A moment later, a red-faced Miss Jones appeared running behind the cat. 'I'm sorry – you know what he's like,' she said. 'He took one look at me and fled. He led me a merry dance all around the courtyard.' She knelt down and guided the cat towards Mrs Woolley's open door. 'Go on, Tom – look, there's your mistress.'

Katharine pursed her lips and the cat came running towards her. She made a series of clicking noises in the ginger tom's ear. The animal responded immediately with a loud purr. She ran her fingers up and down its long tail – a tail encircled by markings that reminded me of those on a rattlesnake – before she bent down and scooped it up into her arms. That unmistakable expression particular to a person strongly bonded to a domestic pet – a mix of utter devotion, pure joy and unconditional love – melted away any signs of anxiety, pain or mania from her features. It was clear that, in this moment, Katharine Woolley was happy.

Unfortunately, I had a feeling she would not remain so for long.

Chapter Six

After a rest I emerged from my room refreshed, ready for Mr Woolley's tour of the site. As I stepped into the living room I was met by Lawrence McRae and his nephew, Cecil, who sat drawing at the table.

'Hello,' said Mr McRae, standing up as he saw me. He introduced himself and his nephew, who refused to meet my eye. 'Have you met everyone on the site yet?'

'I think so – apart from the Archers, who I believe are returning later.'

'And what about she who must be obeyed?'

I pretended not to know who he was talking about.

'Come off it,' he said in his Scottish accent. 'If you've met Katharine Woolley you must know what she's like. I mean, she's hardly what you might call a shrinking violet.'

'She's suffering from one of her headaches at the moment.'

'Oh, is she now? That's very convenient,' he said with a note of bitterness in his voice. 'Just when we could do with her help to document the new finds.'

On the table before him were a number of pencil drawings of earrings, tools, cosmetic jars, robe pins, necklaces, bracelets, rings – and a beautifully realised image of a dagger.

'Are these yours?' I asked.

'Yes, and some are Cecil's – some of the best ones, in fact,' he said, smiling at his nephew who had started to blush.

'I thought you were the architect here?'

'I am. But I'm also called upon to help out with some of the drawings when Mrs Woolley is ... indisposed.'

'Does that happen often?'

'With more and more frequency, I'm afraid,' he said, placing a pencil behind his ear. He was a tall man, with serious grey eyes, striking auburn hair, and pale skin which had a tendency to freckle.

'Is she ill?'

'I doubt it very much,' he said. 'I don't want to sound cruel, but I think most of Mrs Woolley's problems exist in her head. I know you've just arrived here, and of course you must make up your own mind, but let me give you one piece of advice. Don't let yourself be bullied by her. She is an arch manipulator and will flatter and cajole, say anything in fact, to get what she wants. I've witnessed it first hand. I've also learnt it's better to speak one's mind. Of course, she hates me for it with a vengeance. But I didn't come to Ur to make friends.'

'What did you come for, Mr McRae?'

He turned his head towards his nephew, who was concentrating on a drawing. 'Him, mostly,' he said. 'It was important that we make a new start, away from – from everything that had gone before.'

I nodded in silent agreement. Harry Miller had told me that Cecil had lost his parents in an accident and I knew that any further discussion on the subject would only distress the boy. At this moment, Cynthia Jones entered the room and informed me that if it was convenient, Mr Woolley was ready to show me around the dig. She said that she would be happy to accompany me.

'The way Mr Woolley brings the past to life, it really is quite extraordinary,' she said. 'I've heard him dozens of times on the subject, but I learn something new on each occasion.'

As we walked towards the veranda, where we were due to meet Mr Woolley, I stopped and gestured for Miss Jones to step to one side.

'I wanted to ask you something about what you said earlier, about Mrs Woolley.'

'What do you mean?'

'About Miss Bell and what she had told you about Katharine.'

'I don't know what you are talking about,' she said. Cynthia looked around her, terrified someone might over-hear our conversation.

'I just got the impression that it was Mrs Woolley who—'

'I can't say any more,' she hissed. A figure appeared at the door. 'Look, here's Mr Woolley,' she said, her voice sound-ing a little strained as she tried to regain her composure.

'Are we ready for our grand tour?' he asked, his blue eyes shining.

'Would you mind awfully, Mr Woolley, if I didn't accom-pany you on this one?'

'My dear Miss Jones, of course not. Is something the matter?'

'No, just that I realise I'm rather behind with my correspondence with Philadelphia.'

'Yes, we don't want our American friends to feel they are not being kept up to date. Now, Mrs Christie, where would you like to start?'

I looked at Cynthia Jones, whose eyes darted back and forth as if she were seeing some invisible enemy.

'Are you going to be all right?' I asked in a low voice.

'Yes, why shouldn't I be? I'll be perfectly fine. I really must get back to my desk. If you'll excuse me.'

After Miss Jones made a hasty retreat, Leonard Woolley whisked me away. As he began to tell me about the history of Ur and the civilisation of the ancient Sumerians, I began to feel worried for Cynthia. It was clear that she regretted what little she had told me and she was anxious that someone had overheard our conversation. Perhaps Lawrence McRae and his nephew Cecil had heard snatches of the encounter? And earlier, when Miss Jones had first related to me what Gertrude Bell had told her – about how she believed her life was in danger – had anyone been listening then? Certainly, there had been no one in sight and we had kept our voices low, but that did not mean anything in a house such as this. Privacy was in short supply – I had worked out that eleven people, including myself, shared the space – and there were many dark corners where people could stand unobserved and eavesdrop on a conversation.

'You see, many people have assumed that Egypt is the

oldest civilisation in the world,' said Mr Woolley as he led me out of the compound. 'I mean Tutankhamun was all very well, but it was rather showy, don't you think? When historians come to write the definitive account of archaeology in the twentieth century they will realise that this – this!' he said, gesturing towards the land around him and the great ziggurat, 'was the most important find of the modern age.'

I nodded and pulled my hat further over my head to shield my face from the intense sun. Woolley's little legs carried him forwards at great speed and as he dashed towards the site he continued his lecture about Ur. He was nothing if not passionate about his subject.

'As I was saying, what you are about to see here, the tombs and suchlike, date from between 3500 and 3200 BC, the same time the Egyptians were nothing but bar-barians. And when Egypt does manage to forge itself into something like a meaningful civilisation, guess where it gets many of its ideas from – yes, here! From the ancient Sumerians, a great civilisation, one that flourished here, in old Mesopotamia, the land between the two rivers – the Tigris and the Euphrates.'

He stopped on a barren mound and, with his stirring voice, his flashing eyes, and magician-like hands, he began to conjure up the past.

'Imagine, if you can, this place not as a desert, dry and parched, but an island full of greenery,' he said. 'Water was plentiful then, and the sophisticated irrigation channels directed it in an ingenious way. Trees, fruit, vegetation – nature was in bountiful supply. Imagine, too, that on this spot, Abraham – the Abraham of the Old Testament – once walked.' He took my hand and led me towards a section of land that had been excavated, a series of low-lying walls that looked like many ruins I had seen in the past. 'This indeed is a house built at the time of Abraham. One can see, as a result of the excavation, that it was a two-storey structure, with thirteen or fourteen rooms, all arranged around a central courtyard. In fact, it could be the very house where Abraham lived, the great Hebrew patriarch from whom all Jews are descended, a man who was ordered by God to leave Ur of the Chaldees and journey to a place that would be shown to him.'

'How fascinating,' I said, but I'm afraid I did not quite believe my own words. 'And have you found evidence that Abraham actually lived here?'

'The place is mentioned four times in the Old Testament,' Woolley said sharply by way of an answer. It was obvious that I had touched a raw nerve. I wanted to believe in the story, but the cynic in me whispered a snake-like suggestion only I could hear: perhaps Mr Woolley needed to show his belief in Ur's link to Abraham so as to obtain that extra funding from those rich Americans.

'Your guests, the Archers, must be thrilled by your discoveries,' I said.

'Oh yes, they are,' he said, his mood revived by my more positive comment. 'So you've heard of them?'

'Only from what I've picked up since I arrived,' I replied.

'I think Mr Archer's input will make an enormous difference to the excavation,' he said. 'We can take on more people, make the site more secure, perhaps even stage another exhibition, a joint one, organised by the British Museum and the Penn Museum in Philadelphia.'

'That's wonderful,' I said. 'It seems as though you have a very talented team here already.'

'We've been fortunate, yes. Of course, we're missing Max – Max Mallowan, my assistant – but that can't be helped. Appendicitis.'

'And I hear you also received the great support of Miss Bell?'

At the mention of the dead woman's name the light faded from Leonard Woolley's eyes. 'That was a very sad business,' he said quietly.

'She must be greatly missed,' I said. 'Particularly here in Iraq, where she did so much good.'

'Indeed,' he said, looking across the desolate plain. 'We didn't always see eye to eye – I think she found me quite tiresome at times – but we did go back an awfully long way.'

'Really?'

'Oh yes, we knew each other during the war. In Egypt. But those days are best forgotten.' He turned from me and scanned a dusty horizon. 'From looking at it, at this desert, you'd never believe that such treasures could be found underneath the sands. But some of the most beautiful – the most precious and exquisite – objects that I've ever seen in my life have been unearthed here.'

'Yes, I saw some of them featured in the pages of news-papers and magazines back in England. That's one of the reasons why I wanted to come here.'

He turned to me, squinting in the sunlight. Was there also a look of suspicion in his eyes?

'What do you make of my wife, Mrs Christie?'

For a moment I was lost for words. He clearly knew that I had met her. 'She seems like a remarkable woman,' I replied. 'She is talented – I saw the sketches on the walls of her room. She is very beautiful too. But I think it's obvious that her headaches give her a great deal of distress.'

'Do you think they're genuine? The headaches, I mean.'

'I'm sure they are – why?'

'It's been suggested to me by certain individuals on the dig – I won't name them – that Katharine suffers not so much from a physical illness as a psychological one. They believe her presence here to be quite damaging. They accuse her of ... of poisoning the atmosphere here. I've tried to explain: Katharine has her headaches and, yes, she has her moods too. But she contributes so much – her drawings, her models, the recreations of the headdresses. But above all – and this is what people tend to forget – she is my wife. My wife.' He had to stop himself for fear he might break down. He took a deep breath. 'However, I am worried about her. Her problem seems to be getting worse. She's started seeing things, hearing things.'

'What sort of things?'

'Faces at the window. Voices.'

'That does sound worrying.'

'I don't want to bring the doctor in – he's two hours from here and it will only cause further upset in the house. It will give ammunition to the people who think that I should send Katharine back to London.' He turned to me and with pleading eyes continued, 'I know you're not a doctor, Mrs Christie, but I believe you worked as a nurse during the war. Would you do me a favour? While you are here, would you mind spending some time with her?'

'Of course, it would be a pleasure. As long as you realise I'm not at all qualified in these sorts of matters.'

'Yes, I understand. She told me she already likes you a great deal – and that's something she doesn't say about many women,' he said, a touch of humour coming back into his voice now. 'Between you and me, she thinks most women are fools. That's one thing she had in common with Gertrude, with Miss Bell.'

I knew the answer to the question I was about to put to Mr Woolley already, but experience had taught me it would be worth asking: the way someone answered was just as valuable as what they had to say. 'They must have been good friends – your wife and Miss Bell?'

'Friends?' Woolley exclaimed. 'Katharine would have scratched Gertrude's eyes out given half the chance.'

'I see.'

'And Gertrude had no time for Katharine. No, they couldn't bear the sight of one another.'

'Was there a reason for their enmity?'

'I couldn't work it out myself. On paper they were so similar. They both went to Oxford, read History, they shared a passion for the Near East and for archaeology. But they really couldn't stand being in the same room as one another. It was almost like a chemical reaction, I think. Something you read about in novels – a clash of the personalities. That's all I can think. Perhaps there was more to it, but whatever that was remains a mystery. When I asked Katharine about it she chose not to illuminate me.'

'That must have put you in a difficult position? The fact that your wife and Miss Bell didn't get on?'

'You could say that, yes,' he said, striding forwards through the sand.

As we moved nearer to the present excavation I heard a strange chanting, followed by the sight of two hundred or so men digging, sifting, clearing, all singing as they worked. The image reminded me of a colony of ants who worked not singly, but together to achieve their aim. Woolley noticed the expression of wonder and admiration on my face and went on to outline the system under which the Arab men operated. They were overseen, he said, by a foreman, Hamoudi, or to use his more formal title, Mohammed ibn Sheikh Ibrahim, whom Woolley had first met on the archaeological site of Carchemish.

'I wouldn't work anywhere in the Near East without Hamoudi,' he said. 'He taught me the most important lesson there is here: in order to be loved, one must also be feared.'

It was not a motto I subscribed to, but perhaps it was the nature of things here, in this foreign land.

'We've had no trouble with the men as a result – no thieving, no insolence. They rise in the middle of the night, walk across the desert from their homes a few miles away, and work all day shifting sand, bricks and stones in the heat. Of course, we pay them well and give them *baksheesh* if they find anything of value – that's only fair. But Hamoudi keeps a careful watch over them.' A smile broke over his impish face and his blue eyes twinkled with mischief. 'The funny thing is – the person who keeps Hamoudi in check is none other than my wife. The Arab is in awe of her, as are all the servants. There's no doubt that she is the one who strikes fear into their hearts.'

After passing through one of the gates of what he said was the *temenos* wall – a divider between ordinary land and something approaching royal or sacred territory – Woolley led me towards what looked like an enormous hole in the ground. He helped me as we walked down a series of steps dug out of the earth and, with each step, I felt as though I was descending into an underworld. I knew before Woolley told me that I was passing into a burial site. The hairs on the back of my neck rose ever so slightly and, although it may have been my imagination, it seemed as though the temperature dropped a few degrees. This was a site I felt should never have been disturbed.

'What is this place?'

'We're calling it the Great Death Pit,' he said. 'A rather melodramatic name, but it will certainly capture the public's imagination.'

'Indeed it will,' I said, remembering the relish with which I had read of Woolley's discoveries. As I saw the partly excavated site, robbed of its treasures, I felt guilty for regarding the unearthing of the tomb as a source of entertainment. I should have known better. This was a burial site, a sacred place. My very presence here was questionable.

'You are one of the first people, outside the team here, to actually see this,' said Woolley, looking more like a boy than ever. 'When we dug down we discovered the clean-cut earth sides of a pit, sloping inwards and smoothly plastered with mud. As we worked down we discovered the largest death pit in the cemetery, measuring twenty-seven feet by twenty-four feet at the bottom.'

'And what did you find in the pit, Mr Woolley?'

'There were the bodies of six male servants, who lay along the side by the door, as well as sixty-eight women, all closely laid together, found with their legs slightly bent and their hands brought up near their faces. We also discovered four musical instruments, including a silver lyre, and two statues of rams in a corner. But what was fascinating – as we discovered in the tombs of Queen Shub-ad and her husband – was that there was a total absence of any signs of terror or violence.'

'Which suggests that . . .'

'Yes, that the men and women walked down into the pit of their own accord. They were willing victims of human sacrifice. I think it's likely that these men and women took their places in the grave and then ingested some drug or poison – perhaps something like opium – which induced sleep and then death, before the pit was filled in.'

Despite the heat from the late afternoon sun the image made me shiver.

'We've made a map of the pit, showing each of the bodies in situ.'

The comment reminded me of the drawing Gertrude Bell had received just before her death.

'Is this mapping something you do yourself?' I asked.

'Well, it's a group effort. No one single person. Katharine and I noted down the exact position of the bodies, our photographer Harry Miller took some snaps, while Lawrence McRae and his nephew helped with some of the final sketches.'

I thought of the fragment of conversation I had over-heard between Mr McRae and Cecil earlier: 'If I can't have her, then no one will. I can make sure of that.' The boy had been referring to Sarah, the daughter of the rich American I had yet to meet. But if he harboured thoughts such as these, who was to say what he was capable of?

'As you can see, we're still in the process of excavation, but we've already unearthed some treasures. I'll show you when we get back to the house. They're safely locked up in the *antika* room and then they'll be securely shipped back to London, Philadelphia – or Baghdad, of course.'

'How do you go about deciding which items stay in Iraq and which ones go to the museums?'

'An interesting question. When Miss Bell was alive she and I would hold regular meetings and we would just argue our respective cases. Occasionally it would get heated and I seem to remember once we had to toss a coin over who would take possession of an extremely beautiful – and valuable – gold scarab. Much to my annoyance, Gertrude won that one.'

However, Woolley did not look that upset and the idea that he might have killed Miss Bell over such a thing seemed unlikely. Or could Leonard Woolley be that very dangerous thing: a master dissembler?

'Of course, it's only right that certain objects remain in Iraq, but also I have a duty to the trustees of the British Museum, and to the British public,' he said. But, as I had heard from his own lips, he also had a duty to his wife, who he believed was suffering from some sort of mental condition. Was he covering up for something she had done? 'Look,' he said, pointing at the horizon, 'the sun will be going down soon. Shall we walk up to the top of the ziggurat? The view from up there is magnificent.'

'That sounds like a splendid plan,' I said.

'So this is the courtyard of the Temple of Nannar,' said Woolley. 'The whole complex was devoted to the worship of a moon god. We know this because of certain inscriptions found on cylinder seals here.'

'Are they awfully difficult to read?'

'What? Cuneiform? Yes, quite tricky. Of course, they have to be cleaned first, using a diluted solution of hydrochloric acid. If you're interested I'll ask Father Burrows to give you a crash course.'

'Yes, I'd like that very much,' I said.

As we walked up the steep incline of one of the structure's three staircases, Woolley explained a little of the history of the building. The ziggurat was built, he said, by King Ur-Nammu in around 2100 BC. The people who built it wanted to erect a structure so they could feel themselves nearer to the sky and to the god they worshipped. By the time I had reached the first level I was out of breath, but Woolley, as nimble and lithe as a mountain goat, skipped ahead. I took a moment to look out across the empty, desolate desert plain.

'It's extraordinary,' I shouted so that Woolley could hear. 'There's absolutely nothing here, for as far as the eye can see.'

Woolley turned and retraced his steps so that he stood by

me. 'But there, Mrs Christie, you are quite wrong,' he said, smiling. 'Come with me and I'll show you what I mean.'

A few minutes later, and now considerably out of breath, I stood at the top level of the ziggurat. The setting sun was creating its magic across the sands, casting the desert in an ever-changing palette of colours, one moment mauve, the next apricot or rose pink.

'You may think there is nothing, but as with all things it's a matter of adjusting one's perception. In the east,' he said, gesturing for me to turn around, 'one can see the dark tasselled fringe of the palm trees at the river's bank. Of course, the Euphrates has changed its course since the ancient days. Can you see?'

I squinted until I could make it out. 'Yes, just about,' I replied.

'And there, in the distance,' he said, pointing to the south-west, 'you should just be able to see the ruins of the staged tower of the sacred city of Eridu.'

'That's where the Archers have visited today?'

'Yes, that's right. I'm sure they will be full of enthusiasm for it when they return.' He shifted his position so that he was facing north-west. 'And there, although it's difficult to see, you can just discern the low mound of al-'Ubaid, which I excavated a few years back. But yes, apart from this there is nothing but the empty desert.'

I stared across the vast plains that stretched for miles before me. The thought of being so cut off from the world thrilled and terrified me in equal measure. Perhaps I should have listened to the advice of Madge and Carlo when they had told me not to venture to the Near East. And of course, neither my sister nor my secretary knew anything of my real purpose in Iraq.

A series of images came into my mind. Gertrude Bell's terror on receiving those threatening letters and the map of a grave pit with her initials. My nasty encounter with the boy in the backstreet in Baghdad, an incident which could have ended in violence. The mania that haunted Katharine Woolley's eyes, a look suggestive of madness. The fear that gripped poor Miss Jones when she had realised she had said too much. The unsettling, queer song of those Arab workers who sifted through the rubble for signs of a past civilisation. And the dread I had felt when I had descended those steps into the 'Great Death Pit'.

My feelings of unease were not helped by what Woolley said next. 'Do you know the original Sumerian meaning for the Great Ziggurat at Ur, Mrs Christie?'

I shook my head, unable to speak.

'It went by the name of "Etemennigur",' he said, hesitating for a moment. 'Or "Temple whose foundation creates horror".'

Chapter Seven

By the evening Katharine Woolley's spirits seemed to have improved. In fact, she looked an altogether different woman. Her hair was shining and tidy, combed into a neat bob, and she was wearing a stylish dress of a shade my mother would have called *vieux rose*, complete with gloves of the same colour that ran up to her elbows.

She had taken it upon herself to direct a special supper to welcome me to the site. She oversaw the cooking, frequently dashing into the kitchen to supervise the Arab chef and watch the progress of our meal of food I had never tasted before: cheese pastries, fried chickpea balls, a spicy salad, skewers of meat, a rice and aubergine casserole and cooked lentils. She smoothed a white linen cloth over the table and in the centre she placed a small vase filled with yellow chamomiles. Then she set the table for ten with a makeshift collection of plates and not quite matching knives and forks, but she did so with such style that in the end it looked charming.

She had delegated various tasks to Miss Jones, who

busied about the sitting room sweeping and rearranging the furniture, but Katharine seemed to find fault with virtually everything she did.

'Do you really think that the armchair should be placed quite so near the far wall?'

'It seems you've missed a good deal of dust under that side table just there.'

'I can see some smears on those glasses, dear. Would you mind cleaning them once again, just to make sure?'

Miss Jones met each of these requests with grace and as much good humour as she could muster, but I still felt sorry for her. Yet whenever I tried to help Cynthia, Katharine would send me back to my chair with the words, 'You are our guest here, you must do nothing but try and enjoy yourself.'

At six o'clock, Leonard Woolley appeared, dressed in a dark suit and tie, and, at his wife's request, started to mix drinks. Gradually, the rest of the party arrived. There was Father Burrows, who I noticed behaved awkwardly in company, still wearing the clothes I had seen in him earlier; Lawrence McRae and his nephew, neither of whom had made much of an effort to dress for dinner; and the last to join us were the Americans, Mr and Mrs Archer and their daughter, Sarah.

In contrast to Father Burrows and the McRaes, the Archers had pulled out all the stops. Ruth, a small, rather overweight woman, had donned a dark silk dress which unfortunately did her no favours, but she had made up for this with a dazzling array of jewels: a diamond and emerald necklace, complete with earrings to match, and a beautiful diamond and platinum brooch. Her daughter, a slight figure with blonde hair, blue eyes and the face of an angel, was wearing a shimmering silver-beaded dress – which she later told me was Lanvin – and a long string of pearls. The father, Hubert, a huge haystack of a man with a bald head, greying moustache and prominent sideburns, had dressed in a dinner suit and stepped into the room with the confidence and swagger typical of the very rich.

After making sure that the Archers were served with soda water and lemon – the father said that none of them touched alcohol – Leonard Woolley brought the family over to me to be introduced. Hubert, on hearing that I was a novelist, proclaimed that he didn't believe in works of fiction. The only book he and his family read, he said, was the Lord's book – the Bible.

'That is the only truth I need to know,' he said, puffing up his chest. 'Sorry, Mrs Christie, I don't mean to embarrass you or slight you in any way. But I think it's worth setting out one's principles, don't you?'

'Oh yes, I do indeed,' I said, not quite telling the truth.

I listened as Hubert Archer proclaimed his views on religion and the sorry state of the modern world. I shared his opinion on certain subjects – the very real threat that came from evil, for instance – but we differed wildly on others, such as the creation of the world in seven days.

'You may not choose to support my view, but I believe that every word – every word, mind – of what is written in Genesis is true,' he said.

'And we've come to Ur to find proof, isn't that right,

Hubert?' beamed Ruth Archer, her eyes shining as brightly as her jewels.

As the couple continued to talk about Original Sin and the Biblical Flood, I noticed that their daughter Sarah seemed to show more interest in Lawrence McRae than our conversation about the Old Testament.

'Mr Woolley has found evidence of a flood that he thinks could be linked to the flood of Noah's time,' Ruth said. 'That's right, Leonard, isn't it?'

'Without a doubt,' Woolley replied without hesitation, as his wife announced that it was time for dinner. 'As I've told you, we know that the Flood as related in the Book of Genesis is based on the older Sumerian legend and that—'

'What tosh,' whispered Father Burrows to me as I moved away from the group.

'I'm sorry?'

'That nonsense about Woolley finding traces of the Flood here,' said the clergyman, pushing his wire spectacles back up the bridge of his nose. 'He's just saying that to get hold of their money.'

'I see,' I replied.

'Of course, there was a flood, but a local one, nothing to do with Noah at all,' he said, his voice still low. 'And I doubt we'll find evidence of Abraham ever living here. But those fools will believe anything they're told.'

I looked at Miss Archer as Lawrence McRae held the chair back for her. As she sat down a look of triumph came into her eyes. Across the room, sulking in the corner, Cecil McRae watched his uncle, his face burning with anger, his eyes dark with jealousy.

'Yes, please sit anywhere you like,' said Katharine. 'We're very informal here.'

Leonard Woolley sat at the head of the table, with Hubert Archer and his wife next to him; Lawrence McRae took a place next to the American millionaire and opposite Sarah and I pulled out a chair next to the girl.

'I was admiring your dress,' I said, as I sat down. 'It looks so lovely in the candlelight.'

'Thank you,' said the girl, pleased that I had commented on her appearance. She told me a little of a recent shopping trip to Paris and her love of art and music, topics which, she said, were dismissed as frivolous, if not out and out sinful, by her father. In turn, I shared a few memories of my own time in the French capital when I had been a girl. Of course, I had been without her considerable resources and my time at Mademoiselle T's school was a miserable one. I remembered the raw gnawing in my stomach which I gradually came to realise was homesickness: I had never really been away from my mother before and to be parted from her caused me to suffer. Gradually, however, I came to enjoy my time in Paris – oh, the opera, the music, the fashions! I had improved my French – my grammar was awful, but my accent not so bad – and some basics of arithmetic, history and deportment.

'You must have had quite a time there, in Paris,' said Sarah.

'It was wonderful and I made many friends with girls from all over the world,' I replied. 'Particularly American girls, like you. I always enjoyed their breezy way of talking, so refreshing after the rather stuffy drawing rooms of England.'

'Not all Americans can talk quite so freely,' she said, casting a quick glance at her parents. 'The only reason I'm allowed to wear this dress tonight is because I'm about to come into my own money.'

'Your own money?'

'Yes, you see I'm about to be wealthy in my own right,' she said. 'I know the English think it's terribly vulgar to talk about such things, but yes, I will soon have my own money, nothing to do with father or mother. My grandmother, on my mother's side, left her entire fortune to me, due to be made over to me on my next birthday. It drives father crazy because I know he'd like to threaten to cut me off without a cent.'

'So that makes you—'

'Yes, quite the catch,' she said, her eyes sparkling. She turned her attention to the man sitting opposite her – 'Mr McRae, would you please pass me the salad?' – while I started talking to Cynthia Jones, on my right, and Katharine Woolley, at the foot of the table.

'The food's not too bad tonight – nothing's burnt, nothing's too salty or too spicy,' said Cynthia, smiling at Katharine. 'Usually I have to take a glass of water with me when I go to bed. A good decision of yours to supervise Abdul in the kitchen.'

We talked of cooking and the ingredients you could buy in Iraq, and the ones they had shipped in, and various disastrous meals that they had eaten since arriving in the country. Mrs Woolley made us all laugh by telling us how at one dinner she had been presented with a congealed broth made from sheep brains and on another occasion she had taken a mouthful of a dish so rich in chilli spice that it had made her nose stream. The conversation was light and pleasant: jolly, even. From her demeanour you would never guess that only a few hours before, Mrs Woolley, whose eyes were now bright and clear, had been laid low with a terrible migraine.

During a break in the conversation I expressed my gratitude to Mrs Woolley for going to such trouble with the dinner. It was the least she could do, she replied. After all, I had given her hours of pleasure. She looked down the table and addressed her guests.

'I'm sorry to interrupt,' she said, lightly tapping a fork against her glass, 'but I wanted to welcome our guest tonight, the distinguished author Mrs Agatha Christie.'

As the room quietened and all eyes turned to stare at me, I felt myself beginning to blush. I was not keen on being the centre of attention. Katharine saw the look of consternation on my face and she leant over and whispered, 'Don't worry, I won't ask you to make a speech.' She raised her voice and continued, 'Many of you may know that I have – how shall I put it? – particular tastes in literature. I only read what I consider to be *la crème de la crème*. And I can tell you, without a doubt, that Mrs Christie's work is that rare thing, the very best. I'm not going to embarrass my friend any further – for all her genius, she is an extremely modest woman – but please could I ask you to raise your glasses to our very welcome guest, Agatha Christie.'

The assembled guests stood up from their seats and toasted my name. The whole thing was ghastly, but I told myself that it would soon be over and that it had been done with the best of intentions.

'Thank you,' I managed to say. 'That's most kind. Please sit down.' I turned to Katharine and, in order to try and deflect attention away from me, asked her about the digging season.

'Yes, I'm confident it's going to be our most productive yet, don't you think, darling?' she asked, calling down the table for her husband's opinion.

'What's that?' replied Leonard.

'Mrs Christie here was just asking about our hopes for the season. And I told her that we have every confidence that it's going to be extraordinarily productive.'

'Indeed,' he said. 'In fact, after dinner I'll show you, Mrs Christie – and of course, you, Mr and Mrs Archer – some of the treasures we have already unearthed. Needless to say, a few of the very best artefacts have already been sent to Baghdad, London and Philadelphia. But we still have things of wonder here. To see them by candlelight is a most magical experience.'

'I'd like that very much,' I said.

'Is there anything that once belonged to Abraham?' asked Mr Archer.

For a moment the question flummoxed Woolley, but he quickly regained his composure. 'Not yet – but as I said, I'm sure it's only a matter of time. We've dug up so many other beautiful things – lyres, gold cups, head-dresses, necklaces—'

'Finery doesn't interest me,' Archer replied. 'What I want to see is something concrete that once belonged to the great patriarch. As you know, it could be worth a good deal to you – we're talking thousands of dollars, Woolley.'

'That's most kind of you, Mr Archer. In fact, I have one object that should interest you a great deal and which I think may have a connection to Abraham himself.'

'Really?' said Archer, his eyes lighting up.

From across the table Father Burrows leant over to Cecil McRae and whispered something into the boy's left ear. The adolescent, who had remained silent and sullen for most of the meal, snorted at what Burrows had said. All eyes fixed on him.

'Do you find something amusing, young man?' asked Archer, a man who believed his fortune should protect him from mockery. 'Would you like to share it with the rest of the table?'

The boy squirmed in his seat. 'N-no, sir,' he stuttered, his pimpled face reddening.

'Perhaps you find God a subject of humour, is that it?' There was an unpleasant, bullying tone to Archer's voice now. 'Let me tell you, there's nothing funny about it.'

'I think you've made your point, sir,' said Lawrence McRae, who was sitting between the millionaire and Cecil. 'My nephew meant no harm.'

Archer continued to glare at the boy, and was about to say something to McRae but was prevented from doing so by his wife's pleading look, an expression which I assumed she must have perfected over the years through constant practice.

'I believe you were in Paris recently, Mr Archer,' said Woolley, desperate to try and ease the tension in the room.

'Indeed,' said Archer. 'A place full of sinners.'

'Father – you can't say that!' said Sarah.

'Why not? It's what I believe. Some of the people I saw in that city, well—' he stopped himself. 'I couldn't begin to tell you, not with ladies in the room.' He looked at his daughter's bare shoulders with disgust. 'And you, young lady, should learn to cover yourself up.'

Sarah was about to answer her father back when her mother placed a gentle hand on her wrist.

'Why don't we do that tour of the artefacts now?' said Woolley, dabbing his lips with a white napkin and standing up. 'Mr and Mrs Archer? Mrs Christie? Would you like to join me? We could have coffee when we return.'

As we stood up from the table all of us breathed an audible sigh of relief. Woolley took a key from the inside pocket of his jacket and led our little group towards the *antika* room. I cast a backwards glance to the table and saw that Sarah continued to flirt with Mr McRae, while Cecil remained silent. Katharine Woolley excused herself and Cynthia Jones walked over to talk to Father Burrows.

Woolley turned the key in the door of the *antika* room and the four of us stepped inside. The light from his candles immediately caught the gold of an exquisitely crafted cup that he said had been used to store cosmetics. 'No doubt not dissimilar to the kind of stuff my wife uses at night on her face,' he joked. Archer managed a half-smile. 'She has a seemingly infinite amount of jars – I've no idea what's in them.' From a trestle table in the centre of the room Woolley picked up a cylinder seal fashioned from lapis lazuli, together with its impression that had been transferred onto a flat piece of clay. 'As you can see, this shows a banquet scene – again not dissimilar to the kind of feast we have been enjoying tonight.' Woolley pointed out the figure of what he said was a queen or a princess drinking wine from a beaker.

'What's that?' asked Ruth Archer as Woolley's candle passed over a horrible devilish face complete with horns.

'Oh yes, a very curious find,' Woolley replied. 'It's a copper pin, the kind of thing they used to fasten tunics. But I'm at a loss to know why it's got horns – maybe its owner had a mischievous sense of humour. Can you see the work that's gone into this?' Woolley picked up a gold dagger. 'Look how it's been formed, how it's been crafted with such love and attention to detail. Of course, gold wasn't mined at Ur. Rather it was imported from Persia and—'

His words were interrupted by a piercing howl. Each of us momentarily froze, as if the horrible noise had paralysed our nervous systems.

'It's my wife – it's Katharine,' said Woolley, running from the room.

Everyone had been drawn to the source of the disturbance, which seemed to have come from Katharine's bedroom. A circle had formed around Mrs Woolley, necessitating Leonard to push his way through.

'What the hell has happened?' he shouted. 'Katharine? What's wrong?'

Katharine Woolley stared down at the bed as if in a daze. I followed her gaze and there, by the pillow, was Tom, her adored cat. It wasn't moving and there was a small pool of clear liquid mixed with blood on the pillow case by its mouth. Leonard strode forwards, bent down and gently took hold of the animal's front paw. He lifted it with no resistance. He then opened the creature's eyes; all signs of life had been extinguished.

'I'm afraid to say Tom is dead,' said Leonard, turning to his wife. 'He must have died in his sleep. I doubt very much he suffered.'

Tears formed in Katharine's eyes and the pain she felt was visibly etched on her face. She tried to speak, but the words refused to form themselves in anything but an ani-malistic moan.

'Darling,' said Leonard, taking his wife in his arms. 'I know it's an awful shock, but I'm sure it's for the best. He was probably getting on a bit and he didn't seem to like anyone but you.'

Katharine looked up in confusion at her husband's clumsy attempt to comfort her. She pushed him away and came to sit on the bed next to the limp body. She placed her head right down against the cat's stomach and nuzzled the poor creature's fur.

'Tom, dear Tom,' she whispered.

The rest of the party turned their backs and, with a few polite, well-meaning words – 'I'm so terribly sorry', 'It's always such a shock, losing a pet one has loved' – the guests made their way to the main room. I remembered all too well the deep sorrow I had felt after losing each of my dogs – to recall them brought a lump to my throat – and so I went over to Katharine, sat on the bed and took her hand. I knew that there was no need to utter commonplaces at this time. Woolley himself looked embarrassed at this show of feminine feeling and slowly edged his way out of the bedroom with the words, 'Yes, good to have some time to come to terms with it. Poor old Tom.'

Katharine continued to sob, her tears coating the cat's fur with a fine sheen, until finally she could cry no more.

'Thank you,' she said, as she wiped her eyes with a handkerchief.

'It's the least I could do,' I said. 'I know it's no comfort, but I do know what it's like.'

'What I don't understand is why – why now? Despite what Leonard says I don't think he was that old. Of course, we don't know exactly as we found him as a stray in Baghdad.'

'Perhaps he had been ill?' I suggested.

'He hadn't shown any signs of illness,' said Katharine, taking one of his paws in her hands and cradling it as if it were the hand of a baby. 'After getting ready for tonight, I left him sleeping there, on the bed. I gave him a stroke and he started to purr.' At the memory, tears began to appear in Katharine's eyes once more.

'Think of all the lovely times you had together,' I said, squeezing her hand. 'All the pleasure he gave you – and you gave him.'

'Thank you, Agatha,' she said, tears rolling down her cheeks now. 'If you would just give me a few minutes – to say my goodbyes.'

'Of course,' I said, standing up. 'And just think what a wonderful life he had.'

The grief consumed her and I quietly made my way out of the room. I rejoined the others at the table. Even though the cat had not been popular – I remembered Cynthia's words about how some of her colleagues had tried to get rid of it before, much to Mrs Woolley's distress – there was a funereal atmosphere in the room. Woolley had poured out generous measures of brandy for those in the party who consumed alcohol and most of the group sat in respectable silence. Hubert Archer was engaged in a conversation with his wife and as I passed them I heard the millionaire say, 'I don't know what all the fuss is about. It was only a cat! I'd understand if it had been her child,' closely followed by Ruth's words, 'I think it's best if you keep your own counsel, Hubert.'

'How is she feeling?' asked Cynthia.

'She's in shock, of course,' I said.

'She may need a glass of brandy,' said Woolley. 'Yes, that will do the trick.' He poured a generous measure of cognac into a glass and was about to take it into his wife's room when Katharine appeared at the door. There was something unreal about her, and she stepped into the room with the air of an actress making her final appearance at the end of a tragedy. All traces of tears had gone now, but her face had a deathly pallor to it, almost as if she were wearing a mask.

'Thank you all so very much for coming,' she said in a brittle, artificial voice. 'It was a lovely evening and your company was as charming as ever.'

Woolley walked towards her with the glass of brandy. 'Darling, I think you should drink this,' he said, passing the glass to her.

'No, I couldn't possibly,' she said, waving away the

brandy. 'I need to keep a clear head. I'll start on the washing up now. There's no point in putting it off.'

Each of the guests stared at Katharine Woolley with incomprehension and embarrassment. A few minutes before she had been reduced to tears and now here she was pretending to play the part of the hostess to perfection.

'My dear, you've had a terrible shock,' said Woolley, placing a hand gently on her shoulder. 'Why don't you come and sit down?'

Katharine brushed him off and walked over to the dining table. She reached for a plate to take to the kitchen to wash up, but stopped herself.

'If I am to do the washing up I must take off my gloves,' she said.

'Let one of the servants do that, darling,' said Leonard.

'No, it won't take long,' she said, as she started to peel back a glove. 'It's no trouble, no trouble at all.'

'Oh my . . .' said Ruth Archer, her voice trailing off. Her gaze was focused on Mrs Woolley's lower arms.

Underneath each of Katharine's arms, on the soft flesh that ran between elbow and wrist, ran a series of bloody lines that looked like scratches from a cat. It appeared as though the wounds were fresh and no more than a few hours old.

'What in God's name has she done?' asked Lawrence McRae.

The question did not need to be answered.

It seemed there had been a struggle and Mrs Woolley had killed the cat she said she loved.

Chapter Eight

I took Katharine's hand and led her back into her bedroom. Leonard Woolley followed us.

'What about the clearing of the plates?' she asked. 'The washing up?'

'That can wait,' I said, guiding her towards her dressing table. I glanced over at Woolley and, understanding at once, he nodded and went to fetch a blanket. While I tried to distract Katharine with talk of her cosmetic lotions and cleansing creams, he undertook the unpleasant business of removing the body of the dead cat from the bed. Once he had done that – and stripped away the pillowcase with its nasty stain – he left us alone to talk.

'Katharine – can I ask, what do you remember about earlier this evening?'

'What do you mean?'

'If anything ... particular sticks in your mind?'

'Well, I sat here in front of the mirror and then I dressed for dinner.'

'Do you remember putting your evening gloves on?'

She looked down at the gloves which were curled in her lap like a pair of strangely coloured exotic snakes. Something stirred within her – a fragment of a memory, perhaps – and then she gazed in amazement at her arms.

'What happened? Did something happen to Tom?'

'Yes, I'm afraid he's dead. Don't you remember?'

Katharine clutched at me with a desperation I had only seen in some of the men I had treated in the war. They had lost their minds on the front and had returned as empty shells haunted by the violence of their own actions and the violence done to them.

'Dead? But how?' she asked, her hand trailing up to her forehead.

'Where is he? Why is he not on the bed?' Her voice rose in a panic as she stood up and looked for her cat. It was a pitiful sight. 'I left him on the bed, sleeping. He was purring when I left him. Where is he? Where's Tom? What have you done with him?'

'Katharine, you must calm yourself,' I said, placing a hand on her shoulder. 'You've had some bad news. You're in shock. Try to take some deep breaths.' She looked down at her arms again. 'Why do my arms hurt? How did I get these scratches?'

'You can't remember anything of what happened?'

'I don't know what you mean. Why do you keep asking me what I remember?' Her voice had risen to a fevered pitch.

'Please, you must try and stay calm.'

'What happened? What?' The horrible look of panic haunted her eyes; her breathing had quickened and she had started to perspire. The signs told me that unless she was sedated she could be on the edge of a nervous breakdown.

'I'm going to leave you for a minute, but I am going to return,' I said. 'Everything will be all right.'

I left her and went to fetch Woolley, who was talking to Father Burrows. I explained the situation and told him that I thought it would be best to sedate his wife.

'But, as I told you, the doctor is over two hours away,' he said. 'Will that be too late?'

I informed him that I had a supply of a medically pre-scribed sedative which I could administer if he agreed. He nodded vigorously and I returned to my room to retrieve the drug. I opened my trunk and, with my special key that I always carried with me, unlocked the case inside. From there I extracted my precious poisons. I quickly found what I was looking for – a simple barbiturate, similar to the one involved in the death of Gertrude Bell – locked up my case and went to the kitchen for some tea. I returned to Katharine's room, where she was pacing the floor like a caged animal.

'I've got something here which will help you,' I said, stirring the grains of the sedative into the hot liquid. 'Please drink some of this.'

Her eyes were full of fear. 'Are you trying to poison me?' she said, pushing the cup away.

'No, of course not. What gave you that idea?'

'People have been doing strange things to me. Horrible things.'

'Who? And what things? What do you mean?'

She looked like a sullen child. 'You wouldn't understand. You wouldn't believe me.'

'Here, please,' I said, offering the cup to her once more. 'You can trust me. I can assure you that this will not harm you.'

I had some suspicions about Katharine Woolley, but I met her accusatory stare with kindness. It was obvious that she was very ill indeed. She took the cup and sipped its contents. 'That's right, a little more of the tea,' I said. 'Now, let's get you into bed.' I helped her undress, pulled back the covers and made sure she was comfortable. 'You'll feel better in the morning,' I said, not convinced at all that this was the truth. I cleaned the scratches on her arms and I sat with her as the drug started to do its work. Her eyelids fluttered, closed, and soon she was asleep.

When I returned to the sitting room I discovered that everyone except for Leonard Woolley had gone to bed. He was sitting in a leather chair nursing a large glass of brandy.

'How is she?' he asked, as he gestured for me to come and sit by him.

'She's taken the sedative, so at least she'll get some sleep,' I replied.

'I can't believe it – she loved that animal. You should have seen her with it, it was like her own baby. I thought the attention she showed it was a bit pathetic at times. Surely she wouldn't have done anything to harm it?'

But he – like all of us – had seen the scratches on Katharine's arms. 'I suppose it must have been done in a fit of madness. She must not have known what she was doing.'

'Have you seen her like this before?'

Woolley fell silent, his eyes clouded by painful past memories. 'I suppose you might as well know the truth,' he said, taking a large sip of the brandy. 'I'm at a loss to know what to do. I'm not sure if I can endure it much longer.'

'Do you want to tell me what's been troubling you?'

'You've seen what she's like. I've been trying to pretend to everyone here that Katharine suffers from nothing more than terrible migraines. But over the last few weeks I've begun to worry about the state of her mind.'

'Have you witnessed things that warrant such a concern?'

'I have,' he said, sighing. 'So many, I don't know where to begin.' He stood up and checked the doors to make sure no one was listening. 'I've been trying to keep it from the rest of the team here, but that has not been easy. A number of people came to me – poor souls, I should never have asked them to put up with it. Anyway, they felt that Katharine's behaviour had started to become not only eccentric, but erratic. They said that . . . that they no longer felt safe around her. Well, after tonight's incident, who is going to blame them?'

'I think it's best if you tell me everything, Mr Woolley,' I said. 'I'm not sure I'm the right person, but I will do my utmost to help.'

'You will?'

'Why, of course.'

'Would you care for . . . ?' he asked, proffering the brandy. 'No, I forgot you don't drink,' he said, pouring himself another measure. 'Well, I may as well start at the beginning. I'm correct in saying that's what you writers would do?'

I nodded encouragingly. 'That's right.'

'It's not a pretty tale, I'm afraid, but it is one that you

must hear, if you are to understand anything about my wife. And you are sure you won't be shocked?'

'As you know, I was a nurse in the war, Mr Woolley. There is precious little in this world that shocks me.'

'Very well,' he said. 'When I first met Katharine – Mrs Keeling as she was then – in the spring of 1924 she was a widow. She was a beguiling creature, a beauty – well, she's still that. She came out here as a volunteer and almost immediately I spotted her talents. She had a wonderful eye for capturing the essence of an object, almost as if she had a real affinity with the artefacts that we were digging up from the ground. Of course, she suffered from headaches then, but we managed to work around them. Her presence, however, caused some consternation with the heads of the sponsoring museums, and the director of the Pennsylvania University Museum, George Byron Gordon. He's dead now, by the way, had an unfortunate accident at the beginning of last year at the Racquet Club in Philadelphia.'

'So he was against Katharine's presence here?'

'Yes, and he wrote a letter to me outlining his position. There had been some talk – there's always talk, isn't there – about the propriety of having a single woman on the excavation site, here in the middle of nowhere, thrown together with a number of men. It was beginning to bring the expedition into disrepute. Gordon said that he would hate to think that Katharine might become the subject of "inconsiderate remarks" and that her reputation would be sullied.'

'And what of Miss Jones? Was Mr Gordon not concerned about her presence?'

'Apparently not. I think he assumed she was the spinster type.'

'And so you married Katharine?'

'Yes, we were married by my brother, in fact. In Hampshire, in April 1927. But I want you to know that was not the only reason why I asked her to marry me. It wasn't as if I was forced, if you see what I mean. No, I found her a bewitching woman. She could be vivacious, funny, charming. But she was mercurial.'

That word sounded like a euphemism to me. 'Mercurial?'

'Yes, I suppose that's what I found so attractive, to begin with.' He stood and checked again that the doors to the room were closed. 'I'd hate this to get out. You won't say anything, will you, Mrs Christie?'

'I promise I will be discretion itself.'

'Very well. It was on my wedding night that I got a taste of . . . well, of Katharine perhaps not being like other women.'

'In what way?'

'We had arranged to spend the night at a hotel in London. I don't know what I thought would happen – I was a very inexperienced man, I'd devoted my life to my work, I suppose – but I was shocked and, yes, a good deal upset.'

I let Woolley talk in his own time as I could tell that he found discussing this intimate subject difficult and embarrassing.

'She'd already made it known that she didn't want anything to do with my family. That hurt me, but I understood she wanted to start a new life after the death of her husband.'

It was interesting, I thought, that he chose not to share with me the manner in which Colonel Keeling had died.

'And I had certain assumptions – because she had been a married woman.'

'Assumptions?'

'Yes, I thought that she would have a certain knowledge and experience of these things.'

'I see,' I said.

'B-but,' he said, blushing. 'That ... that night, our first night together, she made it clear that she wanted to have separate bedrooms. I told her that this could be arranged. But she meant it in a more permanent sense, if you see what I mean. I pointed out to her that we had just got married, that this was our honeymoon night, but she got more and more upset. I couldn't see any sense in what she was saying, but in hysterics she locked me in the bathroom and refused to let me out until I had promised that I would not make love to her. In the end, what could I do? Of course I gave her my word. Later, we tried to talk about it, but she was adamant. She did not want that from a marriage.'

'Oh dear, that must have been extremely difficult for you.'

'And since then that is how it has been. I did seek out the advice of my sister, Edith, and, if the truth be told, I even considered divorce. But I decided that it would be unseemly to cause a scandal. And my family believed that divorce might well bring about the end of my career.'

As I thought about my own divorce I felt myself beginning to blush.

'I'm sorry if I've shocked you, Mrs Christie, I didn't mean to—'

'No, not at all,' I said. Even though it pained me to say the words, I thought it right that I shared my own experience with Woolley, so as to reassure him that he wasn't the only person in the world to have experienced marital problems. 'You see, my own marriage broke down and before I came out here I agreed to divorce my husband, Archie.'

'I'm sorry to hear that, I had no idea,' said Woolley.

'He fell in love with another woman,' I said.

'I don't think Katharine has fallen in love with another man. Perhaps she just doesn't care for those kind of feelings – with anyone.'

'Yes, perhaps that's it,' I said. 'Some women – and men, I believe – are like that.'

'But I just wish she had told me before we'd married!' There was anger in his voice now. 'I might well have thought otherwise. I could have chosen someone else or perhaps nobody at all would be better than this.' He paused to reflect on what he had just said. 'Listen to me – you must think me an utter beast, talking like this about Katharine. No, I made my vows and I'm going to stick to them. "For richer for poorer, in sickness and in health, to love and to cherish, till death us do part."' He fell silent and pensive. Had his final phrase struck some deeper resonance within him?

'Has your wife shown any signs of violence?'

Woolley looked appalled at the question. 'What do you mean?'

'I wondered if you had seen her inflict harm on others – or herself?'

'No, she's not like that at all. I don't know what you could be suggesting.'

'I think we both know that Katharine killed her cat,

don't we? And I wondered whether there could have been other occasions when she might have lost her temper.'

'She is headstrong, yes, and can be cruel at times. I think tonight was just an isolated event. She must have momentarily lost her mind.'

'Did she ever mention Miss Bell to you?'

'Gertrude?'

'Yes.'

'No, not really. But, as I've told you, my wife and Gertrude didn't get on.'

'Do you know when the two women last met?'

'I've got no idea. I last saw Gertrude in March, I think it was, when I went to Baghdad. Oh yes, Katharine did accompany me then because we were on the way back to England. I spent a few days with Gertrude at the museum, working on the classification of some artefacts.'

'So that was March 1926? And Miss Bell died in July.'

'That's right, but I don't see how this has got anything to do with Katharine. We were both in England during that summer.'

My mind was running ahead with possibilities. Could Katharine have sent those threatening letters and that drawing of a death pit to Miss Bell from England? Did Gertrude Bell know something about Katharine Woolley that might have prevented her marriage to Leonard? Perhaps Miss Bell had some knowledge of what happened to Katharine's first husband.

'I'm just trying to get a few things into some kind of order, that's all,' I said. 'You say your wife was a widow when she met you.'

'Yes, she was married to a Lieutenant Colonel Keeling. That marriage, I'm afraid, was not a happy one.'

'Do you know how Colonel Keeling died?'

'A very sad case. He took his own life, I'm afraid. Katharine told me he shot himself in Cairo, only a few months after their marriage.'

I did not say anything. The silence forced Woolley to fit together the pieces of the puzzle himself. 'You're not . . .'

'What?'

'You can't possibly believe that Katharine . . .'

A look of horror filled his face as he slowly realised the seriousness of the implication.

'Could she have killed her first husband?' he whispered.

'I don't know – is that something you believe she could be capable of?'

'No, not Katharine. She's not that kind of woman. But . . .'

No doubt the events of the night played through his mind. 'But if she could kill Tom, then perhaps she could have done that too.' The articulation of that thought drained the colour from his face. 'No, I refuse to believe it. It's too ridiculous for words!'

'Yes, quite ridiculous,' I said.

But both of us knew it wasn't so far-fetched at all. In fact, it would fit the pattern. The question was: would Katharine Woolley kill again? And if so, who would be her next victim?

Chapter Nine

I was dreading the gloomy atmosphere at breakfast the next morning, but it was enlivened by the return of Harry Miller from Baghdad. The American photographer was full of tales of the exotic city, but thankfully he chose not to regale everyone with how he had saved me from a street urchin. With him he brought back all the supplies. There was typing paper and carbon sheets for Cynthia Jones, new blocks of clay for Father Burrows, a nice set of pencils for Lawrence and Cecil McRae and some religious pamphlets for Hubert Archer, who had set out on an early morning walking excursion with his wife and daughter.

'Where's the queen?' asked Miller with a mocking tone. 'Is she still in bed?'

The room went quiet except for the scrape of a knife across a piece of toast.

'What's the matter? Why are you all looking so odd?'

I had been to see Katharine earlier that morning. She was still extremely drowsy, but it looked as though her symptoms had dissipated. However, I advised her to spend the day in her room. I said I would check on her at regular intervals and Woolley, who was still sitting with her, was extremely grateful for all my help. However, I could tell that he was guarded around Katharine, mindful perhaps of the conversation we had had the night before.

'She's not feeling too well,' I said.

'What? Again?' replied Miller. 'Don't say we're going to have to get the leeches out.'

'Leeches?' I asked.

'Sure. She likes the little bloodsuckers. Says they help relieve tension or something. Sounds like a load of—'

'Well, I'm sure anything that can help Mrs Woolley at the moment can only be a good thing,' I said.

'What – like that time she made Len sleep outside her room with a string attached to his big toe?' said Lawrence McRae in an unpleasant tone of voice. 'So that she could pull it if she needed him in the night?'

'Is that true?' I asked.

Miller nodded and poured himself another cup of coffee. 'She's got that man where she wants him, all right,' he said. He looked around the table, studying each of us in turn. 'But seriously – has something happened?'

Cynthia Jones told the story of what had occurred the night before.

'I don't believe it,' said Miller. 'I know Katharine's a bit odd, she has her moods, and some of the things that come out of her mouth can be cutting, but to kill Tom? She loved that cat.'

'What else can we think?' said McRae. 'We saw the scratches on her arms. She could pick up Tom, but obviously there had been a struggle when she began to hurt him.'

'So what do you think happened?' asked Miller, still mystified.

'The only logical solution is that she must have killed the cat in her room, where it scratched her,' said McRae. 'She finished it off, put on those long gloves of hers to cover up the marks, got ready for dinner and came in here.'

'Would the cat not have made an almighty noise?' asked Cynthia. 'We've all heard it at times making that awful yowling.'

'Did you hear music coming from Katharine's room last night?' asked McRae.

'Yes, now you come to think of it, I think I did hear the sound of her gramophone,' said Cynthia.

'Perhaps that drowned out the cries of the cat,' said McRae.

'But to be so heartless?' said Miller.

'She may not have known what she was doing,' I said.

'She was out of her mind? Is that what you're saying?' asked the photographer.

'By all accounts it seems as though Mrs Woolley has been under a tremendous amount of strain,' I said. 'She may have carried out the attack under the delusion that the cat presented some kind of threat. And she could have blocked out all memory of it.'

'Is that really possible?' asked Miller.

'Yes, I'm afraid it is,' said Father Burrows, who had been silent up to this point. 'Once, just after the end of the war, I saw a case of a poor man who had been convinced that his wife was a German soldier. When he came back from the front he attacked her – he came at her with a hammer – and as a result he had to be taken into hospital. After he'd come round from sedation he had no memory of the incident. Luckily, the wife survived the attack, but of course the gentleman had to be locked away.'

'So she's . . . ?' asked Cynthia.

'Mad, yes,' said McRae, finishing the sentence for her. 'I always said she was unstable. But now it seems as though she might be dangerous.' Gertrude Bell had used the same word to describe Katharine Woolley.

'What are you saying?' Cynthia asked, clearly frightened now.

'I'm just wondering whether . . . whether it's safe to be here, in the middle of nowhere,' answered McRae. 'Who knows what she might do? I'm not concerned for my safety so much – I'm sure I can defend myself against a woman – but for those less able to protect themselves. We've got women to think of – girls like Sarah, and then there's Cecil.'

'You don't need to worry about me,' said Cecil, finishing off his toast and brandishing a knife. There was something unpleasant in the way in which he smiled as he did this. 'I can look after myself.'

'I think it's far too early to jump to any such conclusions,' I said, trying to calm everyone. 'After all, we don't know exactly what happened. I'm going to try and speak to Mrs Woolley again today. Perhaps we'll know more after that.'

Father Burrows stood up from the table and started to clear away his breakfast things. 'Perhaps the woman needs some religious guidance?' he said. 'I could talk to her if you think it would help?'

'Yes, I think that's a very good idea,' I said.

'What utter tosh, Eric!' exclaimed McRae. 'I don't think that's going to make much difference. I don't know if you've noticed, but God is no longer with us, if he was ever with us at all.'

'McRae, really, I think that's taking things too far,' said Burrows.

'Let's all try and keep our heads here,' said Miller.

'Yes, I'm sure there's a perfectly good explanation for what happened,' said Cynthia. 'Perhaps Tom was already ill or injured and Mrs Woolley had to put him out of his misery.'

'That sounds most unlikely,' said McRae, checking the pot on the table for coffee. 'Where's that servant boy – why can't he—?' He stopped, as if something had just occurred to him. 'Now I think of it – what was that story I heard? About Mrs Woolley. Yes, it was something to do with her time during or shortly after the war. Now who told me about it? I can't remember now, but it may have some bearing on this.'

'What do you mean?' asked Miller. 'What story?'

'Something to do with the death of her first husband,' said McRae with fire in his eyes. 'Yes, that's right, it was about a servant in Cairo. You see the servant discovered something about Katharine – I don't know what it was – and he threatened to tell her husband. What was the husband's name?'

'Keeling, I think,' said Cynthia.

'Yes, that's right, a Colonel Keeling. The next thing, Keeling is found dead at the base of the Cheops pyramid, apparently having shot himself. But it doesn't take a genius to work out what really happened.'

'What really happened?' I asked.

'It's obvious, isn't it,' said McRae. 'Katharine shot him and made it look like a suicide. And if she can do that once, it's my opinion she can do it again. Wouldn't you say that's true, Mrs Christie? What do they say in works of detective fiction? Murder is a habit?'

'Well, I—'

'That's all hypothesis and conjecture,' said Miller. 'I mean, do you know that for sure?'

'That Colonel Keeling died?' said McRae, enjoying the attention. 'That he shot himself? Of course, that's a statement of fact.'

'No – that other thing you said?' asked Miller. 'That Mrs Woolley killed her own husband. You said it was a story, something someone told you.'

'Yes, but—'

'But nothing,' said Miller, walking up to McRae with an aggressive swagger. 'I want to know who told you this.'

'I don't know, I can't remember, but I'm sure it was someone—'

'I thought better of you, McRae, I really did,' said Miller, stopping a couple of inches away from the architect's face and looking at him with disgust. 'Unless you've got concrete proof of Mrs Woolley's guilt in the matter then—'

His words were cut off by the entrance of Leonard Woolley into the room. 'Mrs Woolley's guilt in what matter, Mr Miller?' The room fell silent. 'I'm sorry, I seem to have interrupted the flow of your conversation. Please carry on.'

'We were just talking about Mrs Woolley and Tom,

116

the cat,' said Miller, thinking quickly. 'Whether she was responsible for, for . . .'

'For its death?' asked Woolley coolly.

'Well, yes,' Miller replied.

Woolley appraised the group, casting his eye over each of us in turn. Had he been standing at the door for long? What exactly had he heard?

'Really, all this fuss over the death of a stray cat,' he said, laughing. 'It's really quite ridiculous.'

'But if she did kill Tom then how are we—' said McRae.

Woolley cut him off quickly. 'Let me stop the rumours and the speculation right now,' he said. 'My wife didn't kill the cat. I did.'

The confession sent ripples of confusion through the room. 'If you'll just let me explain,' he said, holding up his hand.

'You see, Tom had not been quite himself for some time. I'm sure most of you had noticed it. He'd become increasingly bad-tempered. He wouldn't let anybody pick him up or stroke him even at the best of times. But last night when Katharine tried to stroke him he went for her and injured her badly. You saw for yourself the marks on her arms. And so, I did the kindest thing possible – I put him to sleep.'

'I don't understand . . .' said Cynthia, her face the colour of ash. 'Sorry, I—'

'I think what Miss Jones is trying to say is why did you let Mrs Woolley find him like that?' I asked. 'Why didn't you tell her what you'd done?'

'I did,' said Woolley, looking taken aback that I had dared to question him in this way. 'But I think Katharine

117

didn't want to face up to the truth. She adored that cat. But she knew that as soon as it attacked her then it would have to be put down. I told her last night, before dinner, but she pretended not to hear or understand what I was saying.'

This version of events did not correspond with what Woolley had told me only the night before and I suspected this story to be a lie designed to protect his wife. 'But to leave the cat there on the bed like that for Mrs Woolley to find?' I asked. 'Was that not terribly cruel?'

'Again, Mrs Christie, that's not quite how it happened,' replied Woolley. 'After the cat was dead I covered it in some sackcloth and moved it into one of the storage rooms. I thought I would dispose of its body the next day. But Katharine must have found it and brought it back into her bedroom. She told me she couldn't bear to live without it – even after what it had done to her.'

Each of us stared at Woolley, none of us quite sure about what to say in response to his bizarre explanation. Finally it was Harry Miller who walked over to Woolley and shook his hand.

'I wasn't here to witness any of this, but by the sounds of it you handled a difficult situation well,' he said.

'I wouldn't say that exactly,' replied Woolley. 'I really should have told you all about it last night, but Katharine was in such a state. I wasn't sure how to deal with it. Anyway, I apologise for any upset I caused.'

I had some questions that I still wanted to put to Mr Woolley. Cynthia Jones too looked confused, while McRae seemed possessed by a mix of embarrassment and anger.

'Now, do you think we can get on with some work,' said Woolley; it was not so much a question as a statement. 'Burrows? What's the latest on the inscription of that damaged cuneiform tablet that you've been studying? Can you read any fragments of it yet? McRae, I'd like to take a look over the latest batch of drawings, if you don't mind. Miller – I'm interested in looking again at those photographs of the two sets of gold and lapis lazuli beads, if that's not too much trouble. And Miss Jones, I need to send a memo to the directors of the museum in Philadelphia.' It was clearly business as usual for Woolley and soon the director of the expedition had restored, for the moment at least, a sense of order and confidence in his team.

I slipped away and gently knocked on Katharine's door.

'Come in,' she said.

I entered to find her sitting up in bed, the shutters to the windows still closed. When she saw me she smiled awkwardly; it was clear from her expression that she regretted the scene that she had caused the previous night.

'How are you feeling?' I asked.

'A little tired,' she said.

'That's only to be expected,' I said, as I came to stand near the bed.

'Agatha – I'm terribly sorry about what happened,' she said, gesturing for me to sit down next to her. 'You must think me such a fool, and as for the rest of the team ... I know how they view me, how I must come across. I can hardly bear to show my face today.'

I needed to find out the truth about the cat. 'Don't distress yourself further,' I said. 'It was obviously very upsetting for you. I think everyone understands what must have happened.'

She looked blankly at me. 'What do you mean?'

'Leonard explained everything,' I said, giving her a kind smile. 'About how he had put the cat down.'

There was a long silence during which time I began to feel more and more unnerved.

'Leonard didn't kill Tom,' Katharine said finally.

'But he just came in and told everyone how—'

'Yes, about how Tom had turned on me and how he discovered me covered in scratches? That was just a story Leonard made up to try and protect me.'

So I had been right. Automatically I shifted my position to sit a little further away from her on the bed. Katharine noticed that and also the subtle change in my eyes. I tried not to look afraid, but the more I concentrated no doubt the more frightened I appeared.

'It's not what you think,' she whispered.

'But if your husband didn't kill the cat, then who did?'

There was a pause before she asked, 'Do you think I did it?' There was a coldness to her voice now. 'Is that what you think I'm capable of?'

'I'm just trying to make some sense of what happened,' I replied.

She studied me as if I were an object, a lifeless piece of sculpture in a gallery or a mannequin in a shop window.

'I had higher hopes of you,' she said at last. 'From reading your books I thought you would be cleverer than this.'

'I'm sorry, Katharine, I don't understand what it is you're trying to say.'

She took a deep breath as if she were a teacher in the presence of a particularly stupid child.

'So you're saying Leonard didn't kill Tom?' I asked.

'Yes, that's right. And I know I didn't kill him either.'

'You didn't? But what about the scratches on your arms?'

'I can't explain those.'

Was Katharine to be believed? Was this just a game she was playing?

'Don't you see someone else is behind all of this?'

'Someone else? Who?'

'I don't know,' she said, looking around the room as if there might be someone hidden in one of the corners. 'I've seen things. Heard things.' Was she in the grip of another horrible delusion? 'I can't explain.'

'Perhaps you need another sedative?' I asked, but I was wary of giving her more. After all, she couldn't spend her life on the drugs.

'No, I need to keep my wits about me,' she said, panicking. 'What if someone wants to kill me? Or they might try to murder Leonard.'

'You're scaring me now, Katharine,' I said. 'Why would anyone want to do that?'

'I don't know, it's just a feeling I've got,' she said. 'I know you don't believe me,' she snapped, looking at me with resentment in her eyes. 'You may as well go.'

'I'm not going to leave you until I understand what it is you are afraid of,' I replied. 'Because I can see that you are frightened. Very frightened indeed. I know – let's have some tea.'

'It won't have a sedative in it?' she asked, her voice rising. 'Please – I told you, I can't afford to go to sleep.'

'No, I promise. Nothing but good English tea.'

'Very well,' she nodded.

I called the servant boy and as we waited for the tea to come we talked of other matters such as the weather, the shifting nature of the desert sands, the social scene in Baghdad and, finally, the things the Woolleys missed about England.

'Do you have any family back at home?' I asked, as I poured the tea. There was no response. 'Someone mentioned to me that you were married before. Is that right?'

'Who told you that?' she asked with an accusatory look.

'I think it might have been your husband,' I said.

'Why were you talking about that? What's that got to do with you?'

'I know it was quite presumptuous of me, but I was asking Mr Woolley about you, the state of your health, your headaches and so on, and he must have mentioned it. I'm sorry if I've offended you in some way.'

Katharine fell silent. 'Yes, I was married,' she said finally. 'It didn't last long.'

'I'm sorry to hear that,' I said, before I told her a little of the breakdown of my own marriage and my recent divorce. I explained how my husband had left me for a younger woman. 'You see, I've had rather a hellish time of late,' I concluded.

Something akin to pity came into Katharine's eyes and she began to warm to me again. 'That does sound most unpleasant,' she said. 'What a beast of a man.'

'No, not a beast,' I replied. 'Just utterly, utterly selfish.'

'At least with Bertie I never had to worry about that sort of thing,' she said.

'Was that your first husband? Bertie?'

'Yes, and he was devoted to me,' she said.

I didn't ask the next question that sprang to mind, but instead I let her tell me something of the history of their marriage. 'He was always a very sensitive man, and I think he was left disturbed by the things he had seen in the war. Who wouldn't be? He was in the Royal Engineers and so terribly brave. He was wounded in France in 1916 and sent home, but then returned the following year. He was at the Somme and the battle of Cambrai, and survived both.'

'He sounds like an extraordinary man,' I said. I noticed she said nothing about the intelligence work Davison told me Keeling had done. Perhaps she had not known about that, or was she keeping something secret?

'He was so very clever – he loved the stars, astronomy, and mapping and lots of technical things, triangulation and things like that. Did a lot of work in Egypt, which he loved. He'd just got this new job in Cairo when . . .' Her voice faltered. 'When he fell ill and died.'

'I'm so terribly sorry,' I said. 'Did he require a lot of nursing at the end?'

'No, his death was rather – sudden,' she said, taking a deep breath. 'I suppose you may as well know: Bertie took his own life.'

'Oh my,' I said, pretending to be shocked.

I watched her closely as she related the sad story to me: how she had said goodbye to her husband one morning as he left their home in Cairo for his position as the Director-General of the Survey of Egypt. The next thing she knew was that an official arrived bearing the bad news that her husband had died. It seemed that the Colonel had shot himself. She did not appear to be lying. In fact, it seemed as though her emotions were genuine. However, I knew that people who worked in intelligence, like some murderers, possessed that frightening ability to lie so convincingly that even trained observers were taken in. Perhaps Mrs Woolley was both things: a spy and a killer.

'But why did he do it? Were there any signs?'

'Not that I could see,' she said, blinking. There was something incomplete about her answer, as if she was holding something back.

'I'm sorry I have to ask you this, Katharine,' I said, gently. 'But did you see his body?'

'But why do you ask that?' She looked at me with deep suspicion. 'I don't understand. What do you suspect me of? What have you heard?'

'I wondered whether you had actually identified him?'

'No, the head of the police in Cairo told me that it would be too distressing for me,' she said, with tears in her eyes. 'Even though I didn't see him, I could just imagine the scene . . . the blood, the—'

'Let's not think of that,' I said. 'It's my fault for asking. I'm terribly sorry.'

She got out of bed, opened a shutter of one of the windows and looked out. 'I've seen things out of this window,' she said, straining her neck, patterned by red blotches, and peering upwards. 'Faces. Horrible faces.'

'Please calm down,' I said. 'I didn't want to upset you.'

'Why did you ask about Bertie?' she said, turning away from the window and walking towards me. 'What do you know? And why did you ask all those questions about his body?' She blinked and the expression on her face changed, as if she had just realised some awful truth. 'Oh no, please God, no.'

'Katharine – what is it? You've thought of something, haven't you? What's occurred to you?'

'No, it couldn't be – it can't be true.'

'Tell me – what can't be true?'

She was shaking now. 'What – what if Bertie isn't dead?'

'What do you mean?'

'You asked whether I had seen his body. I hadn't. I believed the police who told me that Bertie had died – that he had shot himself.'

'But surely there was a death certificate. You had that?'

'Yes, but Bertie could have substituted a body to try and fake his own death.'

This was like something I might dream up for a plot in one of my own books. 'That couldn't be possible, could it?'

'All he needed to do was find a victim, shoot the poor man in the face so he was unrecognisable, dress him in his clothes, substitute his identity papers for the victim's and then disappear.'

'But why would your husband do that?'

She stared at me with eyes full of terror. 'A week or so before he died we had an argument. A terrible row. Awful things were said on both sides, and I confess I expressed certain sentiments I now regret. But during that brutal argument he said something to me. He said, with a cold tone of voice I'll always remember, "I could kill you – you know, one day I may just do that."'

Chapter Ten

The next few days passed without incident. Katharine locked herself in her room, convinced that her first husband was still alive and that he might want to kill her. Woolley told the rest of the group that his wife was suffering from one of her headaches and although it seemed that most people believed his story about the cat, an air of suspicion still hung over Katharine. Lawrence McRae continued to feel a certain level of antipathy towards her and, while Woolley wasn't looking, did everything in his power to turn others against her.

I wrote to Davison in Baghdad asking him to find out more about Colonel Keeling's past and who had identified his body. I also wanted him to dig out reports of other deaths of men of a similar age in September 1919. I kept Davison up to date about the events at Ur – adding a number of character sketches – and voiced some of my own suspicions. Was Keeling still alive? Could that really be possible? Had he taken over the identity of the man he had killed and dressed in his own clothes? Even if that was true – and it sounded so unlikely – that did not explain what had happened to Gertrude Bell. I had to remember that I was here to find out who had killed her and so far I was no nearer the truth. Had Katharine sent those letters to Miss Bell? But if she had, why would she go to the trouble of telling Gertrude Bell that her future murderer was based at Ur? Or had Colonel Keeling sent the letters in order to cast suspicion on his wife?

I told Davison that there was no need for him to venture south to Ur just yet; his presence, or that of another offi-cial, would only raise more questions. After all, no crime had yet been committed here, save that of the suspicious death of a cat. There was something about the demise of that ginger tom that unsettled me, but I couldn't quite put my finger on it. Had Woolley killed it so as to try and drive his wife insane? Was he subjecting her to a subtle form of psychological torture?

It was no surprise that, after the events of the last few days, an air of melancholy descended over the expedition. Despite the best efforts of the seemingly eternally cheerful Harry Miller and the youthful exuberance of Sarah Archer, the group's spirits could not be raised. Breakfast was mostly a silent affair, while the evenings seemed to stretch on for eternity, with endless conversations about cuneiform tablets, the minutiae of the archaeological process and the lineage of Abraham. Finally, one night, after a particularly depressing dinner, Woolley declared that, as he knew it would soon be Miss Archer's birthday – she would be twenty-one on the following Saturday – he thought it would be a good idea to celebrate it with a picnic at the top of the ziggurat.

'What a swell idea,' said Miller, slapping Woolley on the shoulder. 'We could all do with something to look forward to.'

'Thank you, Mr Woolley,' said Sarah. 'I was beginning to wonder how long you British could carry on with your long faces.'

'Now, don't be rude, dear,' said Ruth Archer, giving Woolley an apologetic smile. 'What would your father say if he heard you?' Luckily for Sarah, Hubert Archer had stepped outside with Father Burrows for some night air.

'No, Sarah is quite right,' replied Woolley. 'We've been down in the dumps for too long. It should be a jolly event. Your father has given his approval and I've started planning it with the chef and the servants already. There will be food, a nice fruit punch for those of us who partake, and a fresh lemonade with mint too for Mr and Mrs Archer, and you, Mrs Christie.'

'And will Mrs Woolley grace us with her presence?' asked McRae, barely trying to disguise the the sarcasm in his voice.

'I've yet to hear from Katharine on that point,' said Woolley. 'As you know, she has not been quite herself since Tom died. But I hope she will come – she needs cheering up, as do we all. This will give us an opportunity to put that unpleasantness behind us. Then we can go into the season with a renewed vigour.'

On the Saturday of the picnic, Woolley announced that Katharine was feeling a great deal better and had decided that she would, after all, join us on the outing. On hearing this news I went and knocked on her door.

'Come in!' she called from inside.

I opened the door and stepped into what seemed like a different room to the one I had seen a few days before. Mrs Woolley had thrown open the shutters and light streamed in, bathing everything in a delicate yellow glow. She had taken the trouble to order her desk – there was no sign of the papers that she had been working on; the floor appeared to have been swept, the clothes that had previously spilled out of the wardrobe had been tidied away and the pots that had cluttered her dressing table had been arranged neatly. Like the room itself, Katharine appeared to have tidied herself up. She had washed and brushed her hair, her neck seemed free of the blotches that had plagued her complexion and she was dressed in a striking lavender blouse, skirt and jacket, with shoes, gloves and a hat to match.

'Isn't it a beautiful day?' she said brightly, taking a grey silk scarf from the wardrobe and tying it lightly around her neck. As she did so I noticed that she had painted her nails a vibrant red colour and was wearing an elegant gold dress watch.

'Yes, it is,' I replied. 'And you're looking a great deal better.'

'I thought it was about time I faced the world again,' she said.

'I'm very pleased to hear that.'

Katharine walked over to the window and looked out. 'I can't imprison myself in here any longer. And I refuse to continue to be afraid. I told Leonard about my fears regarding my first husband and he convinced me that I had nothing to worry about. In fact, looking back I'm sorry I said such silly things to you. I don't know what came over me.'

'It was understandable after—'

'After what happened with Tom. Yes, I remember now Leonard telling me about what he had done. Yes, the cat did scratch me and my husband took it away to put it down.' There was something artificial about her delivery, as if she were reading an over-rehearsed script. 'Best really. One cannot have an animal like that around. Leonard did me, and everyone else, a favour.'

I knew she was not telling the truth, but I decided not to voice my suspicions as I didn't want Katharine to suffer another relapse.

'And as regards my own safety,' she continued, 'as Leonard said to me this morning, what can happen to me? I'm going to be surrounded by him and the other men, you, all my friends on the dig. Then there are the servants. I've nothing to worry about at all.' She said this as if she were trying to convince herself. 'After being cooped up in here for days I can't tell you how much I am looking forward to the picnic, even though Sarah Archer is not exactly my favourite person.'

'Yes, I don't quite understand her,' I said.

'Oh, good. I was worrying that she and you might become as thick as thieves.'

'No, not at all, she's very young,' I said. 'And also, she's too rich.'

'Too rich? Surely no one can be too rich.'

'You know what I mean,' I said, smiling. 'She's got different values, different ideas about the way life should be.

All that money – imagine! And she told me that she will be soon be very rich in her own right. Independently wealthy, quite apart from her father.'

'I know – aren't some people lucky?' Katharine said. 'That's not to say I have any regrets in marrying a poor archaeologist. There are other benefits, it must be said.'

I recalled the conversation I had had with her husband. The couple, or so Woolley had told me, had never been intimate. Perhaps Katharine was referring to the simple joys of companionship; that was, I knew, enough for many people. It would have been enough for me, but never for Archie. During the dark days leading up to my divorce I had even suggested that Archie continue his relationship with Miss Neele as long as he remained married – in name only, of course – to me. Thank goodness Archie had had the good sense to reject such a proposal.

'Oh, please don't say anything about my aversion to Sarah,' pleaded Katharine, suddenly realising the implication of what she had said. 'If it got back to her, or her father, the Archers might think twice about pledging their financial support. And Leonard would be furious with me.'

'I'll keep it just between the two of us,' I assured her.

'Anyway, I must get going,' she said. 'I promised Leonard I would supervise the picnic. I don't know why he suggested to the Archers that we have it at the top of the ziggurat. What's wrong with the courtyard? Anyway, it's done now. But it does mean an awful lot more work.'

Katharine was not exaggerating. The chef had to draft in extra help from his sons and, throughout the day, Leonard had even been forced to requisition some of the

Arabs from the dig to carry things from the house up to the top of the ziggurat. As was their custom, they sang as they worked, that strange sound that was not so much a song as a monotonous chant. However, by four o'clock they had erected various tents, a couple of tables, a selection of chairs, cushions and rugs, and an hour later they finished bringing all the supplies of food and drink to the top of the ziggurat. We left the house all together in a group, and to outsiders it would have appeared as though none of us had a care in the world. The conversation was superficially light and a delicious sense of anticipation hung in the desert air. Harry Miller took snaps of us all as we walked – he said he wanted to give a selection of the resulting photographs to Sarah Archer as a birthday present – and he made us smile and laugh with his seemingly endless supply of jokes.

'Do you know why photographers can be so nasty?' he asked, a mischievous sparkle in his eyes.

'Nasty? I don't know – why?' asked Sarah, twirling her parasol around in her hands.

'You really can't work it out?' teased Harry.

The girl shook her head. There was no doubt there was a spark that existed between them, something that both McRae and his nephew pretended not to be bothered about.

'Well, here we go. First we frame you, then we shoot you, and finally . . . we hang you on the wall.'

'That's a terrible joke!' said Sarah, laughing.

'It amused you, though,' said Harry. 'You can't deny that.'

'Only because it was so awful,' she replied.

It did not amuse me, however. I also noticed that it failed to raise a smile from Leonard Woolley. Perhaps he was too worried about the smooth running of the picnic, and as we walked he busied himself with last-minute arrangements, talking to various Arab servants in snatches of conversation I did not understand. I could not fail to be impressed by the sound of the language – rich guttural tones that seemed to be wrenched up from the very back of the throat – but there was something quite alien and unnerving about it too.

By the time we had reached the top of the ziggurat, climbing steadily up one of the three staircases that led to the pinnacle, we were all out of breath. We were met, however, by a glorious sight: a series of tents had been erected in one corner of the enormous plateau and a number of servants were ready with trays of drinks. The lemonade was tart, but delightfully refreshing, while the fruit punch was, according to those who sampled it, both delicious and quite potent.

'Just think, Sarah, you could be celebrating your twenty-first birthday in one of the best restaurants in Paris, London or New York,' said Ruth Archer, 'and here you are at the top of an ancient temple in the middle of a desert. Who would have thought it?'

'I couldn't think of anything nicer,' said Sarah, flashing a smile first at Miller and then at McRae. The girl – who was out of earshot of her father – was playing a very dangerous game with this flirtation. Ruth cast a concerned look at her daughter, and tried to engage her in conversation with Miss Jones, but the girl's eyes kept returning to the two men. She never once looked at poor Cecil, who stood by himself in a corner of the tent with a brooding expression on his face and a large glass of punch in his hand.

'Don't you think this is just splendid?' asked Katharine, coming up to me.

'Yes, it was a lovely idea,' I said.

'I feel so much better,' she said. 'How could one not, standing up here, with this view?'

I looked across the desert plain, the setting sun turning the sands the colour of blood. The sound of Sarah's laughter drifted across the ziggurat. She had made her way over to talk to Miller, who said he wanted to take some photographs of her standing at the edge of the ancient structure.

'Be careful, Sarah,' said Ruth. 'There's a terribly long drop there.'

'Don't worry, Mrs Archer,' said Miller. 'She's in safe hands.'

Sarah changed her pose with each click of Miller's camera, clearly enjoying the attention from her fellow American. As she kicked her heels she dislodged a few small stones, sending them over the edge of the ziggurat. Her black net dress, embroidered with chenille, made the near-translucent whiteness of her skin stand out, giving her the look of a ghost.

'I'd really like to have a suntan,' she was telling Miller, 'as it's the fashion now on the Riviera. But Daddy would be furious.'

'Well, I think you look swell just the way you are,' replied the photographer.

'In fact, he would not be at all pleased by some of these pictures,' said Sarah. 'He can be such a prude at times. By the way, Mommy, where is he?'

'Father Burrows is showing him some inscriptions on one of the temple walls,' said Ruth. 'And you're right, your father would most definitely not approve. So I'd make it quick if I were you, Mr Miller.'

Although Woolley was trying to persuade people to sit as the food was ready, Sarah Archer and Harry Miller were reluctant to move away from the precipice. Harry asked Sarah to turn her back to him so he could photograph her silhouette set against the sky.

'Just look at this light!' exclaimed Miller. 'The shadows are something else.'

McRae mumbled something to Cecil and both of them glared in the direction of the photographer. As Sarah pirouetted around she caught sight of her father approaching and signalled to Miller to stop.

'I get the message,' said Harry Miller. 'But I think we got something special there.'

The two walked over and took their seats under the tent as the servants began to plate out the food. The group fell silent as we began to eat a thick tomato stew made with meat which I think was lamb, but may have been goat. Woolley took out a bottle of wine and offered it to the group.

'Yes, I think I would like a glass of that too, if you don't mind,' said Katharine.

'Do you think that's a good idea, my dear?' asked Leonard. 'After all . . .'

'What?' said Katharine under her breath.

'Just that it may not be wise, that's all,' he said gently, trying to prevent the conversation from escalating into a row.

'I think it will do me a world of good – just like this party,' said Katharine, holding out a glass.

Husband and wife locked eyes and finally, after a few moments, a defeated Leonard poured out the wine for Katharine.

'Thank you,' she said.

Miller, glass in hand, stood up and proposed a toast to both Mr and Mrs Woolley for hosting the picnic and to Sarah Archer to wish her happy birthday. We all sang along, and even Lawrence and Cecil McRae roused themselves from their black humour and joined in.

'Thank you,' said Sarah, coming to sit by me. 'I can't believe I'm twenty-one.'

I thought of myself at that age. What had I achieved by then? Absolutely nothing of consequence. 'And what do you have planned when you return to America?' I asked. 'I suppose you must have a great many suitors waiting for you at home?'

'There are some young men who have been picked out for me as being eminently suitable, yes,' she said. She cast a flirtatious glance over at Harry Miller. 'But whether or not I will accept them is another matter.'

As the servants lit a series of long torches, enclosing the group within a circle of fire, Sarah began to tell the story of how, aged eighteen, she had endured a proposal from a man she considered old enough to be her father. Mr Adams, the man in question – whom she regarded as really quite ugly – had taken great pride in outlining what he thought were his considerable financial resources. Sarah had sat and listened patiently to his list of assets: his properties scattered across New York and Philadelphia, his portfolio of stocks and share certificates, his savings and investments. 'So you see, Miss Archer, it would be greatly in your interest if you were to accept me,' proclaimed Mr Adams, 'as I can offer you a personal sum in excess of two hundred thousand dollars.' She told everyone how she had made a pretence of amazement, thanked him for his offer and told him that, as she would only have five hundred thousand to her name, she was sure that he would find her too poor for his consideration. 'Also, I fear that you are too handsome for me, sir,' she added with extra spite.

Although the group laughed at the story and her punch-line, I thought Sarah's comment to be ill-judged. Her aspiring suitor had been crass, yes, but that did not – in my opinion – permit her to be so cruel. It was obvious that her father thought so too.

'You don't do yourself any favours by telling that story,' said Mr Archer. 'In fact, it's not at all Christian of you.'

'Papa, you know as well as I do that Mr Adams would have been wrong for me.'

'That may be so, but—'

'I think you should listen to your father,' said Katharine, interrupting the girl.

Sarah looked astonished that someone outside her family had dared to contradict her. 'Excuse me, I didn't quite hear what you said.'

She was giving Katharine the chance to retract her comment, but it was obvious from what Mrs Woolley said next that she had no intention of doing so.

'I think at times it is wise to listen to those who know

better,' said Katharine. 'Those with a little more experience of life.'

Sarah Archer looked around at the startled expressions of some of those in the group and, emboldened by what she took to be support, placed her fork back on her china plate and said, 'And those who think they know better should sometimes keep their own counsel.'

Katharine's eyes burnt with a dark fury. Leonard reached over and placed a hand on his wife's arm, but she brushed it off.

'In fact, the kind of advice I get from people never ceases to amaze me,' continued Sarah, pausing for full effect, 'especially from those who are not quite all there.'

'How dare you,' said Katharine, standing up.

'Sarah!' hissed Mr Archer.

'What? I'm only voicing what we all think, aren't I?' She looked around for someone to try and back her up. 'Or am I the only one brave enough to say the unsayable?' She turned to Katharine and addressed her directly. 'Very well. Here goes. The truth is everyone thinks you're cracked. That you're crazy. Unhinged. There are plenty of words to choose from – take your pick.'

Mr Archer took hold of his daughter's wrist in an effort to stop her. 'What's gotten into you today, Sarah? I know it's your birthday, but there is no need to—'

'There's every need!' said Sarah, her voice rising. 'We've all been pussyfooting around, pretending everything is fine. When we're living with – I'm sorry to say this – but with a madwoman. Someone who could be dangerous. After what happened to that poor cat . . .'

'But Mr Woolley explained what happened to Tom,' said Cynthia Jones. 'That he had to put him down.'

'It was obvious he was just covering up for her,' said Sarah. 'Weren't you, Mr Woolley?'

Leonard, who looked mortified by the scene playing out before him, did not answer. How could he?

'Are you going to stand there and let this girl talk to me in this way?' Katharine asked her husband. 'Leonard?'

'I – I think it would be best if—'

'Yes, it would be best if Miss Archer here found it in her soul to behave in a more Christian-like way,' said Katharine. 'But perhaps her parents were more interested in studying the good book than in instilling some basic good manners.'

'Excuse me,' said Mr Archer, reddening in the face. 'I really don't think that is called for.'

'Now, Katharine – why don't we all try and calm down,' said Woolley.

'Well, this sure is a birthday I'll remember,' said Sarah as she flung down her napkin. Before anyone could stop her she took a torch from one of the servants and dashed from the tent, across the ziggurat's plateau and into the darkness.

'Sarah!' shouted her father. 'Don't be a fool now. Come back. It's pitch black out there.'

'Yes, the sun does seem to just drop out of the sky,' said Woolley. 'Don't worry, I'll go after her.'

'I'll come with you,' said Miller. 'If we use the light from her torch to guide us we should be able to catch up.'

'So will I,' said McRae, who was followed a few minutes later by his nephew.

'Honestly, what a ridiculous scene,' said Katharine to the rest of us sitting in the tent. She seemed completely oblivious to the part she had played in the drama. 'Had the girl been drinking on the sly, do you think?' she said to me so the Archers couldn't hear. 'I know her father said the family didn't touch alcohol, but I wouldn't be at all surprised if she'd had some of that fruit punch. The men did say it was awfully strong. What else could have caused such an outburst?'

'Birthdays can often lead to tension,' I said diplomatically. The real issue was, of course, the rivalry that existed between the two women. It was clear that Katharine was jealous of the attention Sarah had been receiving from both Harry Miller and Lawrence McRae. And, like a younger sibling who resents the flow of birthday gifts to an elder brother or sister, she had been determined to try and ruin the day for her. Unfortunately, Sarah was not the kind of person to be silenced and had struck back with a vicious verbal blow.

'Well, I think it's time we got back to the camp,' said Mr Archer, somewhat stiffly. 'And then tomorrow we can think about leaving.'

'Leaving?' said Katharine. 'What do you mean?'

'Well, you don't expect us to hang around here and let ourselves be insulted, do you?' he said.

'But what about—'

'And of course the additional money I talked to your husband about is quite out of the question now. Come on, Ruth.' Mr Archer signalled for two servants holding torches to accompany him and his wife down from the ziggurat. 'We've got an early start tomorrow.'

'Please, I know I may have spoken out of turn, but . . .'

Katharine said, as the Archers disappeared into the night. 'Oh dear,' she added, turning to me. 'What will Leonard do when he finds out? He was depending on the Archers' money. He had so many plans.'

'I'm sure Mr Woolley will find another rich donor willing to sponsor the expedition,' I said. 'After all, the project has had so much publicity.'

'Leonard would kill me if they withdrew their support now,' said Katharine. 'No, I'm going to go after them and see if I can change their minds.' She too took a torch and ran across the plateau with a plea for the Archers to stop.

'If you and Father Burrows are happy to make your own way down, I'll go with her and make sure she doesn't come to any harm,' Cynthia said to me.

'Yes, good idea,' said Father Burrows. He called for the remaining servant men to provide us with some more light and ordered them to clear up as best they could; if they took the food and drink back down to the house that night then they could dismantle the tents the following day. 'Well, I'm sorry you had to witness that, Mrs Christie.'

'Yes, it was most unfortunate,' I said. 'But I suppose Mrs Woolley could not endure it for much longer.'

'Endure what?'

'The behaviour from that silly girl.'

Father Burrows looked at me blankly. 'What on earth do you mean?'

'Didn't you notice the way Sarah was flirting with those two men, one minute with Mr Miller and the next with Mr McRae? And then there was that awful thing she had said to poor Cecil.'

'What thing?'

I related what I had overheard that day I had first arrived: how Sarah Archer had told Cecil that he was ugly and stupid and that he should never talk to her.

'It seems as though you've got it all worked out,' he said. 'But perhaps I've deliberately made myself blind to such things. Only way if one wants to cope out here.'

'Yes, I can imagine you're right on that score,' I replied, as I took Father Burrows's arm and two servants, holding torches, led us across the plateau towards the ziggurat's staircases. The sky was covered by a blanket of cloud, eclipsing the stars and the moon, and although it was not yet cold, nevertheless I felt a chill at the back of my neck. Just as I pulled my shawl around my shoulders a scream split the night air. Was it the noise of the animal I had heard when Woolley had taken me up to the ziggurat that first day? No, this was unmistakably a human cry, the cry of a woman.

'What on earth . . . ?' said Father Burrows.

'I think that's Mrs Woolley,' I said.

Father Burrows pressed the servants – who were wide-eyed with fear – to accompany us as fast as possible down one of the staircases to the source of the cries.

'Quickly,' I said. 'Each moment of delay could make all the difference.' But Father Burrows, for all his slim form, was unsteady on his feet, and as we rushed he tripped a few times, nearly dragging me down with him. 'Please, let me go on ahead,' I said. 'Can you ask one of the servants to accompany me and also to let me have a torch of my own?'

He looked dumbly at me. 'Please, Father Burrows,' I

pleaded. 'It could be a matter of life or death.' Fighting to try and regain his breath, he gave the appropriate orders in halting Arabic and I hastened down the stone steps of the ziggurat, the flames from my torch casting a series of grotesque shadows onto the ground. As I approached the bottom of the structure I heard cries of alarm and then the screams of another woman. I saw the backs of Lawrence McRae and Cynthia Jones, peering at something on the ground, and then Ruth Archer, lifeless as a rag doll, being supported by her husband.

'What's happened?' I asked, breathless from both the exercise and the fear that was rising within me.

Cecil appeared out of the darkness, his face white with shock, and pointed at something on the ground, a shape that seemed to be even darker than the black night. I stepped closer to see that there were in fact two figures: Katharine Woolley on her knees and a strange unnatural configuration before her on the ground.

'Don't look,' said Lawrence McRae, trying to stop me. 'It's not—'

I pushed past him. 'It might not be too late,' I said. 'I was a nurse. Let me see if I can help.'

I thrust the torch nearer to the scene before me. The glistening form illuminated by the flames was like something from an unholy nightmare. Katharine Woolley stared at her palms, covered in blood, as if her hands did not belong to her. In front of her, in the ancient sands, lay the lifeless body of Sarah Archer, the back of her skull crushed. Nearby lay a rock covered in blood. Without much hope of finding signs of life, I took her pulse, but there was nothing.

The noise of footsteps diverted our attention for a moment. Leonard Woolley and Harry Miller appeared out of the darkness. 'We couldn't find her – she must have gone down one of the other staircases, but we saw some other—' said Miller, stopping himself as he took in the shocked expression on all our faces.

Leonard Woolley stepped forwards and, as we parted to let him through, he saw for himself the horrors of the scene – his wife with her bloodied hands, Miss Archer's dead body.

'Oh my God, Katharine,' he whispered. 'What have you done?'

Chapter Eleven

There was a moment not so much of calm, but of near silence, when no one spoke and it was possible to hear the cry of birds down towards the river and the whispers of the dry wind across the desert. Then everyone began to speak almost at once.

'Katharine – get up, and come with me,' said Woolley, stretching out his hand.

Ruth Archer looked up from her husband's shoulder and tried to break away from him. 'What have you done to my daughter?' she asked, her voice full of rage. Luckily, Hubert Archer managed to restrain her, otherwise there was a danger she might try to scratch out Mrs Woolley's eyes.

'Sarah – surely she can't be . . .' said Cynthia.

'I'm afraid it's too late,' said McRae, checking Sarah's body.

'But – I don't understand,' said Miller. 'She had only been gone a minute. We were following her torch and then it disappeared out of sight.'

'It must have been a terrible accident,' said Father Burrows. 'She must have slipped and hit her head on a rock.'

'Does that look like an accident to you, you damn fool,' said Hubert Archer. 'Sorry for my language, but the evidence is right in front of us. We can see it with our eyes as plain as day.'

The only person who did not speak was Cecil McRae, who had turned his back on the scene. His uncle went to place a hand on the boy's shoulder, but Cecil brushed it off and ran back towards the camp.

'It's understandable he's in shock,' said Lawrence. 'He worshipped that girl. As I'm sure many of us did.'

Mrs Woolley tried to stand, but as she did so she caught sight of her bloodied hands and fell back onto the sands. She opened her mouth to speak, but no words came out.

'Here, let me help,' said Woolley. He took out a handkerchief from the top pocket of his jacket and offered it to his wife. But, still paralysed by shock, she did not know what to do with it and so let the white square simply drop to the ground. Woolley bent down by his wife and supported her as she stood up.

'If a crime has been committed,' said McRae, looking first at Mrs Woolley and then at the body, 'then nothing should be touched or removed.'

'McRae's right – we need to get hold of the police,' said Hubert Archer, fighting back tears. He was the kind of man who did not like to show signs of grief in public.

'Let's not jump to any conclusions,' said Woolley.

'It's obvious what's happened,' replied Archer. 'How long does it take to call in the authorities here?'

'If we sent a servant now to the nearest town there still wouldn't be anyone here before the morning,' replied McRae.

Archer looked down at his daughter, the blood from her head collecting beneath her like a sinister, dark pool. 'But we can't leave Sarah out here all night,' he said. 'No, we'll sit by her side, won't we, Ruth?'

His wife did not respond, but continued to stare at Katharine Woolley with hatred in her eyes. Woolley gave orders to two servants to ride to Nasiriya, from where they would fetch a doctor and a policeman.

'I can't believe it,' said Cynthia. 'It seems so unreal.'

'I'm sure there's some explanation about what happened,' said Woolley.

'But you more or less admitted it yourself, man!' exclaimed Archer. '"Oh my God, Katharine – what have you done?" you said, unless I misheard you.'

'Yes, but I'm sure it's just a case of—'

'Cold-blooded murder, yes,' said Archer. 'I don't think the authorities look kindly on killers out here, do they Mr Woolley? And I would advise that you lock up your wife tonight – just as much for her own safety as for the rest of us.'

Woolley refused to be drawn into an argument and instead turned to practical matters. 'I suggest all of us make our way back to the house. We need some strong, sweet tea or a spot of brandy. I'll send some over to you and your wife, Mr Archer, together with a couple of camp beds and blankets. Obviously a couple of servants will stay with you all night. Then, in the morning, when the authorities arrive we can try and make some sense of this.'

Woolley led his wife away from the body, the rest of us trailing silently behind them. Back at the house, after making sure Katharine was safe in her bedroom, Leonard took me to one side and asked me to talk to her. 'She's saying nothing to me,' he said, carrying a chair and placing it outside his wife's door. 'I'll sit on guard here all night. I'll make sure she doesn't get out – and that no one else gets in.'

I knocked on the door, gently turned the handle and stepped into the room. Katharine was standing at a wash bowl, scrubbing her hands with soap like a real-life Lady Macbeth. She stared down at her palms, which were now nearly clean of blood, but the water was a reddish-brown colour. A cup of tea, untouched, sat on the dressing table.

'Here, let's get you into bed,' I said. 'You've had a nasty shock.'

I helped undress Katharine and tied a silk dressing gown around her slender form. There was a deadness to her eyes, as if a part of her too had perished out there at the base of the ziggurat. Once I had managed to persuade her she should get into bed I fetched her a small glass of brandy and urged her to drink it.

'Now, that's better, isn't it?' I said, as she sipped the dark liquid.

Slowly she turned her head to me, blinked and said, 'No, nothing will make it better.' She brought up her hands to her face. 'I can still smell the blood. It's seeped into me. I'll never be free of it.'

I took her hands. 'But you've almost rubbed them raw,' I said. 'Why don't you try to sleep and we can talk about this in the morning?'

'You're not going to leave me? Please don't leave me.'

'No, I'll stay with you as long as you like,' I said. 'And Leonard has stationed himself outside your door, so you are quite safe.'

'Do you still have those sedatives?' she asked.

'Yes, I do.'

'I know I said I didn't want them, but now – after this, well . . . if I could just forget about it for a few hours?'

'Of course,' I said. 'I'll go and measure some out for you.'

Outside the door I found Leonard playing the sentinel of sorts; he was sitting in a chair, a candle by his side, making some notes about the dig. When I told him of his wife's request, he looked concerned. 'Will you be careful with the dose?'

'What do you mean?'

'Well, after tonight's terrible events, we don't want . . .'

His voice trailed off, but I understood his meaning. 'You think Katharine might take an overdose?'

'Well, there is that possibility. After all, she's been under an awful lot of strain. What happened tonight could push her over the edge.'

'I see. Can I ask you a question, Mr Woolley?'

'Yes, of course,' he said. We stepped away from the door and walked across the courtyard so that Katharine couldn't hear our conversation.

I looked Woolley straight in the face and asked, 'Do you believe your wife was responsible for the death of Miss Archer?'

The question clearly pained him. 'She's not that kind of woman, she abhors violence of any kind. I know what

I said before, but . . .' He turned away from me, unable to finish the sentence. 'No, I refuse to countenance any suggestion that she was responsible for this. It must have been a tragic accident.'

'If you're so certain of that why would you be worried about your wife taking an overdose?'

'I know her character. She's terribly highly strung, as you've seen for yourself. Her mental state has never been strong.'

There would be time for more questions tomorrow. 'Very well,' I said. 'Anyway, I must fetch that dose of sedative. A good sleep is what she needs. And don't worry, I'll be very careful how much I give her. Could you please order some tea for me? I find that's a good way of taking the drug.'

I collected the barbituate from my room, nodded to Woolley on my return and entered Katharine's quarters again. As I prepared the sedative Katharine, now lying in her bed, studied my every move: the careful measurement of the dosage, the addition of the sedative to the tea, the gentle stirring action as I made sure the grains were evenly dispersed in the liquid. Then she started talking in a strange, trance-like fashion.

'How wonderful it would be just to forget everything,' she whispered. 'Just to go to sleep knowing that all one's worries had slipped away. Do you ever think that, Agatha?'

'I must admit I have on a number of occasions,' I said. 'Now, here's the sedative. If you drink this it will help. It won't provide a permanent solution to your anxieties, but it will guarantee you a good night's sleep.'

Katharine took the cup in her hands and although

the tea was hot, she gulped it down as if she was greedy for unconsciousness.

'I can't stop thinking of the blood – all that blood,' she said, as she laid her head back on the pillow. 'I was rushing down the steps of the ziggurat, trying to find Mr and Mrs Archer who must have taken one of the other stairways, when I slipped and fell. I could feel a warm stickiness on my hands, but I didn't know what it was to begin with. There was a torch on the ground, casting its sideways light onto ... and then I realised what it was. Blood. And then I saw Sarah. I reached out to try and lift her to her feet, but it was no use. Then I saw that where the back of her head should have been there was ...'

'That's enough. You've had a horrible shock. We all have. We can speak of it tomorrow.'

The drug had started to have an effect. Her eyelids were drooping and her breathing had slowed.

'But then everyone gathered round,' she said. 'The things people said. Mr and Mrs Archer, Mr McRae, Leonard even. And the look in Cecil's eyes frightened me. They were all standing there accusing me of killing Sarah.' She paused as a horrific thought possessed her. 'Maybe every-one is right. I know I was angry with her. I felt like putting my hands around her neck and strangling her until she was blue in the face. Perhaps it was me, after all ... Perhaps I am guilty of murder. But I don't remember doing it. Am I going mad?'

As her arm went limp I took the teacup from her hands and made sure that she was comfortable. She would sleep soundly now, I hoped, for at least eight hours. I watched her as her eyes closed, thinking over not only the events of the night, but of the days since I had arrived in the camp. There was so much evidence that pointed to the fact that Katharine Woolley was, as Gertrude Bell had said, a truly dangerous woman. Although I knew she was not in Iraq at the time of Miss Bell's death, she could have sent the threatening letters and that drawing from England to an intermediary in Baghdad. Perhaps she was guilty of sending a poison pen letter, but nothing more? But what of the suggestion that she had played some part in the death of her first husband? And what was I to make of the dreadful events since my arrival at Ur: of the awful atmosphere, the thick scent of suspicion that hung in the air, the demise of Katharine's cat and then the horrible death of Sarah Archer? Yes, Katharine seemed to sit at the centre of it all. And yet . . . there was something not right.

As I sat there gazing at Katharine's quietly beautiful face, I did not feel any fear. Despite what I had witnessed tonight – the sickening image of Sarah Archer with her skull smashed in, the sight of Katharine Woolley with the girl's blood smeared across her hands – I realised that I was not afraid of her. Mrs Woolley probably suffered from some sort of mental illness, but my instinct told me that she was not a murderer.

I walked over to her desk and picked up the incomplete manuscript of Katharine's novel, which she had entitled *Adventure Calls*. I sat down and read for half an hour. It was a strange tale of a girl called Colin, who disguised herself as her twin brother to undertake a dangerous secret service mission in Iraq during the early 1920s. It was an entertaining, if unlikely, story, but then I thought of some of my own novels and short stories, whose plots required leaps of the imagination too. One paragraph from Katharine's manuscript, about the law of the desert, jumped out at me.

> In England, if a man is killed the murderer is pursued by the far-stretched arm of the law; both the detection of crime and its punishment are the duty of the State alone. But tribal law is different. There the family of the murdered man takes upon itself the duty of pursuit and punishment and the guilt extends from the murderer to every man of his house. There is here no jury and no judge, only the avenging kinsman's bullet fired at the first member of the blood-guilty family caught unawares. The feud, a chain of murders, may be drawn out for generations, till the original crime is forgotten and the alternate slaughter goes on senselessly.

As I read this, Katharine's words sent a chill straight through me. What if she was playing an elaborate game? I had been taken in before by the surface sheen of appearances, and, despite my best efforts and past experience, I could no doubt be deceived again. Of course, I knew that one should not use literary criticism as a form of detection. After all, how would I feel if strangers started looking into my novels and short stories for clues to my own inner life? Yet there was something about the words that Katharine had written that made me suspect she was drawing from something very close to home. My imagination started to turn over. Could Katharine be using the mask of insanity as a cover? Could she simply be pretending to be unhinged in order to carry out murders in cold blood? Was she enacting some perverse form of justice that she had dreamt up in her twisted mind, setting herself up as judge, jury and executioner of people who had offended her? If not that, then what was her motive? Was there a pattern? I read the passage again and one phrase in particular jumped out at me: 'a chain of murders'. I knew what she meant by that – after all, one death often sparked off others, like some awful kind of catalyst.

I feared that the death of Sarah Archer was not an isolated event. In fact, I had a terrible premonition that this was just the beginning of a cycle, one which could be long and bloody.

Chapter Twelve

Katharine woke with a start, jolting me from my slumber. My neck was sore from the awkward angle at which I had been sleeping in a wicker chair by her bed.

'. . . Sarah's head . . . something in the dark . . . all sticky,' she said. 'All covered in blood.' She looked down at her hands, as if surprised to see that they were clean. She looked at me like a confused child. 'What happened?'

'There was an accident with Miss Archer,' I said. 'What do you remember of it?'

After taking a few sips of water she began to tell me something of the incident. As she spoke I was conscious of not taking her word as the gospel truth. Her version of events would, I knew, have to be compared to the evidence gathered by the authorities. I assumed that, once the local police had been informed, the message would be relayed to the high commissioner in Baghdad and thence to Davison. I surmised that it would not be long before my friend arrived at Ur.

'I think I said something unfortunate, something that

upset Mr and Mrs Archer,' said Katharine. 'They said they would be leaving, that they would be pulling their funding for the expedition. I wanted to explain and so I ran after them. It was dark, but I followed their torches. But then I think I must have lost them. I saw another light and I started to walk after that, and as I came down the steps I heard what sounded like a girl's cry. I ran towards the noise. I couldn't make out what it was at first – I thought it was a doll, a life-size one, on the ground. But I realised it was Miss Archer. I think I collapsed and as I reached out that's when I felt something wet on my hands. I screamed and the next thing I knew I was surrounded by the rest of the group, all with horrified expressions on their faces. Then the accusations started. I think listening to those – hearing what people thought I was capable of – was worse than the discovery of that body. Even Leonard thought . . .'

'And did you hear or see anything before you heard the cry?'

'I'm not sure,' she said. 'I might have heard the sound of footsteps or some sort of scuffle. And it was so very dark, wasn't it? I was finding it difficult enough to find my way with a torch.' She sat upright and grasped my hand. 'You do believe me, don't you? You're not one of those traitors who would try and get me to hang for something I didn't do?'

'I'm sure it's not going to come to that,' I said, trying to disguise the fact that I wasn't quite sure what I believed. There was a gentle knock at the door. 'Excuse me,' I said, relieved that I could turn my back on Katharine so she could not see my ambivalent expression.

It was Woolley. 'How is she?' he whispered as he ushered

me outside the room and gestured for me to shut the door. 'Has she said anything about last night?'

I informed him of what his wife had told me – that she might have heard the sound of footsteps, or a scuffle, soon followed by a cry.

'So it may not have been an accident, then?' he said, his face darkening.

'No, I don't believe it was,' I replied.

'Just that we've heard from Nasiriya and they won't be able to send anyone over here until tomorrow at the very earliest,' he said. 'Minor uprising or something along those lines.'

'I see,' I said.

'I just told the Archers the news and they blew up in my face,' he said. 'They demand I get someone higher up from the Baghdad police. I've told them of course we will try, but the police are not going to get here any sooner, what with the journey.' He collapsed back down on the chair, looking exhausted. 'Normally, I don't mind not getting much sleep.' He stared at me. 'You probably didn't get much either, eh?'

'No, but it doesn't matter. I thought it was important to sit with your wife.'

'And now Hubert Archer is demanding that I keep Katharine locked up in here. He says that it would be too much of a risk to let her out, that she would be a danger. I've tried reasoning with him, but he's not having any of it.' He ran his hands through his hair and exhaled deeply. 'I suppose if my daughter had been murdered I would feel the same way.'

'I don't mind staying with her,' I said.

'You're not . . . afraid?'

'Should I have cause to be?'

He did not reply. 'Mr Woolley, if there is anything you have kept from me – for example, anything relating to your wife – I think it's only fair that you tell me.'

'Well, I—' Just as he opened his mouth to say something, something that I was sure would prove to be important, Katharine opened the door.

'Excuse me, Leonard, if I could just have a word with Agatha for a moment,' she said. She spoke as if she were at a village tea and she wanted to talk to me about the minutes of the parish meeting or to ask me about a recipe for a particularly delicious sponge cake, not about the ramifications of a brutal murder in which she was the chief suspect. Had she been standing behind the door listening to our conversation? Was she fearful that her husband would say something that would implicate her?

'Of course,' he said, with a strained smile.

Inside, Katharine rushed me away from the door and immediately started to talk in whispers. 'I heard what Leonard was telling you,' she said. So she had been eavesdropping after all. 'That monster of a man, Mr Archer. I wish he'd never set foot in this camp, the moralising idiot. Sorry, but to suggest that I should remain locked up here in my bedroom like a common criminal. It's preposterous!'

'It does seem rather extreme, but then often after a sudden shock some people's reactions can be quite uncompromising,' I said.

'I know the police are delayed and I don't want a cloud

of suspicion hanging over me for longer than is necessary,' said Katharine, colour rushing into her face now. 'So this is what I want you to do. I'd like you to go around to everyone here and interview them about what exactly they saw and what they were doing last night. It's only what the police would do and you may as well get on with it, don't you think?'

'Yes, but—'

'How would you persuade people to talk to you? I've thought of that too. People will be open to you, you've got that kind of face. And if they don't you can say that the police have sent a message asking you to do some ground work for them in preparation for their arrival, or something along those lines. What do you think? You'll do it, won't you? Please say you will.'

Strangely enough, I had already thought of such a plan, as I was sure I could get Davison's retrospective backing to carry out some initial investigations.

'If you think it will help, then yes,' I said.

Katharine threw her arms around me. 'Thank you,' she said. 'I can't tell you how grateful I am.' I remained as stiff as a board and extracted myself from her embrace, looking at her with a serious expression.

'What's the matter? Why are you looking at me like that?'

'Katharine, you must realise that if I'm going to do what you ask, I have to remain impartial.'

'Yes, of course, but—'

'Which means that, if it wasn't an accident, I have to regard you as much a suspect as anyone else,' I said.

'Oh, I see,' she said quietly. She fell silent for a moment

or two before she said, 'Well, you may as well start with me then. Ask anything you like.'

Once again, she went over her movements of the previous night – how she had followed the Archers, how she had lost sight of their torch, how she had heard the sound of footsteps soon followed by a cry, and how she had found Sarah's body at the base of the ziggurat. Katharine made it seem all very plausible, but how did I know that she wasn't lying? Now was the time to ask her some more searching questions, questions relating to her past. This line of enquiry did not come naturally to me – I would have preferred more subtle measures – but I needed to gauge her reaction. The only way I could get through this would be if I imagined myself to be a detective character from one of my own novels.

I took a deep breath and began. 'Have you ever committed a crime, Mrs Woolley?'

Katharine looked startled, appalled. 'What do you mean?'

'Your first husband, Bertie, the one you told me shot himself.' I paused. 'Did you kill him?'

'How dare you!' she exclaimed, her eyes full of fire.

'You said that I could ask you anything.'

'Yes, but I meant about last night, not something that happened years ago.'

'But as you know, the events of the past often have a bearing on the present,' I said, trying to swallow, but my throat was dry. 'I must ask you again, did you murder your first husband?'

'Have you lost your mind? Of course I didn't kill Bertie.'

'And what about Miss Gertrude Bell?'

'What about her?'

'Did you play any part in her death?'

Katharine looked at me as if I were playing some cruel practical joke, almost as if she expected that at any moment my mask would slip. 'I don't understand where this is all coming from,' she said. 'Leonard and I were in London when Miss Bell died.'

'Did you send her any kind of threatening letters or drawings?'

'No, what makes you think I did?' She stopped and looked at me as if seeing me for what I really was: a false friend who had won her intimacy by not wholly honest means. 'I see, it's all beginning to make sense. How could I have been so foolish as to trust you.'

'Katharine, if you'll allow me—'

'Mrs Woolley to you,' she said with ice in her voice. 'And to think that I confided in you. You, who came here under false pretences. Who sent you, I wonder? It doesn't matter, but whoever it was had something they wanted to find out. I'm right, aren't I? They suspected me all along – what, of killing Bertie? And driving Gertrude Bell to her death? And now you think I'm the one behind Miss Archer's death too.'

'Please, if you—'

'Don't waste your breath,' she hissed. That manic look had returned to her eyes. 'A snake in the grass, that's what you are, or rather something more deadly: a snake in the sands.' She stood up and looked towards the door, a clear sign that I should leave. 'There are other less polite names for your kind, but I wouldn't lower myself by saying them.'

I opened my mouth to speak, but there was nothing left

to say. Katharine Woolley was right. I was a snake, yes, worse than that. She had every right to be angry. I had handled the situation badly. I had assumed one could act like a character from a novel. What a fool I had been! It had been wrong of me to accept Davison's offer. I should have stuck to what I knew: sitting at my typewriter, writing stories, spinning tales from the rich swell of my imagination. What had possessed me to think that I had the talents for anything but that? I opened the door slowly, stepped through it, then turned back to her, feeling the sting of tears in my eyes.

'I'm sorry,' I managed to say before Katharine slammed the door behind me.

'Oh dear,' said Leonard, who had already jumped from his seat outside the room. 'Are you all right? You do look terribly pale.'

I nodded, unable to speak for the moment.

'Did Katharine give you a piece of her mind?' he said. 'Yes, we've all been subject to her sharp tongue, I'm afraid. Here, why don't you sit down.'

I took Woolley's chair as I tried to compose myself. 'She – she asked me to try and help by taking everyone's statements about their movements last night.'

'Yes, a very good idea in the circumstances, I would have thought,' he replied.

'She said I could ask anything, anything at all. I thought she wouldn't mind a few probing questions. But I'm afraid I was wrong.'

'Oh, I see,' he said.

'It was my fault entirely,' I said. I recalled the crazed look in her eyes, a look that frankly frightened me. 'I was too clumsy in my line of questioning. I'd like to apologise to her, but she's so angry that she won't hear me out.'

'What we've all learnt – those of us who are close to Katharine – is that she has a temper, a temper so strong that it can burn. She can be charm itself one moment and turn on you the next. She can say the most vicious, the most unforgivable things. Some people have it in their hearts to forgive her. But there are others – such as Mr McRae, for example – who refuse to have anything more to do with her. I do hope you decide you are of the former, not the latter, type.'

I didn't know which camp to place myself in. I only knew that I had seriously misjudged a very delicate situation. As I returned to my room I thought of the blood that had been spilled at the base of the ziggurat and Miss Archer's lifeless body. I feared I had made an enemy of Katharine Woolley. I did not want to be the next victim.

Chapter Thirteen

When I returned to the living room I was informed that Mr Archer had called a meeting to discuss how to proceed following the death of his daughter. Everyone, apart from Mrs Woolley, who was locked in her room, and Ruth Archer, who was still with Sarah's body, had taken a seat around the table.

It was obvious that Hubert Archer, in an effort to cope with the tragedy, had clicked into automatic mode, playing the part of the American businessman and millionaire to perfection. By doing so it was obvious that he was attempting to deflect the shadow of grief that loomed over him.

'Now, top of the agenda as I see it, is to decide what to do about the police, or rather their absence,' he said. 'I can't believe that there is no one around who could help with this matter. Burrows? McRae? Do you know any alternatives?' The men shook their heads. 'Miller?'

'Sorry, but I'm afraid Woolley is right when he says the man from Baghdad will take just as long as the chap from Nasiriya,' replied the photographer.

'So what do you do when there's an emergency?' asked Mr Archer.

'We just wait,' said Woolley, dryly. 'It's one of the hazards of being out here, I'm afraid.'

Mr Archer stared at Woolley in disbelief. 'The next thing we need to decide is what to do about ... about the body,' he said. I surmised that he was using deliberately clinical language so as to distance himself from the awfulness of what had happened to his daughter. 'Obviously, we can't leave it out – there,' he said, glancing towards the ziggurat. 'Ruth is with her now, but she can't do that all day. And even though it's winter the heat of the sun would still ... well, it wouldn't be ideal. I'm sure all of us can agree on that.'

The room was filled with an echo of murmurs and agreements, as if we were voting on the implication of a new card system at the local library or a particular planting scheme in the municipal gardens.

'I'm wondering which is the coolest room in the house,' said Woolley. 'I suppose it must be the pantry. Is that correct, McRae?'

'Yes, indeed,' replied the architect. 'A good few degrees cooler. Of course, we would need to move out the supplies.'

'The pantry?' asked Mr Archer. 'It hardly seems suitable.'

'No, but I'm afraid it is the best place,' said McRae. 'After all, we wouldn't want—'

Hubert Archer cut him off with a nod of the head, before he cleared his throat in preparation for the most difficult question on his agenda. 'And then there is the problem of what to do with Mrs Woolley.'

'In what regard?' asked Woolley.

Archer's reply came as quick and as deadly as a rapier. 'Well, I know my wife for one would not be comfortable sleeping under the same roof as a murderer. I'm sure the same could be said for most of you around this table.'

The statement – articulated and declared with the conviction of a man of standing and means – forced many in the group to acknowledge what they had only been thinking. Cecil, clearly distressed and red in the face, tried to talk, but was stopped from doing so by his uncle who placed a hand on his nephew's shoulder. Father Burrows looked like the saddest man on earth, an Adam being given the news that there was no place for him in paradise. Miss Jones had gone pale and it looked as if she were unable to raise her big brown eyes from a dirty mark on the table, a stain that seemed to hypnotise her. All traces of jollity and optimism had been wiped clean from Mr Miller's face. And Woolley himself appeared inscrutable, for once unable to interpret or predict.

Finally, he stood up and addressed the group, talking quietly at first. 'I know that you all have very grave concerns about what has happened,' he said. 'And I must express my deep sadness and condolences to Mr and Mrs Archer. Although my wife has not been well – many of you have witnessed that – this does not mean that she is the one responsible for . . . for what befell poor Miss Archer.'

'How can you stand there and say that?' said Cecil McRae, standing up with such force that his chair nearly fell over. 'There's only one of us who was found with blood on our hands – Mrs Woolley.'

'That's enough, Cecil,' said Lawrence, trying to calm the boy.

'Everyone's thinking the same thing,' said Cecil, shaking with anger. 'We more or less saw it with our own eyes. First the cat and then this. She clearly took a rock and smashed it on Sarah's head. How could she do that?' The boy's voice broke, and tears had started to stream down his face. 'Maybe she's just mad or evil or a bit of both, I don't know. But I think she couldn't bear the thought of anyone prettier than her getting more attention. And Sarah was younger, much younger, and much more beautiful too. I'm sorry, but if you've snuffed out a life that means, by my reckoning, well, that she's got what's coming to her.'

'Cecil, control yourself!' commanded Woolley. 'I will not have such talk in my camp.'

The boy looked at Woolley with a mix of anger and self-loathing and, unable to bear the contradictory impulses which were raging inside him, he dashed a water tumbler to the floor and ran out. The impact sent shards of glass across the hard floor.

'Cecil – come back!' shouted McRae. He bent down to pick up the pieces of the glass and, in the process, cut his thumb.

'You're bleeding,' said Cynthia.

'It's nothing, don't worry,' said McRae, taking a handkerchief from his pocket and wrapping it around the wound. 'And, Woolley, I'm sorry about that – he's young and headstrong and you know how much he thought of Sarah.'

'Never mind,' said Woolley. 'Now, where were we? Yes, the question of what to do about my wife.'

Just as Mr Archer was about to launch into another speech I thought it best if I tried to calm things down a little.

'May I venture a suggestion?' I said. All eyes turned to me, astonished that I had an opinion on the subject.

'Yes, Mrs Christie?' asked Mr Archer.

'I can see that we are all getting rather heated about the matter,' I said. 'And I can comprehend the very strong feelings. But I do believe it is important to stick by the English principle of innocent until proven guilty.' I said this although I was quite uncertain of Mrs Woolley's innocence in the matter.

'A very noble idea, but in this instance ...' said Mr Archer in a rather condescending manner.

'In this instance, I think it's all the more important we don't let tribal instincts take over,' I said. I thought of the paragraph in Mrs Woolley's manuscript relating to tribal law. 'We can't let ourselves be swayed by outmoded notions of justice. Even though many of us might consider the judicial system in Iraq to be – well, not quite what we would find at home – it's good to remember something of the country's contribution to the history of the law.'

'Indeed, but with the police so far away, and—'

'Exactly so, Mr Archer,' I said, trying to disarm him with a smile. 'With the police so far away we cannot allow this to degenerate.' An image of an ancient relic, an enormous tablet from a bygone age, came into my mind. Something I had been taken to see in the Louvre all those years ago when I had been a carefree girl in Paris. 'Correct me if I am wrong, Mr Woolley – as you will know so much more about this than I – but is there not something called the Code of Hammurabi?'

'You mean the ancient Babylonian law code? Discovered by Jéquier in 1901 at the ancient site of Susa. Translated a year later by Father Scheil.'

I had no idea of the history, but I knew to trust Woolley's expert knowledge. 'Yes, I'm sure that's right,' I said. 'Does that code not lay down what later became the precept of a presumption of innocence?'

'*Ei incumbit probatio qui dicit, non qui negat*?' he replied. 'Which roughly translates as, "the burden of proof lies upon him who affirms, not he who denies".'

'Yes, that's right,' I said. 'It's up to those who are doing the accusing to find the body of evidence.'

'Hold on there, my dear,' said Mr Archer, addressing me as if I were a child. 'I'm sure you can get away with this kind of thing in your books, but unless I'm mistaken you aren't qualified in the law?'

I could feel myself blushing. 'No, I'm not, but I do think—'

'And if you'll forgive me, you're not the only one acquainted with the Code of Hammurabi,' he continued. 'While we were in Paris, I went to the Louvre to see the famous stele for myself. As a scholar of the Old Testament, it was only natural that I wanted to study an ancient relic that spoke of the *lex talionis*. My view, like young Cecil's, is very much an eye for an eye, a tooth for a tooth.'

Leonard Woolley interrupted him. 'I can understand that in a state of shock you may feel like this, but surely it's more Christian to think of the Sermon on the Mount? If anyone strikes you on the right cheek, turn to him the other also.'

'Woolley, Mrs Christie, I'm not here to have an argument about the finer points of theology,' snapped Archer. 'Let me outline the facts as I see them. My daughter has been murdered. Her corpse is lying outside, being watched over by my heartbroken wife. The police aren't due until tomorrow morning. Under this roof there is a murderer. A murderer who has been discovered with blood on her hands. A murderer who is none other than Mrs Katharine Woolley.'

'But Mrs Woolley told me—'

Mr Archer cut me off. 'Told you what? That she was innocent?'

'Yes, that she heard the sound of footsteps or a scuffle and the next thing she knew she stumbled upon your daughter's body.'

'And you believe her?'

'Well, I . . .'

'Why the hesitation? From the look in your eyes it's obvious that you're not completely sure of Mrs Woolley's innocence.'

It was true. I couldn't stand up before Mr Archer and tell him that Katharine had not killed his daughter. Yet neither was I certain of her guilt in the matter.

'Just as I thought,' pronounced Mr Archer. 'Now what I propose is this.' He began to outline his scheme for dealing with the immediate aftermath of his daughter's death. 'Don't worry, I'm not suggesting that we try Mrs Woolley in some kind of kangaroo court. I'm not that ignorant or barbaric. Just that I think it would be best if we removed her from the house. That we contained her somewhere under lock and key while we wait for the authorities to arrive tomorrow. Woolley, is there any such place here?'

Leonard turned his head away from Archer so he could not see the resentment burning in his eyes.

'Woolley, did you hear what I said?'

'Yes, I know a place,' interrupted McRae. 'There's a storeroom just by the outer fence that is used to keep spare tools, building materials and the like.'

'Are you out of your mind?' shouted Woolley. 'You cannot expect my wife to spend the night in that shed?'

'Perhaps we could make it comfortable for her? Could a mattress conceivably be brought in for her?'

'Yes, but ...' said Woolley, the colour draining from his face.

'And it's lockable from the outside?' asked Archer.

'Yes, there's a padlock on the door,' answered McRae.

'There's no way Katharine will agree to this,' said Woolley. 'No way at all.'

'Well, I'm afraid she won't have much choice in the matter,' said Archer. 'As far as I'm concerned she gave away her rights when she smashed that rock down on my daughter's head. Ideally, I would have her removed from the compound altogether, but as the police will want to question her first thing, this is the most appropriate compromise, don't you think?'

Archer looked around the group as if he had just won a particularly tricky game of cards. No one had the energy or the courage to argue with him any more. Even though

I was not totally convinced of Mrs Woolley's innocence, I felt as though the plan was a mistake. What if Katharine was right and someone wanted to do her harm? Even if she was locked up in that shed she would still be terribly vulnerable. The phrase "lamb to the slaughter" came to mind.

'But is this really the best course of action?' I asked.

'What do you propose, Mrs Christie?' Archer's tone of voice was withering now. 'That we all carry on as if nothing had happened? Should we all sit down and enjoy a nice cup of tea? Yes, that would be very English, wouldn't it?'

I ignored his sarcasm. 'Why don't I stay with Katharine in her room as I did last night? You can lock it from the outside if you like. That way Mrs Woolley will continue to be comfortable and you will feel safe knowing that she is . . . contained.'

'You're prepared to spend hours locked up with a woman who could be a killer?' whispered Miss Jones, her eyes shining with the anticipation of future horrors to come.

'I don't think you need be quite so melodramatic, Cynthia,' I said, trying to lighten the mood with a smile. 'I'm sure I will be quite safe, as will the rest of you.'

'No, that's completely out of the question,' said Archer.

'But—' said Woolley before Archer cut him dead.

'Your wife will sleep in the storeroom, which will be locked. And I'm saying this for her safety as much as ours.' Archer walked over and whispered something in Woolley's ear, before he turned to the rest of us and added, 'I'm sure you do understand now that this plan would be for the best, don't you, Woolley?'

'Yes, I suppose under the circumstances you're right,' said Woolley. He no longer had the energy to fight. 'I'll go and talk to Katharine now.'

What had Archer said to Leonard to make him change his mind? Perhaps he had dangled the prospect of not withdrawing the funding from the dig after all. Before Archer excused himself he said that he would go and talk to his wife and tell her about the arrangements for the body of their daughter to be moved into the pantry. His face was inscrutable, as expressionless as an ancient carving; I wondered at what point Mr Archer would allow himself to grieve for his dead daughter. If he was so ardent in his Christian beliefs then the thought of everlasting life would at least give him some comfort. But was there not also a danger in such an attitude too? Did his fundamentalism not blind him to the very real joys and sadnesses of the concrete world in which he lived?

'Thank goodness Mr Archer is taking charge of the situation,' said Cynthia to me after Archer had left the room and the rest of the group began to disperse.

'I only hope he knows what he is doing,' I said.

'What do you mean?' she asked. 'Surely everyone will sleep more soundly knowing that Mrs Woolley is locked away in that shed?'

'Perhaps,' I said. 'But there's something not right about this, something not right at all.'

'In what way?'

'I'm not sure yet,' I said.

'You're beginning to worry me, Agatha,' said Cynthia. Her eyes searched the room for signs of danger. 'When the police arrive they will take away Mrs Woolley and then we can get back to normal, can't we?'

'I hope so,' I said.

'You're not telling me something,' she said, her hands beginning to tremble. I could sense the difficulty with which she formed her next question. 'Y-you think someone else is in peril, don't you?'

'Yes, I'm afraid I do.'

'But who?'

'It's Mrs Woolley herself. I fear someone means to do her harm. I think someone wants her dead.'

Chapter Fourteen

Back in my room I took out my notebook and scribbled down the events leading up to the death of Sarah Archer at the base of the ziggurat.

There she was, the young girl full of life celebrating her twenty-first birthday, kicking up her heels at the edge of the ziggurat, as Harry Miller took his photographs. Her mother told her not to go too near the edge, warning her of the long drop. Harry Miller replied that the girl was in safe hands. Then we all sat down to eat. Sarah told her story about that inappropriate proposal, cruelly describing the man as old and ugly. Her father told her off which then gave rise to the awful clash between Sarah and Katharine Woolley.

The atmosphere had changed from jovial to something more unsettling in a moment. Sarah had grabbed a torch and stormed off into the night, quickly followed by Leonard Woolley and Harry Miller, then Lawrence McRae and Cecil. Mr and Mrs Archer, disgusted by Katharine's behaviour, left the picnic, soon followed by Mrs Woolley and Miss Jones, leaving Father Burrows and myself. Then the night air had been split by that unholy scream and we all clustered round that horrible sight, the figure of Katharine Woolley, her hands covered in blood, kneeling by the body of Sarah Archer.

As I finished reading through what I had written I heard the sound of wailing. I opened my door to see two servants carrying the shrouded body of Sarah Archer on a bier. Behind her trailed her father and mother. Ruth Archer's pale face was a mask of misery like that of a painting I had once seen of Mary after the death of Christ. The sad parade made its way across the courtyard to the pantry, where Harry Miller stood at the door. As the servants manoeuvred the body into the room the photographer bowed his head in respect, leaving the Archers to kneel by their daughter's body.

'That was the saddest thing I think I've ever seen,' said Miller, catching my eye and walking over towards me. 'I can't believe one moment there she was, happy as anything, standing on the top of the world with everything to look forward to. Then the next minute her life had been snuffed out. It doesn't seem right somehow.'

'No it doesn't, there's something very evil at work here,' I said. 'Mr Miller, would you mind coming into my room? I'd like to ask you something.'

'Not at all,' he said, running his hand through his hair.

'I'm just trying to build up a picture of what happened before and around the time of the – of Miss Archer's death,' I said, as I closed the door behind us. 'And I wondered whether it would be possible for me to have a look at some of the photographs you took. Just to see if they might provide some kind of clue.'

'To the murder, you mean?'

'Well, yes,' I said. It seemed that there could be no other explanation for the girl's death.

He shuffled uneasily on his feet and a certain nervous quality came into his eyes. Was there something he wanted to hide? 'I would need a couple of hours to develop the film.'

We stepped out of my room to hear an almighty row coming from Mrs Woolley's quarters. No doubt Leonard had broken the news to his wife of her proposed imprisonment within the confines of the shed.

'I will not agree to it,' shouted Katharine. 'I refuse to be treated like an animal.'

'But Mr Archer insists—' replied Leonard.

'And why didn't you stand up for me? It's always the way with you.'

'If we want the—'

'You're weak, Leonard. You've always been weak.'

'But it's only for one night until the police arrive tomorrow. I'm sure they'll be able to clear everything up and once you've talked to them you'll be free to—'

We heard the sound of something like a vase smashing against the wall and I suggested to Harry Miller that we take a walk around the compound while we waited for the argument to finish. We continued to talk about what he had seen on the night of Sarah Archer's death. He told me once more of how he had accompanied Woolley down from the ziggurat and how they had lost sight of Sarah Archer's torch. Did he remember which staircase he had used as he had made his descent? He thought, he said, that he and Woolley had used the one that ran down the left-hand side of the ziggurat.

'Would you mind showing me?' I asked.

'Of course,' he said.

We strode out of the compound and across the dry landscape towards the great ziggurat. The digging had been stopped for the day – the men had been told not to turn up for work because of what had happened to Sarah Archer – and, instead of the queer low chanting of the Arabs, all I could hear was the sound of the wind as it passed across the desert sands. I recalled Mrs Woolley's fears about her first husband being alive. Here, in this vast open plain, there was nowhere for him to hide. And yet . . . what if Colonel Keeling had disguised himself as one of the Arabs? After all, no one in the household actually paid any attention to the workmen, apart from Hamoudi. I made a mental note to ask Woolley whether it would be possible to have an interview with the foreman to ask if he had noticed anything unusual among his men.

'I can't believe Mrs Woolley would do such a thing, can you?' asked Harry Miller as we approached the ziggurat. 'I know the two women didn't exactly see eye to eye, but even so. To smash the girl's head in with a rock. It seems unbelievable.'

'Indeed it does seem hard to comprehend,' I said. I was careful what I said next. I didn't want to offend Mr Miller the way I had offended Katharine Woolley. 'May I ask you a – delicate question, Mr Miller?'

'Go right ahead. What do they say about Americans? Nothing embarrasses us.'

'I think you were friendly with both Miss Archer and Mrs Woolley?'

'Yes, that's right.' I did not say anything and luckily Mr Miller picked up my meaning exactly. 'I see – you want to know whether anything serious was going on between us?' He broke into a smile and at that moment he seemed more handsome than ever. 'No, it was nothing like that. Sure, Miss Archer and I enjoyed a little flirtation here and there, but it was just a bit of friendly banter.'

I gave him an inquisitive stare.

'Nothing more, I promise!' he exclaimed. 'What do you take me for? Some kind of gigolo?'

The comment made me smile, but I tried not to show that he had amused me. Harry Miller was charm personified – and he knew it – but I didn't want to give him the satisfaction that he had won over yet another woman.

'And I suppose Mrs Woolley did get a bit jealous – not that anything had ever happened between us, Lord, no!' he said. 'I hope you don't think that. Just that her nose was put out of joint by the fact that I was showing Sarah a little attention. Katharine – Mrs Woolley – had been used to having me to herself.'

'I see,' I said.

'But you don't think that's why Mrs Woolley—?' He broke off, unable to finish the sentence. 'I refuse to believe it.'

'It does seem unlikely. Let's go and have a look where it happened,' I suggested.

As we walked I ran through the list of possible killers in my mind. The question a detective would ask would be: who benefits from the death? In this case, I had to assume that the only people to be made substantially richer from Sarah Archer's murder would be her parents. But surely neither Mr nor Mrs Archer would kill their own child so as to increase their wealth. Or would they? When Davison arrived – as I was sure he would when he had learnt of what had happened – I would ask him whether his department could check on the existence and contents of Miss Archer's will.

If money was not the motive behind the murder then what else could it be? Desire? Certainly, there were a number of men at the site who were infatuated by the girl. There was Cecil McRae, whom I had overheard saying that if he couldn't have her then he would make sure no one else could. And what about Lawrence McRae and even Harry Miller? They were men with appetites. What if Sarah had led them on and then had withdrawn her attention? Would they have felt so angry and full of frustration that they would have been prepared to kill her? Yet Harry Miller had come down from the top of the ziggurat with Leonard Woolley and the two men could each verify the other's movements. Could they be working in tandem? But what motive could they possibly have for wanting Sarah Archer dead? Or could Harry have murdered Miss Archer on Leonard Woolley's behalf? Could the archaeologist have a reason to be blackmailing the handsome photographer?

As we arrived at the spot where Miss Archer's body had been found we bowed our heads in a moment of silent respect. Then, when I opened my eyes, I knelt down and examined the ground, a patch of earth still reddish-brown from Sarah's blood. I saw something glinting in the sunlight. Using my handkerchief I eased away the sand. It was Mrs Woolley's gold watch, its face cracked and smeared with blood. Perhaps, just as Katharine had said, she had tripped over the body and in doing so she might have fallen over, smashing her watch in the process. But the other, darker possibility seemed more likely: she had indeed murdered Miss Archer and the watch had slipped off and been damaged in the process.

'What's that you've found?' asked Harry.

'It's a watch – one that I think belongs to Mrs Woolley.'

His face looked grave. 'What should we do with it? Do you think you should leave it there for the police to look at when they arrive tomorrow?'

'Yes, I suppose we should,' I said, letting the watch drop back into the sand.

Next I turned my attention to the rock that appeared to have been used to kill Sarah Archer. It was a large piece of pink sandstone, so substantial that one would need two hands to lift it. Half of it was still covered in Miss Archer's dried blood and, on one of its jagged edges, I could make out a small clump of blonde hair attached to a bloodied patch of skin. Although I had seen some true horrors during the war when I had been working as a VAD nurse, the sight of the bloody rock turned my stomach. As I tried to stand I felt myself feeling faint and, for a moment, I thought I might pass out.

'Are you all right?' said Harry. 'Oh my, no, you've gone

pale. Here, let me help.' He took hold of my arm and led me away from the spot where Miss Archer had been killed. Gently, and protectively, he helped lower me down onto a boulder. His touch lingered a little longer than necessary and I could feel myself blushing at the way he made me feel.

'Thank you,' I managed to say. 'It must be the heat.'

'And the sight of all that . . . mess,' he replied. 'I should never have allowed you to come up here. I thought you wanted to see the layout of the ziggurat, not the actual spot where it happened.'

'I'm feeling much better now,' I said. 'Really I am. Sorry for being such a nuisance.' Using my hat as a protection against the sun I looked up towards the ziggurat. Taking a deep breath, I raised myself to my feet and walked back to the spot where Miss Archer had died. Straining my neck I tried to work out where the rock could have fallen from if indeed it had been an accident, something which seemed increasingly unlikely.

'Let's walk up there,' I said.

'What is it you want to see?' asked Harry, as he took my arm. The lightness of his touch thrilled me. After all, I had been without physical contact of this sort for years now. Of course, I had felt my daughter Rosalind's soft skin against mine and enjoyed the occasional kiss on the cheek from my sister, but nothing – no, nothing – like this.

'Just in case it was an accident – there might be a trace,' I said, coughing, pretending I had a frog in my throat, but of course this was just a ruse to mask my real feelings.

As Harry guided me up the ziggurat's staircase I

concentrated on counting the steps in front of me, studying each stone and brick before me as a way to distract myself from his attentions. The heat stimulated my imagination and I had to blink away the inappropriate images that floated through my mind: his moustache set against his rather full and fleshy lips; his sturdy shoulders that looked as though they were strong enough to carry half a ton of stones; his large and powerful hands, one of which now rested on my arm, hands capable of taking hold of a woman and lifting her towards him.

When I reached the top of the staircase I paused for a few moments to recover my breath.

'Are you sure this is a good idea?' asked Harry. 'Why don't we go back to the house and have a glass of water?'

'No, I'll be all right,' I said. As I looked across the desert the heat beat down onto the plain. I stepped forwards along the edge of the precipice to the point which was, by my estimation, situated directly above the spot where Sarah Archer's body had been found. I kicked the ground around me to see if the action would dislodge any rocks, but nothing more than a couple of handfuls of sand cascaded down. Just as I peered over the edge of the ziggurat I felt the periphery of my vision begin to darken. My legs started to sink from beneath me. Was this how I was going to die? Death from a broken neck or a terrible head injury, my body twisted and broken after falling off the edge of the ziggurat. What was it Woolley had said to me about the etymology of the name? '*Etemennigur . . . Or "Temple whose foundation creates horror"'*?

I tried to step back, but my balance had gone. I had

begun to fall now and I couldn't stop myself. I closed my eyes, hoping that the impact wouldn't be too painful, that my death would be quick. But then, just as I collapsed, just as I saw the ground disappear from beneath me, I felt the strong hands of Harry Miller around my waist.

'What the—' he said, before pulling me back from the edge. 'It's my fault – I knew we shouldn't have come up here. Don't worry, I've got you now.' I felt something soft and cool across my forehead. I opened my eyes to see him fanning his hat across my face. 'Does that feel better? Agatha?'

For a moment I couldn't speak. I stretched out my hands and my fingers felt the sharp ridge of a rock and then the softness of the sand. I took in Harry's face, really the most pleasing face imaginable, and I was conscious of smiling. He had saved me – yet again he had been there when I had been at my most vulnerable. Without him . . . well, I knew what my fate would have been. 'Thank you,' I said.

'You must have fainted,' he said. 'Let's get you back to the house.' His powerful hands raised me upwards and he led me slowly and gently down the staircase. 'That's right – just one step at a time. You're doing great.'

Even though my body was weak, my mind was still working. I was certain now that Sarah Archer's death was not an accident. There were no loose rocks on the ridge of the ziggurat that lay directly above where her body had been found. I had seen traces of blood and strands of hair on the rock by the place where she had died. Miss Archer had been murdered. Yet there was something that still troubled me.

What if the intended victim had not been Sarah Archer but Mrs Woolley herself? Could the murderer have mistaken the light from Sarah's torch for Katharine's? Certainly, Mrs Woolley had seemed terrified of something or someone. What if her 'delusions' – the voices she claimed to have heard, the horrific faces at her bedroom window that she said she had seen – were real?

I was about to share my thoughts with Harry Miller but, as we passed the scene of the crime, I thought better of it. I still had quite a few things to work out and I was not certain whether I would be able to express myself clearly.

'You must have had a shock up there,' said Harry as he led me into the compound.

'Yes, I did,' I replied. 'Thank goodness you were there. I don't know what would have happened otherwise.'

'We don't need to dwell on that. Look – they don't hang around, do they?' he said, pointing towards the shed which was being cleared out by two servants. 'I suppose they'll have Mrs Woolley locked in there by sundown.'

I walked up to the store and inspected the door, which, as described, had a padlock attached. 'Do you know who holds a key to this?' I asked Harry.

'Woolley has one for sure, and I think McRae, and I've got one too, as at one point I stored some chemicals here,' he said. 'To be honest, I'm not exactly sure how many keys there are. It's only been used for holding tools and supplies, nothing valuable. There's never been any need to worry about that before.'

His answer, however, made me worry a good deal. Miller told the servants to make themselves scarce for a moment and we stepped inside the shed. The shifting of the tools to the outside had disturbed a great amount of dust and within seconds both of us started to cough. There were no windows that could be opened and the floor was nothing but bare earth.

'Do you think this is a fit place for Mrs Woolley?' I asked, placing a handkerchief over my mouth.

'Clearly not, but it's what the Archers want,' Harry replied. 'Since they arrived I've learnt that the will of that couple rule. Anyway, you can't worry about that now. We need to get you inside, into your room, where you can lie down.'

'But what if—'

'What if you don't go and lie down as I say? Well, I couldn't be responsible for my actions,' he said, with a mischievous twinkle in his eye. 'No, seriously, Agatha, you really do need to rest.'

'I'm just worried about the keys. You see, if—'

'I know, why don't we make a deal? Why don't I go around and ask for everyone's key to the store? I'll track all of them down and then we can decide where is the safest place to keep them. How does that sound?'

'That would be something,' I said, as we entered the courtyard. 'And could you go and fetch your camera? We need to think about developing the film. Just in case any of the photographs give us a clue to what happened.'

'I'll go and get my trusty Leica now,' Harry said, as he accompanied me to my room. 'Once that is done, you must take to your bed. I wouldn't be surprised if you had a touch of heatstroke.'

'Very well,' I replied, and we parted.

Inside my room I still felt light-headed, not so much from my fainting spell as from the feel of Harry Miller's arms around my body. I poured myself a glass of water. As I passed the looking glass on the wall I caught a glimpse of myself. What a sight! Even though I had been wearing a hat my hair had become loose and was hanging around my head, making me look like a dishevelled fishwife. My dress was dirty, my skin looked all red and blotchy and a fine line of perspiration had formed under my arms and around my neck and breast. What a fool I had been to think that Mr Miller could be interested in me! A divorced woman with a child, a middle-aged woman who was on her way to becoming an old maid. I had made the mistake once before with Archie, forming an attachment with a man who was too handsome for me. What a disaster that had been. No, I couldn't – I wouldn't – let it happen again. Not that Mr Miller had any intentions in that direction. He was probably just being kind to me. Yes, he had taken pity on me, that was it. I blushed as I thought of my girlish behaviour.

A moment later there was a furious knocking at my door. 'Agatha – Agatha!' It was Harry Miller. 'Let me in.'

In that instant, everything I had just told myself slipped from memory, my resolve melting as I felt my heart race.

'What's wrong?' I said, as I flung open the door.

Harry Miller stood there with a pale face and open hands. It took me a while to work out what lay in his palms: a mass of broken, shattered metal, together with a spool of spoilt film.

'It's my Leica,' he said, his voice flat. 'Someone's destroyed my camera and everything that was on it.'

Chapter Fifteen

'Who could have done such a thing?' I asked, as Harry Miller stood dumbfounded before me. He continued to gaze down at the fragments of what had been his camera with a look of sadness so profound it almost broke my heart. 'When did you last see it? Your camera?'

'It was j-just before we heard that argument between Mr and Mrs Woolley,' he said, blinking. 'I put it on my desk in my room and left it there. How long had we been gone? An hour at most?'

'Someone must have gone into your room and smashed it to pieces while we were at the ziggurat.'

'I guess so,' he said. 'But why would anyone want to do that?'

'I can think of one very good reason,' I said. 'The person who did this must have believed that you had captured something on your camera that could have implicated them in some way. It seems very likely that the man – or woman – who did this is the very same person who was responsible for the death of Sarah Archer.'

He turned away from me, walked over to my desk and dropped the pieces onto the surface. He took the crumpled reel of film and lifted it towards the light, before he flung it across the room, swearing under his breath as he did so.

'Is there really nothing that can be done?' I asked.

'No, they're all ruined,' he said.

'Do you remember what photographs you took? Besides those of Sarah?'

'There were some photos of the various artefacts here – pots, earrings, cuneiform tablets. And I took some of the grave pits too.'

'Casting your mind back to the day you took the pictures of Sarah – do you remember seeing anything that struck you as unusual?'

He fell silent for a moment, then shook his head. 'I don't think so. It was such a jolly occasion to begin with, wasn't it? Everyone in such good spirits. I took some shots of us strolling towards the ziggurat, people chatting, the food, the servants.'

I thought back to Mrs Woolley's wild theory regarding her first husband and the incredible suggestion that he might actually still be alive. 'Can I ask – have you noticed any of the Arabs behaving in a peculiar manner recently?'

'In what way?'

'I don't know – perhaps a man who seemed out of place or one who looked like he might be in ... disguise?'

'Disguise? What do you mean?'

'I know it sounds extremely unlikely, but when I was talking to Mrs Woolley she mentioned her fear that her first husband might not be dead. That he could be here, on this dig, perhaps disguised as one of the workers?'

'But that sounds absurd!'

I had to agree. 'I wanted to mention it in case you had seen anyone acting strangely.'

'Sorry, but no, nothing springs to mind.' Harry returned to the desk and looked down at the shards of his Leica. 'What I want to find out is who is responsible for smashing up my camera. I don't know what to do.' He gazed at some pieces of metal in the vain hope that they might meld together and somehow miraculously form a camera once more. 'Do you think we should go and ask the others if they saw or heard anything suspicious?'

'Yes, it might help,' I said. Although the culprit was unlikely to declare himself, perhaps it might be possible to gauge the reactions and pick up traces of guilt. I told Harry that if he broke the news of what had happened I would watch the faces and the bodies of the rest of the group for any hints or clues. He picked up the remains of his camera, together with the ruined spool of film, and walked into the main room, where he threw everything onto the table.

'Oh my, what's happened?' asked Cynthia Jones. She was about to lay the table for lunch.

There were similar cries of astonishment from Father Burrows and Lawrence McRae, while Cecil remained sullen and silent. Harry explained how he had found his broken camera and voiced his suspicions about a possible motive behind the destruction. And, as I thought, no one admitted they had seen or heard anything untoward.

'And no noise at any time?' asked Harry, as he stared at each of the group in turn. 'Nothing? It seems extraordinary that none of you heard the sound of smashing from my room while I was gone.'

'Perhaps you need to inform Mr Woolley?' suggested Cynthia.

'Don't you think he's got enough on his plate at the moment?' sneered Cecil. 'What with having a wife for a murderer?'

We all ignored the boy's blunt remark. 'Yes, I'll ask him if I can have a word,' said Harry, before he left us in uncomfortable silence. I studied the figures in the room in earnest; none of them seemed to be showing any sign of guilt. But then when did those with a criminal bent ever show signs of culpability? They were, I knew, masters of deception, well-practised in the arts of duplicity.

A minute or so later, Harry Miller returned with Woolley. The photographer had already explained what had happened, but Leonard wanted to see the evidence for himself. He walked over to the table and picked up the shattered pieces of Harry's camera.

'Damn shame,' said Woolley, turning the fragments over in his hands. 'Nice piece of equipment it was too. I used to own one similar, older model though.'

'Do you think Mr or Mrs Archer heard anything?' asked Harry. 'Where are they now? In their room?'

'No, they're still with their daughter's body,' replied Woolley. 'I don't think now is the best time to ask them.'

'How is your wife?' I asked.

'As well as can be expected,' he said.

'She didn't take the news of her imminent move well, I imagine?' asked Lawrence McRae.

Woolley ignored the question and encouraged the group to get on with their lunch as normal. Soon he would accompany his wife out to the shed, which he believed should now be ready for her.

'Mr Miller and I were just talking about the keys to the padlock,' I said. 'We were wondering how many copies exist and who has them?'

'Yes, it's something I was thinking about too,' he replied. 'Probably a good idea to gather all of them together, considering what's happened.' He turned to address the group. 'If you could search through your things and see if you have any small keys that might fit a padlock, I would be so very grateful.' It was difficult to know whether Woolley was being sarcastic or whether he really was apologising to his colleagues for putting them to this trouble.

The group started to check their pockets and some of them went off to their rooms to search their drawers and desks. A few minutes later three people – Leonard Woolley himself, Harry Miller and Lawrence McRae – returned to the main room and placed their keys on the table.

'That's all of them?' asked Woolley. 'Miss Jones? Cecil? Father Burrows?' Each shook their head.

'And who is going to guard the keys?' asked McRae.

'Well, I thought I might,' said Woolley.

'But what if the desire to liberate your wife suddenly came upon you?' persisted the architect.

'What would you suggest, Mr McRae?' said Woolley

icily. I could tell he was doing his very best to keep his temper under control.

'Perhaps Mr Archer should keep them?'

'Oh no, I don't think that would be a very good idea,' I said. All faces turned to me.

'Why ever not?' asked McRae.

'I was thinking that we shouldn't place that burden on his shoulders at such a difficult time,' I replied. The truth was that I did not trust him or his wife with the keys.

'Yes, that's a good point, I suppose,' said Woolley. 'So why don't we let Father Burrows take charge of them.' He turned to the emaciated priest, who had taken his spectacles off and was in the process of cleaning them with his handkerchief. 'Eric, what do you think? Would you be prepared to do that?'

The Jesuit paused for a moment, looked through his glasses, placed them back on the bridge of his nose and nodded. 'Yes, I don't see why not. I've got a strongbox in my room. They'll be safe enough there.' He was indeed the best choice for the job; after all, I knew that he had not killed Sarah Archer because father Burrows had remained at my side as he accompanied me down from the top of the ziggurat.

'Well then, what are we waiting for?' said McRae, addressing Woolley. 'You may as well take her over there.'

The archaeologist began to protest. 'I think she should be allowed to enjoy a few more hours in her room. After all, lunch—'

Just then, two pale figures appeared like ghosts in the corner of the room. Mr Archer's demeanour, once so confident and impenetrable, seemed to have crumbled: his eyes were red and raw from grief and his shoulders had slumped forwards as if all life had been sucked from him. His wife looked as if she was almost deranged: her hair hung about her swollen, tear-stained face and it was clear that she had not changed her clothes from the day before.

Mr Archer cleared his throat. 'I think it's best if she went into the shed straight away, don't you?' he said weakly.

How could anyone argue with a parent who had just spent hours sitting next to the body of their murdered daughter?

'Very well,' said Woolley in a quiet voice. 'I'll take her in there myself.'

Woolley left the room and we fell once more into silence. Cynthia Jones went over to Mr and Mrs Archer and tried to encourage them to eat some lunch, but the couple refused. I suspected that none of us had much of an appetite for food at that moment.

'At least come and sit down and have a cup of tea,' urged Cynthia.

'No, thank you,' replied Mrs Archer, her head angled towards the door and her eyes full of fury.

'But you must be feeling so weak,' continued Miss Jones. 'You have to keep your strength up. Mr Archer, please try to persuade your wife to have something. We don't want you to—'

At this point, Ruth Archer broke away from her husband and ran out of the room. A moment later we heard her screaming as though she were demented. Each of us sprang up from our seats and ran towards the courtyard, where a nasty scene between Mrs Archer and Katharine Woolley was taking place.

'You murderer!' shouted Ruth, her face now red with rage.

'Please, contain yourself, Mrs Archer,' urged Woolley as he led his wife from the house.

'Why did you do that to my daughter?' shrieked Ruth Archer, her voice breaking. 'Just tell me why? What did she ever do to you?'

Katharine opened her mouth to say something, but Leonard pulled her towards him. 'Try to ignore her – she's suffering, one must try and pity her,' he said.

'It's you I pity,' hissed Mrs Archer as she lurched forwards.

'Ruth, that's not helping matters,' said Hubert Archer, placing an arm around his wife and drawing her away.

'It's you I'd like to kill!' she shouted at Katharine.

'Ruth, now that's enough!' continued Mr Archer. 'Try to control yourself, please.'

At this, Ruth Archer burst into tears and ran back into the house; the sound of the slamming of the door echoed through the hot, dusty air.

'You can understand how distressed she must feel,' said Mr Archer.

'She's still in shock,' said Woolley. 'Totally understandable in the circumstances. I'm sure Katharine won't—'

Katharine Woolley stared in astonishment at the two men, before she interrupted her husband. 'Well, I'm pleased that you both feel able to talk for your wives. And, Leonard, if you were about to say that I wouldn't mind, well, you're wrong. I do mind. In fact, I'm not only angry but deeply hurt by the whole saga.' She turned to the rest of us watching the drama unfold and with a deep and theatrical voice said, 'As for you – all of you! Look at you gawping at me as if I'm some common form of entertainment you might see in some end of the pier show. You should be ashamed of yourselves.'

'Come on, let's get you into your new quarters,' said Leonard, trying to calm her down.

'You make it sound as though I'm moving into a hotel. I don't know if it's escaped your notice, Leonard, but I'm being imprisoned, against my will, in a dirty, decrepit shed. Is that how little you think of me?'

'Let's not make another scene now,' he said. 'I'm sure all this will be sorted out when the police arrive tomorrow. It's only a temporary measure.'

Katharine was about to answer back, but clearly thought better of it. Instead, she sighed in frustration and allowed herself to be led by her husband down towards the shed. She cut a pathetic figure, with her smart lavender hat positioned so neatly on her head and her mauve scarf blowing in the dry wind, almost as if she had dressed ready to take a pleasant jaunt in a motor car. As she entered the hut she turned to the rest of us, her face full of dismay, hurt and betrayal. It was the kind of look a sick horse might give you in the moment before it was shot.

Chapter Sixteen

Nobody cared for lunch and so each of us drifted back to our rooms. I went through my notes, assessing the case from the very beginning. Despite hours of reading and thinking, I could not find any link between the death of Gertrude Bell – which was, after all, the reason I had been despatched by Davison to Ur – and the murder of Sarah Archer.

Were there two different killers at work here? It certainly seemed that way. After all, Miss Bell had died from an overdose in her bed at her home in Baghdad; even if she had a premonition that someone wanted to kill her, her death would have been a peaceful one. It looked as though Miss Archer, however, had been hit over the head with a rock – a nasty, brutal, painful murder. But could the two killers be working together? If so, what had they to gain from the deaths? Could Katharine have employed someone to murder Miss Bell? Leonard Woolley himself had told me that life was cheap in Iraq. What was it that Gertrude Bell had written in that letter to her father? Yes, she had

ANDREW WILSON

been talking about a perilous journey to Hayyil, where she had been taken prisoner: *'There I heard it said that in that place murder was considered so normal it was likened to the spilling of milk.'*

Did Mrs Woolley harbour a secret resentment towards other women? Was it a simple case of madness? After all, if Katharine was that insane then there really would be no need to examine the case for evidence of a complex motive. Perhaps, possessed by a blind rage or in the midst of some sort of deranged frenzy, she had indeed taken that rock and bludgeoned Sarah Archer to death. I knew that it was also perfectly possible for such murderers, those classified as mentally unhinged, to erase violent acts from their memories. Maybe Katharine was guilty of the crime, but believed herself to be innocent.

Yet, as I read my notes, the thing that struck me was how Mrs Woolley had been regarded as an oddity from the very first, almost as if someone was trying to prejudice opinion against her. I had heard her described as everything from strangely attractive, demanding, manipulative, jealous, eccentric and unreliable through to hysterical, erratic, unhinged, insane. After this, perhaps it was only natural to attach that other deadly label to her personality, that of a murderer, especially when she was discovered sitting next to a body with blood on her hands.

The more I thought about it the more I became convinced that there was someone behind all of this, someone who was desperate to cast Katharine Woolley in the role of a killer. But could this be a case of a very clever double bluff? I remembered the way Katharine told me that she had admired my novel *The Murder of Roger Ackroyd*. I also thought back to the plot of my first published novel, *The Mysterious Affair at Styles*, which featured a killer who was the most obvious suspect. Sometimes it was better for a murderer to hide in plain sight.

And what was I to make of her wild surmise that her first husband might actually still be alive? It sounded preposterous, but nevertheless it needed to be ruled out. I intended to try to investigate this matter further by asking Woolley whether it would be possible to set up a meeting with Hamoudi. I doubted that the foreman spoke much English and so Woolley or one of the other members of staff would have to be present to translate.

I found Leonard in the *antika* room studying a set of precious stones lying on a sheet of white paper.

'Come and take a closer look at these,' he said, ushering me over. 'They were found scattered in one of the death pits, and this set here belonged on one necklace, I believe.'

'What are they?' I asked, as I peered at the delicate beads, some coloured a dull red, others a startling bright blue.

'A mix of carnelian and lapis lazuli, which would have been interspersed with gold, using a delicate filigree technique. Can you see?' Using a pencil he pointed out tiny fragments of the precious metal. 'The carnelian may have come from India and the lapis from Afghanistan. The person buried in this grave wanted to look beautiful – even in death.'

I asked after Katharine, but it was clear that Woolley did not want to talk about his wife. I then brought up Mrs Woolley's suspicion that her first husband may not have died back in 1919 and her fear that he was the one behind the recent attack.

'But how could that be possible?' he asked. 'And why would he want to do that?'

'I know it sounds rather extraordinary, doesn't it,' I said. 'But I suppose Colonel Keeling could be out to exact some sort of revenge on his wife.' I explained the possibility that he could have disguised himself in some way among the workers and asked whether it might be possible to put a few questions to Hamoudi. Although he acknowledged that the scenario seemed most unlikely, he agreed to send for the foreman.

'I think Keeling spoke good Arabic, but even so I'm sure Hamoudi would have noticed if anything like that was going on under his watch,' he said, before a look of alarm came into his eyes, shortly followed by disgust. 'But that means – no, surely not.' He could hardly bring himself to say the words.

I knew what he was thinking: if Keeling were still alive that would make Woolley and his wife bigamists. 'Don't worry, I'm certain it won't come to that,' I said, not wanting to increase his fear by articulating it.

'I'll go and find him now,' he said, placing the beads back in their box. 'You're very welcome to stay here though. I'm sure you're not going to run away with any of our precious treasures.'

I told him that I would wait for him to return, secretly pleased that I had the opportunity to explore a room that I had only seen once before, in candlelight and only for a few minutes. As I stood in the dusty storeroom the sound of

Katharine's distress when she had found the body of her cat that evening seemed to echo through the air. The thought that the corpse of Sarah Archer lay only a few feet away, in the pantry, unsettled me. Despite feeling on edge and anxious I thought it only right to search the *antika* room for any possible clues. After all, it was a locked room, out of bounds to most of the group.

Ranged around the room were a series of shelves on which lay dozens of brown boxes. A label had been attached to each one, but on close inspection these did not give me much of a clue, as they were written using a particular specification with a code of letters and numbers. I picked up a box at random, carried it over to the trestle table and eased off the lid. The box had been packed with tissue paper, which as I searched through it whispered to me of secrets of the past.

At the very bottom of the box lay a small but beautiful dagger, ornamented with gold. Delicately, I placed my finger on the very tip of the blade, which felt as sharp as if someone had whetted it yesterday. I wondered who had been the last person to carry the weapon. Had it been used to kill? My imagination stirred and I saw a man greeting his best friend, offering him a seat at his table, asking him to consume the lavish feast his servants had prepared, before walking behind him and stabbing his companion in the back. I saw it all play out before my eyes: a man who killed his friend because of the discovery of his wife's infidelity. Although cultures and customs had changed over time, I doubted human nature did; the same problems, desires, wants and passions of people today existed thousands of years ago just as they did today.

I packed up the dagger and took out another box. This one contained a group of exquisite small bowls and shells, each fashioned from bright gold, which looked as though they might once have contained cosmetic powders. The sight made me think of the array of jars and receptacles on Katharine's dressing table. Her husband too had made reference to them. What was it he had said? We had been standing in this room and he had been showing us a pot used for the storage of a primitive kind of make-up. Yes, that was it. 'She has a seemingly infinite amount of jars – I've no idea what's in them.' My mind began to work. *What if someone* . . . yes, that would fit very nicely indeed. That would certainly explain it. But I stopped myself. How did I know that this theory, hardly even formed in my brain, was not some half-baked product of my fancy? My imagination had already conjured a scenario around an ancient dagger. Supposition was one thing, evidence and proof quite another.

I covered the shells and bowls with tissue paper and replaced the lid of the box, but in doing so I knocked to the floor a pencil that had been sitting at the edge of the trestle table. As I bent down to pick it up I noticed a small circle of wax on the ground. I picked up the disc and ran it through my fingers. I recalled the time when Leonard Woolley had given us a brief tour of the *antika* room. Yes, he had been holding a pair of candles because he said the treasures looked even more magical in the soft light. Perhaps, in the panic and chaos that had followed Katharine's cries, Woolley had upset the candle holder and a spot of liquid wax had dropped onto the floor where it then solidified.

I heard the sound of men's voices approaching and a moment later Woolley walked into the *antika* room, together with Hamoudi. The foreman was an imposing character, a tall man with a long face, a pronounced nose and a slightly protruding jaw; in fact there was something animalistic about him. He was wearing a long, flowing tunic or *thawb* fashioned from a light fabric, together with an *agal* and *keffiyeh*, or traditional head covering. His eyes were small and dark and shone with intelligence. Woolley talked to him in Arabic and the foreman replied in a series of deep, harsh-sounding pronouncements.

'He's saying that he cannot believe that my wife would do anything wrong,' Woolley said. 'You see, he holds her in the highest possible regard.' Hamoudi continued to talk over the archaeologist. 'He thinks it is against Allah's will that she be locked up in the shed. There is talk among the men about it. Even rumours of some kind of protest or rebellion.' Hamoudi's voice quickened and rose in pitch. 'Yes, Hamoudi, if you could just slow down, I will tell the lady,' he said, addressing the Arab. 'He wants me to tell you that the Shaytan has come to Ur. The devil. There is a force here that is like a poison festering and killing a healthy body. My wife is innocent and only greater evil can come from keeping her locked up like a prisoner.'

When the speech came to an end both men looked at me for a response. I started slowly, first of all asking Woolley to convey my thanks to the foreman for agreeing to meet. I too believed that no good would come from locking up Katharine Woolley in the store room. I then asked Woolley to put my question to Hamoudi: had he seen anything suspicious among his men? At this the foreman looked at me with distrust.

'He wants to know what you are suggesting?' Woolley said. 'He's not happy with the question.'

'I'm sorry,' I said to Woolley. 'Let me try and rephrase it.' I said that I was not implying that either he or his men had done anything wrong. Rather, I wondered if he had seen any of his workmen behaving in a way that might suggest something odd: that one of them might in fact be a European man in disguise. The question was met with a laugh so loud it reverberated around the room. Hamoudi's little eyes squeezed shut and his mouth revealed an array of broken and discoloured teeth.

'What does he find so funny?' I asked.

Each time Woolley tried to answer Hamoudi split the air with his laughter and soon tears were rolling down his worn, sun-spoiled face. Finally, the foreman explained himself to Woolley.

'He doesn't mean to be rude,' said the archaeologist. 'But it's the idea that a white man – a Westerner – could fool him or the other men by dressing up in that manner. He says the intruder would be discovered in a matter of hours. The man would stand out like – well, he uses an indecent expression that is probably best not translated. Better to say that the Westerner would indeed be rumbled quickly.'

'I see.' I felt myself blush a little. It was best to bring the conversation to a close. 'Well, perhaps you could ask him to keep his eyes open for anything suspicious,' I added.

'Indeed,' replied Woolley, dismissing the Arab in another guttural interchange. When he had gone the mask of joviality worn by the archaeologist fell away and he slumped into a depressed silence; no doubt the interview with Hamoudi only strengthened the suggestion that Katharine was indeed losing her mind. He walked over to a canvas chair in the corner of the *antika* room, sat down and put his head in his hands. He looked like a man broken by circumstance. I knew, from what he had told me, that his marriage to Katharine was not intimate. That was hard enough for any man to endure. But now he was being forced to acknowledge that his wife, at best, might be clinically insane and, at the very worst, a murderer.

'Maybe it is better if Katharine is taken away tomorrow,' he said. 'Perhaps she does need some serious medical help.'

'What do you think will happen to her?' I asked.

'I'm not sure,' he sighed. 'I suppose it depends on what the police decide. As you know, the rule of law is extremely unforgiving in the Arab lands.'

'Would she not get any kind of special treatment? You must be able to appeal to the British authorities here for a certain amount of clemency?'

'I'm not sure,' he said again. 'But I'm afraid to say that I fear the very worst.'

As he said those words a thought wormed its way through my mind. What if it was Leonard who was behind all of this? Could he be trying to drive his wife mad? After all, his marriage was a sham. Divorce would ruin him. If he could arrange it for Katharine to be taken away to some discreet institution where she remained for the rest of her life then surely that would solve all his problems. The marriage could be annulled. He could get on with his work without the burden of having a neurotic wife. He could be free to marry again. Was there another woman on the scene? I knew from personal experience that some men grew tired of their wives and hankered after younger models. And if he had been unable to enjoy personal relations with Katharine then perhaps it was only natural for him to look elsewhere. Did Leonard have a mistress back in England? Or was there even someone in Iraq, in Baghdad perhaps, or in the camp itself? I couldn't imagine that he would find Miss Jones very alluring, but one never knew. Or did he have another kind of secret?

'I see from your expression that you fear the worst too,' he said.

'Well, let's hope it doesn't come to that,' I said, dissembling.

'It's bad enough to think of Katharine locked up in that shed, but ...' he said, his voice breaking. He swallowed a couple of times and then continued, 'How will she cope with a prison cell in Baghdad? I'm afraid it would kill her.'

'Don't think about that,' I said, trying to sound as if I had his welfare at heart. 'You need to remain strong, if only for Katharine's sake.'

He coughed, ran his hands through his hair and stood up. His mouth twisted itself into a smile, perhaps hoping that the gesture would be enough to raise his spirits, but as he did so I noticed that his eyes remained devoid of any spark of happiness. 'Yes, you're right,' he said. 'You know, I do wonder what people back at home would think if they really knew how we lived out here.'

'What – the basic conditions, the dust, the heat and so on?'

'Not just that,' he said. 'Can I speak honestly to you?'

'Yes, of course. You know you can say anything to me.'

He studied me for a moment, as if assessing whether he could trust me. 'There's something about the desert that strips away all the pretensions of human nature. It's a shock at first to find out what lies underneath the veneer of respectability.'

I was careful what to say next. 'Is that something you've experienced yourself?' I asked gently.

'I was just thinking about Carchemish,' he said vaguely, as if his mind was beginning to travel back in time. I knew that he had worked on that site with T. E. Lawrence and Hamoudi. 'Particularly Jerablus. I was there, it must be seventeen years ago now. You see the archaeological remains at Jerablus fell under the control of the Kaimakam, or governor, of Birajik.'

These other names meant nothing to me, but I nodded my head in encouragement.

'It was a place famous for the beautiful black ibis which winters in Sudan and returns to Birajik to nest on the castle wall. We discovered a very fine mosaic floor, dating to about the fifth century AD, with an image of the glossy ibis – an indication, you see, that even in Roman times the bird always flew back to the Euphrates in order to nest.'

As I was beginning to wonder what this had to do with his previous point Woolley sensed my slight impatience. 'Sorry to digress, but it's important that you understand what we were doing there,' he continued. 'There was some kind of mix-up about whether we were allowed to dig there. We feared there might be trouble and Lawrence and I armed ourselves with revolvers, hoping that we wouldn't have to use them. But when we were refused permission by the Kaimakam to dig I felt so angry that I took my gun out and placed it against the governor's left ear. I was shocked when I heard myself say, "I shall shoot you here and now unless you give me permission to start work tomorrow."'

'And did you?'

He paused for a moment, before he said, 'Luckily the governor leant back in his chair and said, with a wintry smile, that he could see no reason why we couldn't start the next day. So you see, I believe many of us could be capable of murder, given the right circumstances.'

'I don't doubt it,' I said, thinking about some of my own terrible experiences.

'Indeed, a murderer can often be a bonus out here.'

The boldness of the statement surprised me, but again the rules of the desert were very different to the ones that existed in the drawing rooms of England. 'In what way?'

'The cook and general factotum employed on the dig at Carchemish was Hajji Wahid, a man Miss Bell found most charming, I believe,' he continued. 'But Hajji was also a murderer. Apparently he had been rather too keen on a local young woman and had ignored the requests of the girl's brothers to leave her alone. The situation got so out of hand that eventually Hajji killed four men in a skirmish. In due course, Hajji was sent to prison, but Campbell Thompson, who was in charge at Carchemish, thought that the man was just the sort of chap he was looking for. And so he proved. When he came out of prison Hajji was the most loyal of employees. He never left Thompson's side, always standing by with a gun, guarding him. And later, when I took charge, he became my bodyguard. Anyone who set foot into the site without permission was risking his life, and for the most part it was a very safe camp because everyone knew that Hajji would not hesitate to shoot them dead.'

'Yes, I can see that,' I said.

'It's a shame he's not here now,' he said. 'Hajji would not put up with any nonsense. In fact, I doubt any of this would have happened if he had been here.'

There was something I wanted to know. 'Can I ask, do you have a gun in your possession?'

'Yes, of course,' Woolley replied. 'It would be madness to live out here and not have any means of protecting oneself and also the very valuable treasures that we've dug up.'

'And do you keep it on your person or in your room?'

'It's kept in a strongbox in my bedroom. Why do you ask?'

'I wonder if you'd mind checking to see if it's still there.'

Woolley's eyes narrowed as he looked at me. 'What is it you suspect?'

I did not, of course, tell him the whole truth. It was enough to reveal only part of what I surmised. 'I'm worried that tonight, or at least before the police arrive tomorrow, someone may try to make an attempt on Katharine's life.'

He couldn't quite take the words in and his eyelids flickered as he tried to comprehend what I had just told him. 'I'm afraid that your wife may be about to be murdered,' I added.

As he contemplated the thought of losing his wife, a look of almost unimaginable grief crossed his face. If he was the one behind the plot to drive Katharine mad – if that was indeed what was going on here – then he certainly was a very fine actor. He opened his mouth to speak, but uttered nothing but a low moan. Finally, he asked me to follow him to his room, where he would show me the gun. Woolley took his time to make sure everything in the *antika* room was safely stored away. I noticed that as he turned the key in the door his hand was shaking slightly.

'I think I've just finally cracked the translation of that particularly difficult tablet we unearthed the other day,' called out Father Burrows as we stepped into the bright light of the courtyard.

'Not now, Eric,' replied Leonard.

'It's just that it is ever so exciting,' the Jesuit persisted. 'You see, I was reading it all wrong. I had been thinking that—

But Woolley cut him off. 'I said not now! For God's sake, man, can't you see I'm busy!' I had never seen Woolley lose his temper – he always seemed so composed, so controlled, even in the most stressful of situations. He pushed on, across the courtyard, with a determination and an energy that reminded me of an express train. He did not look around to see Burrows's puzzled and somewhat hurt expression.

A moment later we were in his room, which contained only the most minimal amount of furniture: a rather sad single bed, a desk piled high with books and papers, and a wardrobe that contained barely a change of clothes. It reminded me of the quarters of a bachelor or a student rather than a married man.

'It's all kept under lock and key,' he said, as he bent down to retrieve his strongbox from underneath the wardrobe. His fingers reached out and took hold of the metal box, presenting it to me with something of a flourish. From the pocket of his trousers he took out a keyring and selected a key. 'In fact, it's the same gun I was telling you about earlier, the one which I used – or at least I was prepared to use – when I was in—' he said, as he tried to turn the lock. 'What's this? Why is this open?' His hands pushed into the box but seemed to turn over nothing but empty air. 'I don't understand – where the . . . ?'

I knew what he was going to say next.

'The gun's gone – somebody's taken my revolver. And the bullets too.'

Chapter Seventeen

Each of us sat around the table trying to hide the suspicion in our eyes. Woolley had called the group into the main room and related the fact that his gun had gone missing. The news acted like a small explosion, sending waves of disbelief and confusion through the camp.

'But when did you last see it?' asked McRae as he stood up.

'I'm not certain,' said Woolley. 'I think it was when I—'

'You can't be vague about it,' interrupted the architect.

'Let me see,' said Woolley, blinking. 'I remember going to look in the strong box to check if there were any extra keys to the padlock on the shed. Yes, that was it.'

'And the gun was there then?' continued McRae in his aggressively interrogative manner.

'Yes, yes, it was,' replied Woolley.

'And you're sure you locked the box?'

'Well, I'm almost positive I would have done,' said Woolley in a way that did not inspire confidence.

'Damn it, man – sorry, ladies,' McRae said, addressing

Miss Jones, Mrs Archer and me, before he turned back to Woolley and continued, 'Can't you see?'

'See what?' said an exasperated Woolley.

'Couldn't Katharine have taken it?'

'I don't see how or why she—'

'But she may have had the opportunity?'

The interchange between Woolley and McRae had taken on the air of a compulsive and ghoulish spectacle.

'She can't have done,' insisted Woolley. 'She was locked in her room.'

'How do you know for certain? Perhaps she picked the lock, crept into your room and took your gun. You see, all of us, apart from Mr and Mrs Archer, were here in this room at the time. No one would have known if that's what she had done.'

For a moment the room was quiet before Mrs Archer, who had been twisting her fingers in her lap, started mumbling to herself.

'She killed my daughter, she killed Sarah, she murdered my beautiful girl,' she said. 'She took a rock and smashed her head in. Sarah was so proud of her hair, her lovely hair. Now it's all matted. I tried to make it look nice, I tried to smooth it out, but the comb got stuck. All that dried blood.'

Mr Archer, who had reverted back to his steady, composed self, placed a hand on his wife's wrist. 'Hush now, dear. You've had a shock. We've all had a shock.'

She broke away from her husband and turned to him, her voice full of venom. 'A shock? Is that how you see it? Is that how you'd describe how you feel after the death of our daughter?'

'Ruth, please . . .'

Hubert Archer bent down and lowered his voice to say something the rest of us could not hear. She started to resist his quiet entreaties, before he took hold of her hand again and appeared to apply pressure to her wrist. Finally Mrs Archer closed her eyes as she winced in pain.

'That's right, come with me,' he said, leading her from the table. 'Let's go and have a rest.'

Just before she reached the door she turned to the table and said, in a series of gasps, 'Let me tell you something – all of you. If you don't do something – something about that – that woman . . .' It was clear Mrs Archer wanted to call Katharine Woolley something else entirely. 'You're all going to be . . .' She forced out the final word in one final exhalation. 'Murdered!'

Mr Archer bundled the weeping woman away, but her cries continued until she had reached her room.

'I feel so desperately sorry for her,' said Cynthia Jones. 'I can't imagine what it must be like to lose a daughter like that.'

'Yes, it's awful to witness,' said Woolley.

'Forgive me for saying, but what she says does have an element of truth,' countered McRae.

'That's right,' said Cecil, his face flushing. 'Who's to say we won't all end up like . . .' The boy's eyes filled with horror as he no doubt recalled seeing Sarah Archer's body. 'End up being killed.'

'So what do you suggest?' asked Woolley.

All eyes turned to Lawrence McRae. 'As I see it, there is only one option and that is to search Mrs Woolley and that shed to make sure she is not in possession of the gun.'

'Very well,' said Woolley with barely disguised contempt. 'And who would you suggest as a suitable candidate to carry out such an investigation?'

McRae looked around the room before his eyes settled on me. 'I don't know ... what about Mrs Christie here?'

There was no reason not to agree. The search would give me an opportunity to talk to Katharine – if she would let me. After all, the last time we had spoken she had been so angry that she had dismissed me with contempt in her voice, calling me nothing but a snake, a deadly serpent that slithered through the desert sands.

'Now, you do know what it is you are looking for?' asked Lawrence McRae in a somewhat patronising tone as he and Woolley led me out of the room.

'Yes, I think I do,' I said. The sweetness of my delivery disguised my true feelings. McRae's insistence of Katharine's guilt had made me more than a little suspicious of him.

I followed the men to the Archers' room. After McRae knocked gently on the door a shaken-looking Mr Archer appeared. He was clearly in no mood to talk and simply nodded in agreement as McRae outlined the plan to search Mrs Woolley's quarters. From there, we went back to the main room and retrieved a key to the shed from Father Burrows.

'I can assure you that my wife is not in the best of moods,' said Woolley. 'So pay no attention if she's rude to you.'

I had an idea. I thought back to something Miss Jones had said when I had first arrived at Ur. 'I know – why don't I take her some coffee, as a peace offering?' I suggested.

'If you like, yes,' said Woolley, vaguely. I left the two

men and went into the kitchen. I made the strongest, most pungent, sweetest pot of coffee I could, the kind of thick brown liquid that could keep you up all night – which was just the effect I wanted to achieve. When it was ready I took a sip of the dark bitter brew. After a few moments my head started to spin and I felt adrenaline begin to course through my veins. That should do the trick, I thought.

As I carried out the steaming coffee pot on a tray I asked the two men whether it might be possible to go by myself to the shed.

'But what if Katharine tries to attack you?' asked McRae.

'Nonsense, I doubt Mrs Woolley will do anything of the sort,' I said.

'What do you think?' asked McRae, addressing Woolley.

'Katharine might throw some cruel things Mrs Christie's way, but I doubt that she'd hurt her,' replied the archaeologist.

'If you feel afraid you must come back to the house straightaway,' said McRae.

I left the men under the shade at the front of the house and proceeded on my own to the storage shed. After placing the tray on the floor I eased the small key into the padlock. The lock gave a satisfying click. I picked up the tray and paused for a moment on the threshold. I strained my eyes to make out the shapes and forms before me. Katharine lay curled up on the mattress, completely covered with a white sheet and, as I stepped closer, I struggled to make out whether she was breathing. Was she asleep or . . . ? I felt panic rise in my chest, as if a small bird had got itself trapped inside my ribcage and threatened to push its way up through my throat. I put the tray down on the earth floor and edged forwards, fearful of what I might discover. I turned my head to the half-open door and the line of light that partly surrounded its frame. It was tempting to rush out of the shed and call for Woolley and McRae, but I swallowed my fear, stepped forwards and knelt down. I stretched out my hand and, unsettled by the sight of my trembling fingers, steeled myself to pull back the top of the sheet.

'Katharine,' I whispered in so soft a manner I wasn't quite sure whether I could be heard.

Just then the sheet moved and a hand shot out to grab me. I reeled backwards, but the grip on my wrist tightened.

'Katharine – it's me, Agatha,' I said, trying not to raise my voice. After all, I did not want Woolley and McRae to come running.

She stared at me with wide, terrified eyes, the look of a woman who was in fear of her life.

'I'm not here to hurt you,' I whispered. 'I'm here to help. Katharine, please.'

The grip on my wrist lessened and finally she fell back onto the mattress.

'Listen, we don't have much time,' I said, taking a deep breath. 'I've been told that I have to search you for a gun that has gone missing. I doubt very much you have it, but—'

'A gun?' she said, looking confused.

'Yes, Leonard's revolver.'

'What do they think I would do with a gun?' she said, brushing her dark hair away from her face.

'I'm not sure,' I said. 'But the house is in a state of panic. People want to blame you for . . . well, for everything that has happened.'

'But you know that I didn't do any of those things? I could never hurt Tom, my dear, dear Tom.' Tears came into her eyes at the mention of the dead cat. 'Lying here I've been thinking about him. The way he used to curl up on my bed. The feel of his soft fur against my cheek. I wish he was here with me now. And although Miss Archer could be irritating, I would never do such a terrible thing as . . . as murder.' She pronounced this as if she were fearful of the word itself. 'You do believe me, don't you?'

That was the question. Did I? The situation was so very complex that I wasn't quite sure what I believed. Yet I did not have time to explain everything.

'Yes,' I said, even though I wasn't quite sure of it myself. 'Now, before we go any further, would you mind if I searched the room? For the gun?'

Katharine responded with a shake of her head.

'It won't take long and then I'll explain what is going to happen next,' I said. There were few places to hide a gun in the room and after I had searched the floor and walls, looking for any recently dug holes or gaps, and then the mattress, there was only Katharine's person that needed to be checked. 'I think this is why McRae chose me for the task,' I said lightly, as I related what I needed to do.

Her face darkened at the request. 'That's completely out of the question,' she snapped. 'I'm not going to be frisked like – well, like a kind of woman I'd rather not mention.'

Was she hiding something in her clothes?

'I understand that all of this must be terribly distressing,' I said. 'But I really do have to know that you are not in possession of that gun.'

Katharine stared at me, that dark energy burning in her eyes. 'Very well,' she said finally. She stood up and raised her arms as I ran my hands over her slim, boyish body.

Even though I discovered nothing, part of me wished that she had indeed stolen the gun, if only to protect herself.

'I hope you're satisfied now,' she said.

'I'm sorry I had to do that,' I said, trying to soothe her nerves. 'I know it may not seem like it, but I'm actually trying to help you.'

'Help me?' Her tone was sarcastic. 'Somehow, I can't see it myself.'

I walked over to the door and looked outside to check that nobody was listening. I picked up the tray and brought it back over to the mattress.

'Katharine, I'm going to tell you something that is going to alarm you,' I said.

She turned her head to the wall like a small child who had decided to withdraw their affection from one of their elders.

'I am fearful that something may happen to you tonight,' I continued. 'I think someone may be planning to do you harm.'

'Harm? What do you mean?' she said, turning back to look at me.

'I'm not certain, but I want you to make sure that you try and stay awake,' I said. 'I've brought you a pot of very strong, very sweet coffee – not to drink now, but to take later. It will be cold and not very nice but it's important you do as I say.'

'I don't understand,' she said.

'I need you to be on your guard,' I explained. 'I suspect someone will steal into the shed tonight and try to ...' It was difficult to say the words.

'To what?'

I took another deep breath. 'To make an attempt on – on your life.'

Katharine looked around at the walls of her prison. 'And you want me to stay here?' She immediately jumped up and tried to make a dash for the door. I grabbed her by the wrist. 'Listen – please,' I urged. 'It's the only way we are going to find out who is behind it all. If you run now you'll only put yourself at greater risk.'

The muscles in Katharine's arms tensed as she considered what to do.

'If you want to discover the identity of your real enemy then it's the only way,' I said. 'I know it's risky, I know that you'll be afraid, but I've got a plan. I'm going to stay awake too and station myself outside. At the first sign of any trouble I'm going to come running.'

'But what about the others? Wouldn't it be better if you shared your plan with them? At least then they might be able to protect me.'

'I'm afraid I'm not sure whom to trust,' I said. 'If I revealed everything then we might lose our opportunity. Waiting any longer would be a very bad idea. When the police arrive tomorrow they will probably arrest you, after which anything could happen, perhaps the very worst. You see, I believe someone wants you out of the way. And I think someone has been trying to drive you mad.' I was conscious that I was speaking too quickly and so I made an effort to slow down. 'Do you remember what you told me about the faces at the window? The voices you said you heard?'

'Yes, it was awful,' said Katharine, the painful memories casting a shadow across her face.

'Well, I've got a theory about all of that. I can't go into it-just yet, but I hope perhaps I can share it with you once tonight is over.'

Katharine nodded. She looked at me with a certain amount of respect and admiration. 'Very well, now tell me,' she said, taking a deep breath. 'What do you want me to do?'

Chapter Eighteen

Throughout dinner we all tried to pretend that everything was perfectly normal, but it proved an impossible task. Our nerves were in a delicate state and jangled with each scrape of a plate or bang of a glass onto the table. Conversation was a stilted affair and we made an effort to avoid the subjects that preoccupied us: the body in the pantry, the destruction of Miller's camera, the imprisonment of Mrs Woolley and, most terrifying of all, the missing revolver.

When I had reported that I had found no trace of the weapon, either in Katharine's meagre quarters or on her person, the room had fallen silent. The inevitable question hung in the air like a nasty gas: if she had not taken the gun then who had? It had to be somebody in the house: Mr or Mrs Archer, Harry Miller, Lawrence or Cecil McRae, Father Burrows or Cynthia Jones. Of course, I wasn't without suspicion and at various points during the meal I caught sight of people casting dark glances in my direction. There was also the possibility that Leonard Woolley was not being entirely honest and he remained in possession of the revolver after all. If so, he could be playing a very clever game of double bluff.

After the food had been cleared away Father Burrows asked me whether now would be a good time for him to show me the basics of cuneiform. I was hardly in the mood, but it would have been rude to refuse as Woolley had already told him that I was keen to learn. Perhaps it would provide a little distraction from the endless play of dark thoughts that continued to circle through my mind.

Father Burrows went to get a sheet of paper and a pencil and with boyish enthusiasm started to explain the principles of the ancient writing system. He told me how the Mesopotamians would use a stylus made from a piece of reed to inscribe what they wanted to say on a fresh piece of clay, clay which had been gathered from the banks of the Tigris or the Euphrates. It was not so much an alphabet, he told me, but an elaborate series of syllables and words. On the piece of paper he began to make a number of marks – there were, he said, vertical, horizontal and oblique wedges – signs which he then asked me to copy onto my piece of paper. The lesson went on to take in more complex aspects of the system – it was, he said, impossible to write a consonant on its own and certain sounds such as 'j' or 'c' simply did not exist – but, despite my best efforts, I could not concentrate as my mind was shadowed by the horrors of the night to come.

'Now, why don't you try to copy this,' he suggested, pushing a sheet of paper towards me. 'After you've had a little practice we can get a piece of clay and use that.'

My handwriting was messy at the best of times and no matter how hard I tried to copy the series of strange signs my efforts were clearly disappointing.

'Don't worry, if you put your mind to it you'll be a dab hand in no time.' Burrows glanced across the room at Cynthia. 'When I first met Miss Jones she was a complete novice like you and yet she picked it up very quickly. Quite the natural, weren't you, Miss Jones?'

'Oh, I wouldn't say that,' said Cynthia, blushing and turning her back towards us. I suspected that, like me, she felt uncomfortable with compliments.

'It's fascinating,' I said, 'but I'm afraid I'm rather tired. Would you mind if we took it up again tomorrow?'

'Not at all,' said Father Burrows. 'Yes, it has been quite an exhausting day. However, I'm pleased that you'd like to learn more about cuneiform. Most people show no enthusiasm for it whatsoever. Once you've mastered the basics then you can go on to read some of the great texts. Oh, the wonders of the library of Ashurbanipal! Do you know that?' He did not stop for an answer. 'The last great king of the Neo-Assyrian Empire. A collection of thousands upon thousands of clay tablets. Discovered by Layard in the mid part of the last century. In fact it was Layard's assistant Hormuzd Rassam – himself quite an intriguing figure – who unearthed the most famous ancient Mesopotamian text of all, the Epic of Gilgamesh, dating from the Third Dynasty of Ur. Yes, really quite fascinating.'

I tried to say goodnight to Father Burrows, but even as I made moves to step away from him he continued with his monologue.

'There are many parallels between Gilgamesh and the Bible, particularly when it comes to the great Flood,' he said. It was obvious that Burrows had no intention of drawing his lecture to a close and although I did not want to appear rude, I also did not care to stand and listen to him for the next half-hour or so.

'I could listen to you talk about this into the early hours, but I'm afraid I do have to go to bed,' I said. 'Will you forgive me if I say goodnight and you can tell me more tomorrow? I have the whole morning free, but only if you have time – I wouldn't want to take you away from your work.'

Father Burrows's eyes lit up at the prospect of hours of uninterrupted discourse on the literature of the ancients. 'I would be delighted,' he said.

I doubted whether he would get the chance, as I was certain that the next day would be taken up with much more serious matters. I said my goodnights to the rest of the group and, feeling relieved, retreated back to the quiet of my room. I knew I would have a long night ahead of me and so, after using a match to light a couple of candles, I lay down on my bed and closed my eyes for a few moments. The events of the last few days flashed before me, unpleasant memories that, like bloated corpses in a river, refused to settle and disappear. The terrible sight of Katharine's cat, that stain next to him on the bed. The image of all those pale, shocked faces, patterned by deep shadows cast by the torchlight, gazing on something unspeakable. The body of Sarah Archer, the girl's head smashed in. The horror in Katharine's eyes as she looked down at her hands streaked with blood. The hatred on Ruth Archer's face as she lashed out and tried to attack Katharine Woolley.

I felt too unsettled to rest and so took up my notebook in an attempt to make sense of it all. Davison had sent me here to look into the death of Gertrude Bell, yet since arriving at Ur her murder seemed almost peripheral. There was something else at work at the camp, something dark and base and evil, that did not look as though it had a connection to the death of Miss Bell. My mind started to work, teasing out the various possibilities, and as I tried to record the myriad conjunction of motives and hidden designs my pen could not keep up with the fast flurry of my thoughts.

As the night drew on I listened for the now familiar sound of the occupants making their preparations for bed until finally there was silence. Before I left the room I made sure that I was sufficiently prepared for what might happen. I did not want to use my arsenal of poisons, but I felt a little more secure knowing that in my handbag was a syringe filled with a fast-acting drug that could put a man to sleep in minutes. I took a deep breath and quietly opened my door, hoping that I could slip out of the house without being noticed. If I were to confront anyone such as Mr Woolley, who I knew did not need much sleep, then I would tell him that I was going to step outside to look at the stars, that I intended to refresh myself with a little night air.

I blew out the candles, picked up the box of matches and let my eyes adjust to the darkness. I edged my way across the courtyard, stretching a hand out before me as I did so. I tried to keep my breath quiet and steady, even though the primitive instinct of fear did everything in its power to close up my throat. I felt my chest beginning to tighten. Was there someone watching me in the dark? The old terror of the Gunman – that nightmare I had had since childhood, the sense of an omen of ill fortune – threatened to return. I remembered how the spectre of that figure had frightened me as a girl and how later, driven to the edge of despair by Archie's infidelity, I had thought it had stolen into my husband's body. How easy it was for someone you loved to turn against you; a face that you had once gazed upon with adoration became possessed by something else entirely, something unfamiliar and strange. I was certain there was someone in the house who had donned a mask of respectability but who, in effect, was the embodiment of evil.

I took a couple of deep breaths and tried to put such thoughts out of my head. I walked slowly and as inconspicuously as I could down towards the shed, making sure that I kept to the pockets of darkness not illuminated by the stars. I came to settle on a pile of sandbags hidden behind the store, from where I would be able to hear anyone trying to open the door of the shed. I thought of Katharine inside the wooden shack: I hoped that she had followed my instructions and had drunk the strong coffee which would keep her senses primed for any signs of an intruder. I didn't need such a stimulant; despite my deep breaths and my repeated mantra to calm down, my heartbeat was already racing.

I tried to make out the constellations, tracing imaginary lines in the sky in an attempt to bring some kind of order to the seemingly random pattern of stars. I loved the names that populated the celestial sphere, made up of mythological characters, animals and objects and, as I tried to pick them out of the sky – where was Hercules? – I wondered how many people had sat here before me, looking up at the heavens, unable to sleep or driven from their beds by thoughts of their own mortality. We were all so small, so insignificant; the thought was a well-worn one, something of a cliché, but still there was a reason why throughout history many of us, when gazing up at the stars, had pondered the nature of it all. We were all the same, I thought. Despite culture, background, status and sex, each of us wept, each of us laughed, each of us, if cut or shot, would bleed; indeed, in the end, each of us would die. If this was the case, what was the point in me trying to prevent the death of Katharine Woolley? After all, I knew that the woman – whom I had only met quite recently – would die one day. The answer was that I believed in the sanctity of life; it was, I felt, something supremely important. Yes, I would do anything in my power to stop this murderer.

Just then I heard someone approaching. I dared not peek out from my hiding place behind the shed, but the sound of footsteps was unmistakable. I clasped my handbag closer to my chest and tried to ignore the rasp of my own quickening breath. The person approached the front of the shed and stopped. Then I heard the faint sound of a padlock spring open and the creak of a door. I knew that I would have to act. I eased myself slowly forwards and felt my way towards the front of the shed. My steps were slow and as soft as I could make them. As I turned the corner I noticed a dark shape surrounded by the frame of the open door.

The figure had their back to me, but there was no doubt that it was a man and that he was raising his hand. I had not a moment to lose.

I moved quickly and took out a match and lit it. Although my fingers were trembling I knew I had only one attempt at this. I concentrated as I pressed down with the match onto the box. The touch, I knew, had to be firm and yet somehow graceful, a difficult combination to master when one was trying to prevent a murder.

As I struck the match a spark turned into a flame and the figure turned round. I lifted up the match and the small light was enough to illuminate the frightened face of a boy. Cecil.

'Stay away,' he said, swinging around to look at me.

'He's got a gun!' shouted Katharine from inside the shed.

'Shut your mouth,' he hissed, turning back towards Mrs Woolley. 'You've done enough damage as it is.'

I had to think quickly. 'Cecil, is this something you really want to do?' I said in as calm and as soothing a voice as I could manage. 'What would your uncle say? Or your poor parents?'

'Don't you dare speak of them,' he said, as he continued to point the gun towards Katharine. 'They have nothing to do with this. She's the one that has got to pay for what she did. For what she did to Sarah.'

'How do you know for certain it was Mrs Woolley?' I asked, conscious that my match would only burn for another few seconds. I knew that I could light another one, but in those moments of darkness that would inevitably follow I had to acknowledge that anything could happen.

Cecil might panic, he could shoot into the shed or fire the gun into the night. I or Katharine, or both of us, might get injured – or worse.

'Everyone knows it was her,' said Cecil. 'We all saw her. Her hands were covered in Sarah's blood.'

I took a step forwards. 'Now why don't you put the gun down and we can—'

'Don't come any closer,' he said, spitting the words out. 'I warn you.' As he swung round towards me I noticed that his hand was shaking, the metal of the gun glinting in the starlight.

'I know that you were very fond of Sarah,' I said, the match singeing the ends of my fingers. 'It's terribly sad that she—' As the flame finally died the sensation of burning was too much and I had to drop the match.

'That she was murdered by – by this . . .' said Cecil, his voice trailing off in disgust.

The sky may have been full of stars, but the loss of the light from the match seemed to plunge the scene into darkness. It took a moment or two before my eyes adjusted to the night. I quickly tried to light another match, but this time my fingers felt clumsy. I saw a spark, but it died as soon as it came alive. Then the second flared up too quickly, burning itself out in a moment. By the time I had successfully lit a third match Cecil had moved inside the shed to stand over Katharine, who was cowering in the corner of the mattress.

'Please, please, no,' she begged. 'I swear to you I had nothing to do with—'

Cecil leant forwards and placed the gun on Katharine's

right temple. Although it was obvious that the boy was nervous – the revolver shook in his grasp – he would still kill her if he pulled the trigger. I had to act. With my free hand I unzipped my handbag, placed it on the floor and took out the syringe. I moved quietly and stealthily across the shed, ready to plunge the syringe into his back. But just as I was about to reach out he turned towards me.

'What the . . .' he said, as if he could not believe what he was seeing. He grabbed my wrist with his left hand and twisted my fingers, forcing me to drop the syringe and the box of matches. Again the match extinguished itself, again I felt the ends of my fingers burn, but this was nothing compared to the pain I felt in my other arm, as Cecil forced my hand around my back.

'Are you in this together?' he spat. 'Is that what this is all about?'

It was my turn then to feel the unmistakable cold caress of the gun against my temple. Fear closed up my throat, preventing me from speaking.

'If that's the case I should finish you both off,' he said, pressing the revolver harder into my skin. 'Two sad old maids, jealous of our youth. Was that it?'

At thirty-eight years of age, I was far from old, but I knew in Cecil's eyes I was like an ancient relic. In that moment, I thought of all the things I wanted to do with my life. I didn't want to die, not yet. I would never see Rosalind grow up, never see her children. And what about my books? After the rather dry spell, when I had suffered from that awful paucity of ideas, now I found that inspiration came easily. I had so many stories swirling about my head that I wanted to tell. One bullet dislodged by Cecil McRae would be the end of all of that, the end of me.

I took a deep breath and although every cell in my body was telling me to fight back, to struggle, to do anything to stay alive, I closed my eyes and forced myself to relax. Perhaps if I went as lifeless as a dead fish the boy might loosen his hold on me. I imagined myself as a child in my mother's arms. I was safe, at my family home, Ashfield in Torquay. Nothing could harm me. However, as soon as I opened my eyes, I realised the real danger which faced me. Was I experiencing my very last moments?

Just then, as I felt Cecil tighten his grip on me, I heard an almighty, high-pitched scream. Katharine jumped up from the mattress and bore down on Cecil. She tried to stab the boy – she must have found the syringe on the floor – but as she reached out Cecil grabbed her hand to stop her. In doing so, he had no choice but to free me and I fell back and away from him. Somewhere on the floor lay the box of matches that I had dropped. I moved across the ground like a crab, stretching out my hands in the hope of finding the box.

'Oh, no you don't,' said Cecil as he bent back Katharine's fingers, forcing the syringe from her hand. 'I'll take that.'

'Agatha, help me,' gasped Katharine, the pain audible in her voice.

'What's in here?' he asked. 'What's in the syringe? Tell me!'

'Don't hurt her,' I said. 'It's a strong sedative, I promise, nothing more.'

'If that's all it is then I'm sure Mrs Woolley here wouldn't mind a little something to ease her misery,' he said. 'I've always wondered about a gunshot wound to the head. How much pain you would feel. They say it's more or less instantaneous, but you must feel something. Although I don't like to think of it, I'm sure Sarah felt pain in the moments before her death.'

As I made another large circular movement with my right arm I felt the familiar rattle of the matches in their box on the ground. I grabbed it and despite my shaking hands I managed to light one. Cecil had bent over Katharine: in one hand he held the gun, which he pointed at her chest, in another he brandished the syringe. It seemed as though Katharine's fate had been sealed.

'You know Miss Archer was really quite taken with you,' I said.

There was no response from Cecil.

'Yes, she told me that she liked you, but that she was too shy to tell you,' I continued.

'I – I don't believe you,' replied the boy.

'Oh no, it is true,' Katharine managed to say, understanding what I was up to. 'The girl more or less said as much to me too.'

'She had to pretend to be cruel to you, because she didn't want you to know the truth,' I said, as I used a match to light a candle. The soft light cast its amber glow across the enclosed space of the shed, sending amorphous shadows across the walls.

'The truth?' His voice was soft now, gentle almost, and he lowered the gun slightly.

'That she loved you,' I replied.

'Then why did she go and spoil it all then?' said Cecil,

the anger rising within him once more as he addressed Katharine. He aimed the revolver squarely at Mrs Woolley's throat, his finger on the trigger.

'Cecil, I don't think Sarah would have wanted you to do this,' I said.

'And what makes you such an expert all of a sudden?'

'She came from a Christian family, didn't she? Surely that must count for something. She wouldn't want you to suffer.'

'Suffer?' he said, blinking as if seeing the gun in his hands for the first time.

'I'm not talking of the sweet – or not so sweet – hereafter, I'm afraid,' I said. 'No, I was referring to the here and now. As soon as that gun goes off you're going to have everybody from the camp running from their beds to find out what has happened. Your uncle will see what you've done. I believe he's tried his very best to give you some stability since the death of your parents. And then it's a question of the police. You'll almost certainly be put to death for your actions. Is it worth all of that?'

In that moment, Cecil looked like a confused little boy who had just woken up from a nightmare.

'If you put that gun down I'm sure we can all agree that this was nothing but a boyish mistake,' I said. 'We needn't mention it ever again.'

He looked from me to Mrs Woolley. 'Are you sure?'

'Oh yes, quite certain, aren't we, Katharine?'

Mrs Woolley reluctantly nodded her head, but I could tell that she did not believe it for a second. She knew that as soon as the revolver was out of his hands we would raise the alarm and he would have to face the consequences of his actions.

'And, if you felt like it, you could even help us find the real murderer of Miss Archer,' I suggested.

He paused for a moment, before he raised the gun.

'It wasn't me, I tell you,' whispered Katharine, trying not to show her fear. 'It wasn't . . .'

I had tried everything I could, but it looked as though Cecil was determined to take his revenge on Mrs Woolley. If I did anything to stop him I knew he would not hesitate in turning the gun on me. I prepared myself for the very worst. I could not bring myself to witness the death of Katharine at such close quarters and so I closed my eyes. I had taken the coward's way out.

However, instead of the click of the trigger and the blast from the gun I heard the soft whimper of Cecil's cries. I opened my eyes to see the boy, now red in the face, throw the revolver and the syringe across the shed towards the door.

'Damn it!' he shouted, as he got up. 'I'm pathetic. My father always said so and so did Sarah. They were right. I'm weak-willed. I can't even kill the person who murdered the girl I loved.' He kicked the mattress as he walked past, but then as he was about to leave the shed, he bent down and picked up the gun, before he hit the door with his fist and disappeared into the night.

My immediate instinct was to run into Katharine's arms. We didn't say anything, but just stood there silently, as we replayed the terrible events of the last few minutes in our minds.

Finally Katharine stood back, looked at me and in a low voice said, 'Do you think . . . ?' I knew from her inference what she meant: the question whether Cecil might take his own life.

'I'm not sure,' I said. 'He may do. He's in an awful state. I'll go back to the house and see if I can find his uncle.'

'Don't leave me,' she said. 'I'm not sure whether I could stand it.'

'I'll only be a few moments. And then I promise I'll come back and I'll stay with you.'

'You won't lock me in, will you?'

'I think it's best – if only for your own safety,' I said. 'You never know, Cecil might be lurking outside, waiting for me to leave.'

Katharine looked at me with an expression of utter misery. 'I just don't understand. How long can this go on? I don't think I will be able to endure it.' She raised her hand to her head. 'Everything feels so . . . I don't know. I can't explain.'

'You can tell me,' I said gently.

'I'm scared, Agatha. Even more frightened than before. Not so much for my own life. After what happened I feel that could be taken from me at any moment. It might come as a blessed relief.'

'Don't say that,' I said, placing a hand on her arm.

'What happens if they put me away? I'd rather be hanged or shot than spend my life in some kind of institution. I've seen what goes on in those places.'

'I'm sure it's not going to come to that.'

'Maybe it's best if I confess to the crime?' The look of

mania that frightened me so had returned to her eyes. 'Yes, I could tell the authorities that I had taken a rock and smashed it over Miss Archer's head. That I relished the sound of the stone hitting her skull. That I did it over and over until she was dead. Would that do it? Would that convince them? They'd take me away, I would be sentenced without so much as a hearing and put to death in some primitive, utterly barbaric manner.'

'Don't talk like that,' I said, quite firmly. 'Now, listen. I'm going to leave you for a moment or two. I'm going to lock the door. I will go and find Lawrence McRae and tell him what has just happened. Then I'm going to come straight back here with some extra blankets. I can sleep on the floor, next to your mattress. And then, in the morning, we will clear all of this up. Also, I have every expectation that my friend, John Davison – he works in a division in the Foreign Office . . .' I thought it best to keep details vague. 'Anyway, I think he'll soon turn up with the police. He's got a good head on his shoulders. The kind of man who can see clearly through a difficult situation. He'll make sure nothing untoward happens to you.'

Katharine did not respond, but as I walked towards the door I saw that her eyes were beginning to dart around the walls of the shed. She watched the shadows cast by the candle with a mix of fascination and fear. Had this later incident with Cecil finally pushed her over the edge?

I turned the key that Cecil had left in the padlock. I paused by the door, listening for signs of the boy. With the light from a candle, I made my way back to the house and across the courtyard until I reached Lawrence McRae's room. I knocked on the door and a moment or two later I heard the sound of the architect stirring from his bed.

'Who is it?' he said from behind the door.

'It's Mrs Christie,' I said.

'Is there something the matter?' he asked, as he opened the door, fastening a Paisley-patterned dressing gown around him.

'It's your nephew, I'm afraid,' I said. I took a deep breath before I explained what I had witnessed in the shed. 'I fear Cecil may do something – stupid. He took off with the gun, you see.'

'Yes, you were right to come and tell me,' he said. I turned my back as he quickly put on some clothes. 'As you might have guessed, he's not been right since the death of his parents.'

'What happened to them?' I asked.

McRae did not answer. Instead he picked up a flashlight and asked, 'Where did you say he went?'

'I'm not sure,' I said, as I accompanied him out of the house. 'He ran out into the darkness. I came as fast as I could. I left Mrs Woolley locked up in the shed.'

'If you could rouse a few of the other men and tell them that Cecil has gone missing,' he said. 'I'll go down and start searching for him now.'

As McRae started to call out the boy's name, I ran back into the house and knocked on the doors of Leonard Woolley and Harry Miller. Mr Archer had been through too much already and I thought Father Burrows would be more of a hindrance than a help. I gave a quick synopsis of how Cecil had tried to threaten Mrs Woolley before taking the gun and disappearing. Disturbed by the noise, the other occupants soon opened their doors to find out what was happening. Cynthia Jones appeared, looking an absolute fright in her old-fashioned nightgown and bed hat. Father Burrows emerged blinking, battling with his wire-framed spectacles. The Archers, the poor grief-stricken parents, stepped into the main room looking as though they had had their spirits sucked out of them.

'Please, I think it's best if you go back to bed,' I said.

'What happened?' asked Mr Archer.

'Please, please not – not that woman,' said Ruth Archer, with fear in her voice. 'She hasn't – escaped?'

'No, nothing of that kind,' I said. I didn't want to add to their worries. 'She's still locked up in the shed. It's Cecil. He seems a little ... unbalanced.'

'When will this nightmare end?' asked Mr Archer. 'There's something rotten about this place. In fact, I don't believe it's where Abraham was born after all.'

'I wish we'd never come,' his wife replied, starting to cry. 'Why didn't we stay in Paris? Sarah was so happy there.'

Miss Jones and I watched them walk arm in arm back to their room. Cynthia asked me what had happened to Cecil and, as I repeated what had occurred, she looked taken aback and more than a little shocked.

'But don't worry, I'm sure he's not a danger to anyone but himself,' I said. 'I must get back to Mrs Woolley. I promised her I wouldn't be long.'

'Please take care,' she said, her eyes full of concern as she handed me her flashlight. 'After all, I'd hate it if anything happened to you as well.'

Miss Jones's words echoed through my head as I made

my way first to my room to get some blankets and then back to the shed. Just as I was about to turn the lock and open the door I heard a call from Lawrence McRae.

'Here's over here, by the spot where – where we found Miss Archer,' shouted the architect.

I saw the beams of light from the men's torches change direction as they ran through the darkness towards the bottom of the ziggurat. I hoped they would be in time to save the poor boy's life. He was in a very bad way, close to complete nervous collapse. As I opened the padlock and stepped into the shed I heard a gunshot split the night sky.

Chapter Nineteen

'Quick, push something against the door,' said Katharine.

'But what about Cecil?' I asked.

'If the stupid boy wants to go and kill himself I say let him,' she replied. 'After all, he hardly behaved like the perfect gentleman, did he?'

The statement was true enough, but I ignored her cruel remark and told her that I felt duty bound to see if I could help. My nursing training had not left me and I still felt a moral compulsion to relieve any suffering I might encounter. If Cecil had indeed shot himself he could still be alive and, if so, his life might be saved.

'Now, where's that syringe?' I asked. 'From my reckoning it should be still in this corner where Cecil threw it.' I bent down and, with the light from the torch, searched for the syringe. I looked in dusty corners, under the edge of the mattress and even outside the door, but it was nowhere to be found.

'Perhaps Cecil took it as he went out, when he picked up the gun,' said Katharine, as she saw me scrabbling on the ground.

'Yes, I suppose he must have done,' I said. 'Anyway, I must go and see what I can do.'

'And you'll lock the door, won't you?' pleaded Katharine. 'I don't want that boy making a second attempt on my life.'

I did what she asked and then, with the torch, ran through the darkness towards the men's cries and calls emanating from the base of the ziggurat.

'Who's that?' It was Miller's voice.

'It's Agatha,' I said. 'What's happened?'

I tried to make sense of what I saw. Cecil lay prostrate on the ground, on the exact spot where Sarah had died. In the sand by his right shoulder lay Woolley's gun. He wasn't moving, but as I ran my flashlight across his body – from his head, neck, torso, towards his waist, lower body and limbs – I couldn't make out any traces of blood or any obvious wounds.

'Is he dead?' asked Miller.

McRae stepped forwards, kicked the gun out of Cecil's reach and bent down. 'No, he's still breathing,' he said. 'Cecil, Cecil, wake up,' he said, shaking him.

'He's probably injected himself with a sedative,' I said. I moved forwards, crouched down by the boy and used my torch to illuminate the ground. There, half buried in the sand, was the syringe. I picked it up and, as I held it towards my flashlight, saw that it was empty. 'He must have had the gun in his hand, but quite wisely decided not to end it all. Instead, he decided to use this,' I said, showing the men the syringe. 'And the gun must have gone off as he injected himself. He's just going to have a very long, very deep sleep, nothing more.'

'But where the hell did he get hold of that?' asked McRae.

'It's mine,' I said.

'And do you make it a habit to go around carrying drugs of this kind?' There was an unpleasant tone to McRae's voice. 'I hope you haven't got anything else stashed away.'

I did not reply.

'Let's get the boy back to the house,' said Woolley. 'We can take it in turns to watch him and then we can inform the police of everything when they arrive tomorrow.'

'What do you mean?' asked McRae. 'Surely we don't need to tell them—'

'I'm afraid they need to know everything that has gone on here,' replied Woolley.

'But it won't look good, not with the boy's history,' said McRae.

'His history?' I asked.

There was an awkwardness that hung in the night air. The flashlight that shone on McRae gave him an unhealthy, sallow appearance.

'I suppose you may as well know now,' he said. 'It's about the boy's parents. The reason why I've been so protective of him. I didn't want to say anything before.' He fell silent before Woolley prompted him to continue. 'You know that Cecil's father and mother died in an accident. That is true enough. But what you don't know is the manner in which they were killed. You see, one day Richard, my brother, was showing the boy how to use a gun. They lived deep in the Scottish countryside and they were going to hunt some rabbits. But that day – this was when Cecil was only fourteen – the boy . . .' He cleared his throat and

243

continued. 'We still don't know the exact circumstances of what happened, but Richard and his wife, Elizabeth, died by gunshot wounds.'

'So you're saying that Cecil killed his own parents?' asked Miller.

'I'm sure it wasn't deliberate,' replied McRae. 'Well, the coroner seemed to be very understanding. He ruled an accidental death. Of course, I had to pull a few strings – luckily I had some contacts in the local force and I had to promise that he would live with me until he came of age.'

'And you brought him to this camp?' said Woolley.

'Yes, but I was sure that—'

'You do know you put everyone's life here at risk, don't you?' Woolley's voice was cold and full of anger. 'How am I going to explain this to the directors of the museums? What will I say when I tell them the truth?'

'I don't see why they need to know.'

'Are you insane?' said Woolley, his voice rising. 'For all we know, Cecil here may be some kind of unhinged killer. What if he were the one responsible for Miss Archer's death?' His brain worked quickly as he tried to make sense of how this new piece of information fitted into the larger picture. 'What if her rejection of him led him to bash her over the head with a rock? And then he planned on framing my wife for the death? Perhaps he was going to make the shooting of Katharine look like a suicide. After all, the facts are undeniable: he stole into the shed, armed with my revolver, with the aim of shooting her.'

'That's nonsense and you know it,' said McRae. 'Cecil would never dream of hurting Sarah. He loved her. And as regards to what just happened in the shed I suspect he was just blowing off a bit of hot air.'

'Hot air indeed!' exclaimed Woolley. 'Had it not been for Mrs Christie's clever intervention I suspect all of this could have turned out very differently. No, I'm afraid that there is no question: the police will have to be informed of all the circumstances. And if I have my way, the boy will be tried and found guilty. He'll certainly be taken away and, even if he does escape with his life, I can tell you that he'll never set foot within the camp again.'

'But—'

'We've stood here in the dark long enough as it is already,' continued Woolley. 'There's nothing more to be said. We need to get Katharine out of that awful place and back into her own quarters. God knows she's suffered enough as it is. But first, let's get this boy inside. Come on, Miller. Will you give me a hand?'

The men took Cecil by the shoulders and legs and carried him into the house, leaving McRae and me to trail behind. I was curious to know more about Cecil and his mother and father. I could understand how the boy might have accidentally shot one of his parents, but not both. Surely after he had seen a bullet go into one of them – say his father – he would have realised what an appalling thing he had done and then would have dropped the weapon. But, if he truly did not mean to carry out the act of murder, why would he then continue and swing around and take aim at his mother? What was it that made him hate them so much? It did not make sense. Woolley's theory – that Cecil had killed Sarah and then intended to frame Katharine and, after shooting her, make that death look like a suicide – seemed much more plausible.

'I understand this must all be very distressing for you, Mr McRae,' I said.

'Yes, indeed it is,' he said, as he stared straight ahead, refusing to meet my gaze.

'I wondered if I could ask a question,' I said. I did not wait for his response. 'Was Cecil a happy boy? As a child, I mean.'

'Happy?'

'Yes – was he a cheerful little boy, without a care in the world?'

'No, I don't suppose he was. But that doesn't make him a killer, Mrs Christie.'

I fell silent for a moment. 'No, of course,' I said. I was conscious of the words that McRae had used in regard to Katharine Woolley: *'Murder is a habit'*, he had said. At the time he had been trying to persuade us that Katharine had killed her first husband, and that, since then, Mrs Woolley had been possessed by some kind of urge to murder again and again. But what if it was not Mrs Woolley who had a taste for murder, but Cecil? And what if his uncle had known this all along and was doing everything in his power to protect his charge? Did that make him an accessory to the crimes?

'I know it's none of my business, but I was just thinking ... about the accident involving Cecil and his parents,' I said.

'I'd really rather not talk about it any more, if you don't mind,' he said.

'You see I can't understand why Cecil didn't drop the

gun after the first shot went off,' I said. 'I must be being terribly stupid, but—'

'Yes,' said McRae, the implication being that I was being very stupid indeed.

'But it seems to me that if it was an accident then the most natural thing in the world would be to throw the weapon from one's hands. Or to simply stop shooting.'

McRae turned to me and put a hand on my shoulder. 'I really don't think it's any of your business, do you?' In the darkness I could sense the anger burning in his eyes. 'In fact, I would suggest that you stop this line of enquiry right now. Am I making myself clear?' I felt his fingers dig into my skin. 'After all, I think we've had enough accidents for the time being.' The pressure on my shoulder intensified for a moment, before he released his grip and stormed away back into the house. McRae was definitely hiding something, something which would not remain in the shadows for long. I would ask Davison, when he arrived, if he could send off and request records of the case, including information about the beneficiaries of Richard and Elizabeth McRae's will.

I had so much to tell Davison that I didn't know where I would start. In truth, I couldn't wait to see him so I could, at least, speak freely of some of the things I had observed since I had arrived in Ur. The phrase 'wheels within wheels' came to mind; I sensed that so much still remained obscured from me. But, as I walked back to Katharine, I had to acknowledge that this was more like a case of 'murders within murders'. I pictured a Venn diagram of sorts, a series of interlinked circles, their edges outlined in blood.

Chapter Twenty

'Mrs Christie, Mr Davison of the Foreign Office tells me he is already acquainted with you, is that correct?' asked Woolley as he led my friend into the room where we were just finishing breakfast.

'Yes, indeed,' I said, as I stood up and shook Davison's hand. The look we exchanged – a rather formal, but courteous expression of recognition – concealed not only the real nature of our purpose in Iraq, but also the depth of our friendship.

'And this is Captain Forster, a representative of the Baghdad Police, who has been dispatched to try and sort out this unholy mess,' added Woolley, as he introduced a sandy-haired young man who looked barely old enough to shave, never mind lead a murder investigation. Clearly, he had been sent out as part of the British administration in Iraq and I wondered how much he really understood about the country.

'I've told Mr Davison and Captain Forster the details of what occurred, but obviously both of them will want to interview each of you in turn,' said Woolley, addressing the group at the table. 'I hope that won't put you to too much of an inconvenience.' Murmurs of agreement came from Mr Miller, Miss Jones and Father Burrows; Mr and Mrs Archer and Katharine Woolley were not present as they had decided to take breakfast in their rooms. 'They are going to question Cecil McRae first of all, who remains locked in my room.'

'The whole thing is bloody ridiculous,' said Lawrence McRae under his breath.

'You may think so,' replied Woolley, who had heard the architect's whispers. 'But nevertheless the process is one that has to be followed.'

'You know as well as I do that the boy would not hurt a fly,' said McRae.

Woolley was implacable. News of Cecil's past had already circulated through the camp and so Leonard felt able to speak freely. 'I've informed Mr Davison and Captain Forster of the boy's background,' he said calmly, 'and they will contact the relevant authorities back in Scotland to find out the exact circumstances of his parents' deaths.'

McRae pushed his plate away from him and stood up from the table, finding it difficult to contain his anger. 'I always suspected you cared more for the dead than for the living, but now I know this to be the case,' he said.

'Mr McRae, I do think that's quite—' said Woolley just as McRae stormed across the room towards him. The architect lunged at Leonard and just as he was about to strike and punch him in the face, Davison and Forster were quick to restrain him.

'If anything happens to that boy I know who to

blame,' spat McRae, as he tried to free himself from the men's grip.

'Really, I think that's—' said Woolley.

'I think it would be best if you calmed down,' said Davison. 'Come on, Forster, let's take him outside for a breath of fresh air.'

As the two men ushered McRae towards the door the architect turned to Woolley and said, 'If that boy dies for a crime he didn't commit you're the one that should be punished.'

An embarrassed silence descended on the room and was only broken by Miss Jones, who asked if anyone would like another cup of tea. 'I do think it would help,' she said.

'I'm sorry about that,' said Woolley. 'I suggest, if you're all in agreement, that we get back to work as soon as possible. Not only are we behind schedule, but we could all do with something to take our minds off, well, off the rather distressing events of the last few days.'

'Indeed,' said Burrows. 'In fact, there was something I wanted to talk to you about.' As Father Burrows moved over to discuss the intricacies of a certain cuneiform tablet with Woolley, Mr Miller came over to take his place at the table.

Since the incident on the ziggurat, when I had half fancied that I had felt a certain amount of tenderness towards the handsome photographer, I had deliberately tried to distance myself from him. I dreaded the rejection and couldn't bear to let myself be humiliated.

'I heard what you did last night to save Katharine from Cecil,' said Harry. 'It must have taken a great deal of courage.'

'It was only what needed to be done,' I said coolly, pretending to butter a piece of cold toast.

'Agatha – I hope I'm not speaking out of turn, but I wondered if, well, if I had done anything to offend you.'

'No, not at all,' I said.

'I know the Americans and the British have very different attitudes when it comes to friendship,' he said. 'No doubt you think us Yanks are vulgar and over-familiar.'

'You're forgetting my father was from New York,' I said.

'Yes, I remember you saying,' he said, trying to warm my spirits with a flash of his winning smile. 'Anyway, whatever it is I've done, I'm sorry. Maybe someone here told you something about me, and well, if you want to believe them there's not much I can do about it.'

As Miller stood up to leave the table I realised I had to say something. 'I'm sorry, it's nothing you've done,' I said, my resolve softening. 'The truth is that I've been rather shocked by everything that's gone on here. Sarah's murder, the awful way in which she was killed. And then all that business with Mrs Woolley and Cecil McRae. It's shaken me to the core.'

'That's hardly surprising,' he said. 'So I've done nothing to upset you?'

'No, and I'm sorry if I gave you that impression.'

He gazed enquiringly at me, as if to test the veracity of my words, and his look was so disarming that I had no choice but to smile back. It was at this point that Captain Forster and Davison returned to the room. As my friend entered I noticed that he had observed the silent interchange between Miller and me, but of course he was too discreet and well-mannered to say anything. Instead, he asked Woolley whether there was a room that he and Forster could use to carry out their interviews. Leonard ran through a list of spaces: the *antika* room was free, but hardly suitable as it contained some highly valuable objects; his own room was being used for the containment of Cecil, while Katharine, after her ordeal, really should not be disturbed.

'The other guest rooms really are too small, which leaves only this space, I'm afraid,' he said, gesturing at the table still laid with breakfast things. 'We could get the servants to clear up and have someone stationed at the door so you won't be disturbed.'

'How does that sound to you, Captain?' asked Davison.

Forster did not look impressed, but agreed that it would have to do. 'If I could ask all of you to return to your rooms, we will call you one by one so you can give us your statements,' he said, rather stiffly. 'I need not remind you that this is a murder investigation. This is a very serious matter indeed.' A few of the group bristled, no doubt feeling that they did not need to be patronised in this way, and indeed the overall effect of the unfortunate Captain Forster was that of a head boy at a minor public school who was intent on giving orders to his superiors. 'Where are the Arab boys?' he said, clapping his hands. 'Let's get this mess cleared up as soon as we can.'

Our little group dispersed and a moment later Harry Miller and I found ourselves in the courtyard.

'Shall we take a walk?' he asked.

'Yes, why not,' I replied. 'I could do with some air.'

'I suppose we can't go far ... we don't want to get into trouble with the Captain,' he said, smirking.

As we walked towards the ziggurat, where a chain of Arab men and boys had resumed their backbreaking work, Miller asked me about the events of the previous night. I went over the horrible scene once more.

'Do you really believe the boy's the one behind it all?' he asked, as we started a gentle circuit around the base of the structure. 'Woolley seems to think as much. That he killed Sarah out of spite, because she rejected him, and then he tried to frame Mrs Woolley for the crime?'

'I'm not sure,' I said. Since last night I had reassessed the situation.

'So you don't believe that he intended to shoot Katharine and then make it look like suicide?'

'I don't think Cecil has that kind of calculating mind,' I said. 'He's a hot-headed adolescent, not a cold-blooded killer. No, the person behind all of this is driven by something altogether different. It's the work of someone with an ability to schematise and plot.'

'You mean somebody like yourself?' said Miller, as we came to stop at the bottom of one of the ziggurat's grand staircases.

'I beg your pardon?' I said, somewhat taken aback. 'I'm sorry, but I don't think that's a very amusing joke.'

'You don't?' said Harry, his eyes glinting with mischief. 'I'm sorry if I offended the great lady novelist. But if I'm not mistaken I suspect you and the killer share more than a few characteristics.'

His comment was so outrageous that I had no choice but to laugh. 'I do hope you're not being serious, Mr Miller.'

'Oh, but I am,' he said, as he moved a step closer to me. 'In fact, I do believe, Mrs Christie, that if I applied a little more pressure you may actually confess to the crime.'

I could feel my face reddening and my breath quickening. 'Indeed?'

'Oh yes,' he said.

Just then, as his face moved an inch or so towards mine, I heard something or someone above me on the ziggurat. I looked up, but as I did so a few grains of sand fell onto my face. Harry followed my gaze and instinctively pulled me towards him. For a moment I thrilled to the feel of his strong arms around me, but then I came to my senses. I was about to object – how dare he presume to act in this way? – but then I realised that he was simply trying to protect me.

'Who is that?' he shouted, straining his head to look up towards the top of the ziggurat. 'Who's there?'

We heard the scuffle of footsteps, but although Miller dashed from the base of the ziggurat towards the open ground he said he couldn't see anyone.

'Whoever it was disappeared in a cloud of dust,' he said, returning to me. 'Are you all right? Here.' He passed me his handkerchief to clean my face. 'Did you swallow any sand?'

'No, I don't think so. But my mouth does feel a little gritty.'

'Let's get back to the compound,' he said, 'where you can have a drink of water.' He turned his head as we walked in an attempt to spot a glimpse of whoever had been perched on the top of the ziggurat. 'I wouldn't worry, I'm sure we weren't saying anything particularly revelatory. Probably just a couple of the Arab boys.'

'Yes, perhaps,' I said, biting the corner of my lip. No doubt they found it all rather amusing, the sight of a middle-aged English woman making a fool of herself with a handsome American man. I really would have to make sure nothing of the sort happened again. What had I been thinking?

We walked back to the camp in silence. I let Miller believe this was because of the disagreeable sensation of having sand in my mouth, but the reality was that I felt eaten up by shame.

'I'd better knock before I go in,' said Harry, as we stood outside the door to the main room. 'I don't want to walk in on an interview.'

'Come in,' said Forster.

Both of us stepped into the room to see the Captain seated at the long table studying his paperwork. Miller explained that he wanted to get some water for me, but the young officer seemed oblivious to the request.

'Go ahead, we haven't started yet,' he said, waving his hand in the direction of the kitchen. 'Davison said he was going to go on a recce of the ziggurat to see where the murder took place.'

'Davison's at the ziggurat?' I said, hoping that I had misunderstood Forster's words.

'I've been here before – one hot summer, when you lot were back in good old England – so I know the layout of the structure, but Davison said he would find it helpful to go and have a look,' he said, not looking up from his papers.

The revelation that it had been Davison who had evidently been spying on us left me feeling confused and a little dizzy. What had he been doing? What had he seen? What had he heard?

'But . . .' I said, gripping the back of one of the chairs.

'Here, sit down,' said Miller. 'Are you sure nothing hit you out there?'

'No, just a few grains of sand, nothing more,' I said, dropping into the chair. 'I think I should go and have a lie down.'

'A good idea,' said Miller. 'You won't need Mrs Christie right away, will you?'

The Captain did not bother looking up from his papers. 'No, no,' he said, sounding distracted. 'As soon as Davison gets back we're going to question Cecil McRae. We'll let you know when we want to speak to both of you.'

I pushed myself up from the chair and although Harry tried to take my arm to accompany me, I brushed him off. I could not show him any further signs of encouragement. But just then I stumbled, either from the effects of dizziness or from a slight ruck in the rush matting, and Miller grabbed my arm and prevented me from a nasty fall. It was then, at this moment, that Davison opened the door and walked into the room.

'Oh dear, has something happened?' asked Davison, rushing to help.

'We've got everything under control,' said Miller, holding up his free hand in a rather possessive manner. 'Mrs

Christie just felt a little faint, I think, and I'm taking her back to her room.'

As I stared into Davison's hardened eyes I saw a hint of something approaching . . . what? Jealousy? Surely it could not be that. Suspicion, yes, but not of me. Perhaps it was some kind of warning?

Chapter Twenty-one

I waited in my room for the gentle tap on the door that surely must come. I ran across the room and opened the door to a grave, serious-looking Davison.

'What was that all about?' I hissed, as soon as he had stepped into my quarters. 'Watching me from the top of the ziggurat? What were you thinking?'

'I know you're angry, but I can explain,' he said.

'I do hope you can,' I said.

'I see you're not wasting any time on being civil,' he said.

The comment stopped me in my tracks. Although I was showing all the outward signs of anger, I realised that my behaviour was very much a performance. I trusted that Davison would always have my best interests at heart. And if he had felt a need to spy on Miller and me, I suspected that there would be a very good reason why. That did not mean, however, that I was ready to give him an easy time.

'Why should I be polite to you after what you've done?'

'If you let me tell you what I know, then perhaps you'll see,' he replied.

'Very well,' I said, looking at him with a frosty expression.

'It's clear that you've become friendly with Mr Miller,' he began. 'I realise he's a very attractive man, but you may not know the whole truth about him.'

'You mean that he's got a reputation as something of a ladies' man? And that he had a relationship with Miss Bell?'

'Yes, and certain other things too,' he said.

I waited for him to go on. He walked to the window that overlooked the courtyard and made sure the shutters were closed.

'Well,' he said, clearing his throat. 'There is the not insignificant matter of his name.'

'What do you mean?'

Davison took a small notebook out of his jacket pocket and read from it. 'Mr Harry Miller, photographer, born in Philadelphia on the seventh of October 1889, died in an automobile accident in that city on the twenty-fifth of May 1925.'

I understood the implications straight away. I blushed as I thought about the intimate talks I had enjoyed with the man I knew as Harry. I remembered the way I had felt when he rescued me, first on that backstreet in Baghdad and then when I had been at risk of falling from the top of the ziggurat. 'So if Mr Miller is not who he says he is, then who is he?'

'I'm afraid we don't know that at the moment,' replied Davison. 'It seems as though he assumed the identity of the dead man, a man who was without family. How he got hold of his papers and his passport, we're not sure.'

My mind started to work quickly and I began to think

out loud. 'But why would he bother to do that? What was he running from?' I said. I paused as I forced myself to think the very worst. There was no use in being sentimental, of thinking of what might have been. Thank goodness I had forced myself to draw back from him before we had become too friendly.

'Perhaps he had committed some crime in America and needed a new start,' I said. 'But what if something – or someone – stood in the way of him achieving that? The man we know as Harry Miller had enjoyed a close relationship with Gertrude Bell. What if she had discovered his real identity? Perhaps he could not risk the exposure and so had no choice but to finish her off.'

'Yes, my thoughts exactly,' replied Davison. 'Now, I think you'd better tell me everything that has happened here, don't you?'

Slowly, but with as much precision as I could, I related the events that had occurred since I had first arrived at Ur. Katharine's odd, paranoid behaviour. The death of her cat. The strange atmosphere among the group. The rivalry between Mrs Woolley and the young Miss Archer. That awful row at the picnic. The discovery of Sarah's body.

'And then there was the thing with Miller's camera,' I said.

'His camera?' asked Davison.

'Yes, Harry Miller – well, the man pretending to be him – told me that he had discovered his Leica in bits. As if someone had deliberately smashed it. I had thought that this was because someone had wanted to destroy the film inside the camera because it might contain a clue to the identity of Sarah Archer's murderer.' I paused. 'But what if "Miller" had broken up the Leica himself?'

'It seems highly likely,' he said.

'I feel such a fool,' I said. 'To think that I was taken in by him.' I told Davison what had happened that day in the alley off Rashid Street. As I related the events the realisation hit me with such force that it left me reeling. 'Do you think that it all could have been staged?' I asked, even though I knew Davison could not give me an answer. 'That this man pretending to be Miller paid that young boy to try and grab my handbag so that he could then ingratiate himself with me?'

'I don't know, but we may have to face up to that possibility,' said Davison.

'But this – this impostor wasn't here when Tom, Katharine's cat, was killed. He was in Baghdad.'

'Or so he told you,' said Davison. 'But he could have stolen into the compound and done that.'

'So what do we do now?'

'Of course we can't let this chap know we suspect anything,' he said. 'It may be hard for you, but it's important that you behave as normal. Continue to enjoy your ... friendship with him.'

I knew exactly what Davison was trying to imply. What was the best way of putting him right? 'At one point, I was foolish enough to believe something may come of our friendship, but fortunately pride and common sense told me otherwise. As you know, I have not had the best luck when it comes to such matters. And I doubt whether I will enjoy such a close relationship again.'

'I see,' said Davison.

'So you don't need to spare my feelings when it comes to Mr Miller, or whatever his real name is,' I said. 'I didn't expect anything from him.' This was not quite the whole truth, but by saying it I hoped to make the sentiment real. 'I wonder what his game is? I'm sure it will come to light soon enough. Now, what else did you discover? Yes, what about Katharine Woolley's first husband, Colonel Keeling, who supposedly killed himself in September 1919. Did you find out who identified the body?'

'Yes, it was one of his work colleagues, Mr Thurley,' replied Davison. 'He told the police that the man on the mortuary slab was Colonel Keeling.'

'But how would this Mr Thurley know for certain if the man substituted for Keeling had suffered extensive head and face injuries? And what about other men of a similar age who disappeared at the same time?'

Davison consulted his notebook once more. 'Let's see – Albert Morrison, a civil servant, who on the eighteenth of September 1919 walked out of his house in Cairo after a row with his wife and who was never seen again. He was born in 1882, so that would make him two years younger than Keeling. And then there was Patrick Deller, a man of independent means, who disappeared from his home on the fifteenth of the month. He was older, forty-five at the time of the disappearance, and lived alone. There is some suggestion that he may have been linked with – well, with something unsavoury.'

'Such as?'

'Do you really want to know?'

'Yes, of course I do. My grandmother always said a true lady can never be shocked nor surprised. Now go on.'

'Well, it seems as though Deller took an unhealthy interest in young girls.'

'I see. So that disappearance could have been an act of revenge by one or more of the justifiably angry parents or relatives? And no trace of a body?'

'None,' he said, beginning to close his notebook.

'And did you find out anything else about Colonel Keeling?'

'I'm still waiting on a few pieces of information to come through,' he said. 'If anything significant turns up, I've asked for it to be delivered here.'

'I see, but it does seem strange that the whole thing could be linked to two separate cases of swapped identity – first Colonel Keeling and then Harry Miller,' I said. 'If Katharine Woolley believes her first husband may still be alive we have to take that suspicion seriously, even if it does not sound at all credible. Unless . . .' I said, almost under my breath. But it was best to keep my thoughts to myself for the time being. 'And there are a couple of other things I wondered if you might be able to help with.' I outlined what I needed to know: who, in each case, benefited from the wills of Miss Archer and Cecil McRae's parents.

'That shouldn't be too difficult to find out,' said Davison. 'I'll send a message to the department back in London.' He looked at his watch. 'Right, I'd better be getting back to Forster.'

I raised an eyebrow. 'He's very young, isn't he?'

'Yes, and between you and me he hasn't got a clue. It's a mystery how he got the job.'

He asked what I thought of Woolley's theory.

'I doubt very much Cecil is the one behind all of this,' I said, walking with Davison towards the door. 'However, he may well be the key to it.'

'What do you mean?'

'Would you mind asking Captain Forster to step inside for a moment?'

'What's going on, Agatha? I can see you've come up with some kind of scheme. What is it?'

As I outlined my plan a real smile, one with verve and mischief and delight, began to form itself across Davison's face.

'I say, that's splendid,' he said. 'Highly risky, of course, and it remains to be seen whether Forster will go for it, but there's a touch of genius there. Yes, only you could have thought of that. I'll go and fetch the boy wonder.'

A moment later Davison returned with an irritated-looking Forster. 'Now, what's all this about?' he barked. 'We really do need to start questioning Cecil McRae. I'm not in the mood for any idle chit-chat.'

'If you could just listen for a moment to what Mrs Christie has to say,' said Davison.

'About what?' asked Forster.

I did not have the time, or the inclination, to reveal everything I knew about the complicated case, and so I told Forster something of my suspicions relating to Cecil and how I did not believe that he had killed Sarah Archer. Then I outlined how he might be able to help.

'No, it's completely out of the question,' said Captain Forster. 'It would be tantamount to interfering with the evidence. No, I couldn't allow it, I'm afraid.'

'But can't you see Mrs Christie's point?' said Davison in as calm a manner as he could manage.

The young Captain looked at me and with a wave of the hand turned away from me. 'To be honest, I cannot,' he said, his face reddening. 'I need to establish the facts of the case for myself. I can't have the evidence messed about with in this manner.'

Davison tried to interrupt, but Forster cut him off. 'I wouldn't expect a lady to know the correct protocol of such an investigation,' he said, addressing me. 'To be honest, I'm rather surprised at you, Davison. You really should know better. In fact, I may even have to mention this to my superiors back in Baghdad.'

Would Davison reveal his hand and tell the pompous young fool of his real status? That, instead of being a mere civil servant in the foreign office, he worked for the Secret Intelligence Service?

'I'm sorry you think that,' said Davison, remaining composed. 'But perhaps you're right. Maybe it was an oversight on my part.'

'We've got to do this by the book,' said Forster, checking his watch. 'Let's get on with it. That boy should have come round now.'

'Indeed,' said Davison, taking a small card from his inside jacket pocket. 'But before we go and question him there is something I should show you.'

'What is it now?' snapped Forster. 'Really, Davison. My chief said he was sending someone with me who would help and assist me, not bother me with all these unnecessary details. It really is quite—'

The sight of what was written on the card – Davison's name and title at the service – stopped Forster in his tracks. He coughed in a half-hearted attempt to swallow his words. 'Well, I – I mean,' he blustered. 'If I'd only . . . then, of course, I—'

'Not to worry,' said Davison, smiling gently. 'You weren't to know. And I'd appreciate it if you kept that information to yourself. Now why don't you listen carefully as Mrs Christie here explains what's going to happen.'

Forster blinked back his astonishment as I turned towards him and began to outline in more detail what I wanted him to do.

Chapter Twenty-two

We had all been told to wait in our rooms until we were called for questioning. My own session with Forster and Davison was done very much for show, as we had discussed certain aspects of the case earlier in my room. The interrogations continued until four o'clock when we were informed that we were invited to return to the living room to take tea with Captain Forster, who wanted to share with us some important information. I was the first to arrive and as I entered Forster jumped up and found a chair with a thick cushion for me at the table.

'Are you sure you are comfortable?' he asked, fussing around me like an old maid. 'Would you like some tea? And how about a cake?' He snapped his fingers and a servant brought over a little plate of Arab delicacies. He lowered his voice so nobody else could hear. 'If you do get a moment to talk to Davison, I would be enormously grateful if you could extend my apologies for earlier. Really, I had no idea. If I had known, obviously I would never—'

At that moment the door opened and Father Burrows

entered with a nervous-looking Miss Jones, soon followed by Lawrence McRae, the man who went by the name of Harry Miller, Mr and Mrs Archer, Leonard Woolley and then Davison. Each of them took their places at the table and helped themselves to tea, the ritual accompanied by the familiar sounds of the clink of teaspoons on china and the low murmur of polite conversation. On the surface the occasion appeared an utterly civilised one, but underneath the sheen of respectability there was something ugly, something evil.

'Now who is it we are waiting for?' asked Forster, looking around the table. 'Oh yes, Mrs Woolley.'

'Katharine still hasn't recovered from her terrible ordeal,' said Leonard Woolley, taking a sip of tea. 'I'm afraid she won't be joining us.'

'Mr Woolley, I thought I had made myself clear that everyone had to attend the meeting this afternoon.'

'Is that strictly necessary?' replied Woolley, his eyes glinting.

'I must insist, there can be no exceptions,' said Forster. 'If you'd be so kind as to go and request her presence.'

Woolley placed his teacup on the table, stood up and with as much grace as he could muster walked out of the room. The conversation resumed as if nothing had happened. Father Burrows started with a monologue about the intricacies of certain cuneiform tablets with Miss Jones, who did not seem to be paying him very much attention. Harry Miller informed Lawrence McRae of his intention to go to Baghdad as soon as he could to order a new Leica camera – the question now was which model he should order? – but the architect remained silent and preoccupied. No doubt he was worried about the fate and wellbeing of his unfortunate nephew. Mr and Mrs Archer talked quietly among themselves, so softly in fact I couldn't make out their words.

A few minutes later, Woolley returned with Katharine trailing behind him. At her entrance, all heads turned, all eyes were directed towards her, as if she were some kind of lodestar. She was dressed smartly not in her customary shade of *vieux rose*, but in black. She held herself with a dignified posture – her expression spoke of stoicism and unspoken suffering – and as her husband pulled out a chair for her she reminded me of one of the great classical actresses of the stage. Ruth Archer opened her mouth to speak – was she about to apologise for some of the awful things she had said? – but her husband placed a hand on her arm.

'Let's wait to hear what the Captain has to say,' said Mr Archer, before he looked up and addressed Forster. 'The sooner we get this over with the better. We have a funeral to organise. A daughter who needs to be buried.'

'Yes, I understand,' said Forster. 'And I do appreciate the fact that you've delayed your departure. As I said, we can help with any arrangements regarding the body and so on.'

The Captain's reduction of Sarah Archer to a mere 'body' startled Ruth Archer and, quite understandably, she took out her handkerchief and pressed it to her mouth. I only hoped that the inexperienced officer would not make any more blunders.

'Now, as I suggested, we have some important information that I would like to share with you,' said Forster.

A silence descended on the group as we waited for the Captain to impart his news.

'As you know, on Saturday night the body of Sarah Archer was found at the base of the ziggurat. She had sustained a number of serious – in fact, fatal – head injuries,' he said. Again Mrs Archer winced at Forster's words. 'It looks as though she had been hit with a rock.' The Captain must have caught a glimpse of Mrs Archer's appalled face and after clearing his throat began again. 'I must thank Mr and Mrs Archer for their patience and for their help at this most distressing of times. And I would like to thank each of you for your co-operation. I know it must not have been easy to relate the details of what you saw and heard.'

Mr Archer shifted impatiently in his seat. 'Thank you, Captain Forster,' he said. 'And we are grateful to you too for coming down here. However, I'm conscious that time is of the essence. I don't want to appear rude, but we do need to sort this out as soon as possible so as to make the necessary arrangements.'

'As I was saying, we have concluded our investigation into the death of Miss Archer.' Forster puffed out his chest slightly, 'And I am satisfied to tell you that we have secured a confession from—'

A low, excited murmur ran around the table.

'From Cecil McRae, who—'

At the mention of his nephew's name, Lawrence McRae stood up, tipping over his cup of tea as he did so.

'Oh my, dear me,' said an agitated Miss Jones, taking out her handkerchief and dabbing it across the table.

'That wicked, wicked boy – how could he have done that?' exclaimed Ruth Archer.

'This is absurd!' shouted McRae. 'Cecil is the last person who would have hurt that girl.'

'If I can ask you to sit down, Mr McRae,' said Forster. 'I can explain to you—'

'No, I will not sit down. I demand that you take me to see him.'

'All in due time,' said Forster.

'What did you do?' said McRae. 'Beat a confession out of him?'

'I can understand why you feel distressed,' said Forster, 'but really that's quite unnecessary.'

'Please, Mr McRae,' said Davison, coming to stand by the architect. 'If you could let the Captain explain.'

McRae glared first at Davison and then across at Forster, no doubt thinking that he'd very much like to punch them both, before he audibly dragged his chair back across the floor and sat down. 'I won't believe a word of it, not until I hear it from the boy's mouth myself,' he said.

'Thank you, Mr McRae,' said Forster. 'As I was saying, we have obtained a confession from Cecil, a confession that was given voluntarily and entirely without duress.' He took out a notebook from his pocket. 'This is not strictly proper procedure, but I can see how the crime has distressed you all,' he said, watching the stricken faces of the Archers and the nervous hands of Miss Jones flutter across the table as she continued to dab away at the last traces of the spilled tea. 'In order to put all your minds at rest I can supply you with a few more details, in Cecil's own words.'

He cleared his throat once more and, from his notebook, began to read out the boy's statement.

'*I wasn't looking forward to the picnic. I didn't want to go and I thought the whole thing was a stupid idea. As soon as I saw Sarah flaunting herself in that black dress I felt my blood begin to boil. Part of me wanted to tell her to go and change into something more modest. Another part of me – well, let's just say I wish she could have been nicer to me. None of this would have happened if she had been a bit kinder. Not so cruel. Why did she have to go and say those things? That I was ugly and stupid? Why did she tell me that I couldn't talk to her?*

'*After walking to the top of the ziggurat I helped myself to a couple of those fruit punches. I suppose they must have gone to my head. I was watching Sarah and Harry Miller laugh and joke about. He was taking her picture and she was standing at the edge of the ziggurat. You should have heard them. It made me sick to my stomach. And then the next thing I knew she was looking at my uncle in the same kind of way. She never once cast a look in my direction. I saw her dancing on the edge, kicking her heels up, in that dress, a dress that was nearly transparent. If she'd never worn it things might have been very different.*

'*My uncle saw the way I was looking at Sarah and told me not to worry. One day I'd forget her, he said. I didn't want to forget her. I wanted to be with her, forever. Then Sarah's father came, everybody sang happy birthday and*

272

she started telling that stupid story about the man who proposed to her. She got into that horrible row with Mrs Woolley and stormed off. "This is a birthday I'll remember," she said, or something to that effect.

'I didn't mean to do it. It was a terrible accident. You've got to believe me when I tell you that I didn't mean Sarah any harm. I just wanted to go and make her feel a bit better. She was upset after that argument. So I went after her. I found her running down the path. I called her name, asked her to stop but she told me to go away. I reached out and grabbed her. I just wanted to talk to her. But she got me all wrong. She said some cruel words to me and I caught hold of her. She went to slap me, but when I reached up to protect myself she fell backwards. It all happened so quickly. She must have banged her head. I shook her, told her to wake up. That I didn't mean her any harm. But then I thought what would happen if she did wake up. She would tell people that I had tried to attack her. What happened with my parents would get dragged up again. I would get locked away, or worse. And so I took hold of a rock and hit her over the head. I know it was a bad thing to do. I can see that now. But I wasn't thinking right.

'That's when I heard footsteps. I suppose it must have been Mrs Woolley. I didn't meant to hurt Sarah. I hope that one day Mr and Mrs Archer will forgive me. This completes my statement – Cecil McRae.'

Captain Forster looked up from his notebook to a sea of bewildered faces. Mrs Archer wept into her handkerchief and her husband's face looked ashen. Lawrence McRae seemed as though he was on the point of storming out again and Miss Jones stared at her tear-stained handkerchief with something approaching horror in her eyes.

'We should pray for the boy,' said Father Burrows.

'I may be a Christian man, but I refuse to pray for him,' said Hubert Archer. 'In fact, I'm going to see that boy hangs for what he did. Come on, Ruth.' He stood up and held out his hand for his wife. He turned to Captain Forster. 'We don't want to bury Sarah out in the desert. We'd like to take her back to Baghdad, if that's possible.'

'Yes, of course,' said Forster. 'I'm sure that can be arranged, can't it, Davison?'

'Indeed, sir,' said Davison, playing the role of the subservient civil servant. 'I'd suggest the British cemetery. I know your daughter was an American, but it does seem the most fitting and distinguished place. After all, it's the same cemetery in which Gertrude Bell was buried.'

The Archers seemed honoured that their daughter had been compared to the famous traveller and archaeologist who had helped to found the new state of Iraq. The mention of the name had other effects too: Harry Miller got up from the table and turned his back on the group; Leonard Woolley smiled fondly to himself and said that she was greatly missed, while his wife remained unmoved; and Miss Jones asked to borrow a clean handkerchief from Mrs Woolley, which she then used to dab her eyes. I still could not work out whether the death of Miss Bell was connected in any way to the murderous goings-on at Ur. An image of an ancient pot lying in pieces flashed into my mind. I couldn't see how it would be possible to fit all the differently shaped shards back together.

'We've all had quite a shock,' said Miller from behind me. I turned around to see the handsome photographer. 'Who would have thought that Cecil was the murderer? Are you all right? You look a little pale,' he said to me.

'I do feel a little shaken by the news,' I said. I had to do everything in my power to stop myself from asking him to explain himself. If he wasn't Harry Miller, then who was he? What had he done that was so terrible he had had to flee America? And what was the truth about his relationship with Gertrude Bell? 'I think I'm still suffering from shock,' I said instead.

'Why don't we take a walk?' Miller said in a gentle voice. 'That might help calm your nerves.'

'Very well,' I said in a louder voice. 'A little fresh air might be just the thing I need. Where shall we go? Perhaps to the ziggurat?'

As Davison saw us leave the room he shot me a look of warning.

'How extraordinary,' said Miller, as we stepped into the courtyard. 'It seems too unbelievable for words.' As he continued to talk about what Cecil McRae had done to Miss Archer I fell silent and studied his profile. Certainly there was nothing in his physiognomy to suggest that he was hiding anything; in fact, his open expression and handsome features suggested nothing but a good, old-fashioned American wholesomeness, the kind I had encountered many times both at home in England and on my travels. Yet I knew that appearances counted for nothing.

'Sorry, I should stop talking about it as I can see that it's distressing you a good deal,' he said, as we walked towards the ziggurat.

'Well, I am worried about what will happen to that young boy now,' I replied. 'Do you think he's the one who smashed up your camera?'

'I guess he must have been. Perhaps he thought that I had caught him looking at Sarah with an angry expression, or maybe he was just so cross about my flirtatious interchange with her on the top of the ziggurat.'

'It will all come out when Forster questions Cecil again back in Baghdad,' I said. I wondered how I could ask Miller a little more about his background without raising his suspicions. 'But jolly annoying for you, having to buy a new camera.'

'Yes, I'll have to go to Baghdad to put in an order,' he said. 'It will probably take weeks to arrive.'

'Have you always been a photographer?' I asked as we started to climb the stairs up towards the first level of the ziggurat.

'I don't think I'm fit for purpose to do anything else,' he said, without a moment's hesitation.

'How did you learn?'

'I got the bug as a kid,' he said. 'I pestered my folks to buy me a Box Brownie and from that first Christmas I was hooked. I must have been about ten or eleven. I started to take shots of our dog, my grandparents, the trees in the park in winter, and then when I was sixteen I got a job on the local paper, the *Middleton Bulletin*. That was great training, let me tell you. I covered everything – crime, sports, personalities, local politics. It was a fast life, but a hard one.'

He didn't seem to be lying, but I knew that practised or pathological liars sometimes came to believe their own falsehoods.

'It seems a bit of a jump from news photography to taking images of artefacts found in the desert sand,' I said.

'Let's just say I had my heart broken,' he said, turning away from me. 'Anyway, I've talked too much about myself.' He stepped closer to me. 'I'd like to get to know you a little better.'

'Ask away. I'm an open book.' That was not entirely true.

Miller began to ask me some questions and I told him a little about my writing, my childhood and my first marriage. He did not, thank goodness, ask about my disappearance in December 1926, an event in my life which had been on the front page of every newspaper. I also shared with him the feelings of guilt I felt about Rosalind: she blamed me for the breakdown of the marriage. It was important for me to show my vulnerable side to him if the plan was going to work.

'Sometimes I feel so alone,' I said. 'Of course, I have my writing, but often I think I only do that to tell myself stories because there's no one around to talk to.'

'You don't have to be – alone.'

'I know it's partly my fault. I'm not a terribly social being, I'm afraid.'

'No, I didn't mean that exactly,' he said. 'You ... you could always marry again.'

'I don't think so,' I said, laughing. 'After all, who would want an old maid like me?'

Miller reached out and placed a hand on my shoulder. The touch was like a current of electricity through me. Although my mind felt detached – I knew exactly what I was doing – my body was a different matter. The photographer saw the effect his gentle caress had on me and, after looking into my eyes, he felt emboldened to take a step closer. It had been so long since I had experienced anything like this. There had only been Archie. And then he had gradually absented himself from our bed. He had turned away at night, pleading tiredness. Then he had started to spend more time away from home. There was business, he said, there was the golf. In truth, all along it had been the other woman in his life, Miss Neele, a woman who was now the new Mrs Christie.

'Don't put yourself down – you're no old maid,' whispered Miller. His hand moved from my shoulder to the small of my back and he drew me towards him. The sensation was a delicious one. I had to tell myself, once more, not to fall for his charms. There was something I needed to accomplish – to keep Miller away from his room long enough so Davison could search it. Davison and I had agreed on that much, but we had had different opinions on the actual methods involved. I could not risk telling my fellow secret agent what I intended to do in case he ruled against it. Davison assumed that we were going for a walk, nothing more than that. Of course, it was a risk, but I thought it was one worth taking.

Miller moved his head forwards and bent down to kiss me, but I stepped away from him and looked around to see if anyone had spotted us. My reticence was all too real, and he simply interpreted my actions as those of a woman of my class and background.

'I'm sorry,' he said. 'I overstepped the mark.'

'No, not at all,' I replied.

'Perhaps we should get back to the house,' he said, looking at his watch. 'No doubt they'll be wondering what's happened to us.'

I knew Davison would still be searching Miller's room. I had to try and delay him.

'I'm rather out of practice, I'm afraid,' I said, blushing. 'Since my husband left me, well, it's been . . .'

'You don't have to explain,' he said, moving towards me again. He reached out and, with the tip of his thumb, touched my lips. He took hold of my hand and led me into a dark, shaded corner of the ziggurat where no one could see us. 'Look, there's nobody around. You don't have to worry.' As his face neared mine I inhaled traces of his expensive-smelling cologne. The aroma was deliciously heady and transported me away from the desert sands, but I had to remind myself that Miller was using it to mask something. The scent wasn't his true smell, his name wasn't even his real one. What exactly was he covering up? I hoped Davison would be deep into the search of his room now.

'You seem distracted,' he said.

'No, I was just thinking about—'

And with that he stepped away from me. 'I really must be getting back,' he said, coughing in embarrassment.

'Oh, dear,' I said, starting to panic. 'I hope I didn't give you the impression that I was . . .' I couldn't quite finish the sentence.

'No, not at all,' he said, trying to smile.

'Perhaps we can go to the top of the ziggurat? We could admire the view from there.'

'I'm not sure,' he said. 'I can see that the recent troubles have taken their toll. I wouldn't want to add to that.'

'What do you mean?'

'Agatha, if I may speak plainly?'

'Of course, please do.'

'It seems to me that you're not ready to embark on a new romance,' he said. 'I think what happened with your husband is still preying on your mind.' He started to walk away.

'But I do find your company very appealing,' I said, feeling a flutter in my breast. 'I can relax with you, something I haven't felt able to do with ... well, with anyone since my husband.'

'That's good,' he said. 'But I think we need to take this – what is it they say in novels? – at a slower pace, if only for your own sake.'

Despite the evidence that seemed to suggest he was covering something up – perhaps something as evil as a murder – he was behaving like a gentleman.

He presented me with his arm and asked, 'Shall I walk you back?'

It appeared I had no choice. I would have to think of some other way to slow his return. 'Yes, that would be very nice, thank you,' I said, politely.

On our walk back to the house I asked him what he missed about America. He tripped off a list of seemingly trivial items – ice cream, the movies, the energy of a big city – before he mentioned one thing that I thought could be significant. 'I suppose the idea that you can be who-ever you want to be.' The comment was a throwaway one and he soon began to talk of skyscrapers and base-ball and the grandeur of railway stations, but the phrase stayed with me.

I slowed my pace as I began to talk about my youthful dreams of being an opera singer, a concert pianist, a sculp-tor. I took my time to describe each of these ambitions in turn, outlining the various hurdles that had stood in my way: my shyness, my horror at performing in public, my lack of talent. Miller did not want to offend by cutting my stories short, but I could tell a certain dullness had stolen into his eyes. As I waffled on, he stood there making appro-priate facial gestures, until finally he looked at his watch and said, 'I'd love to talk more about all of that, but I really do think that—'

'Of course, I'm sorry if you find my stories boring,' I said, pretending to be hurt by his decision to cut our con-versation short.

'No, it's not that at all – just that—'

'I did think we had something in common,' I said, trying to bring tears to my eyes.

'Oh no, I've upset you,' said Miller, stepping closer. 'That's the last thing I wanted to do. Whatever it is I said or did not say, please forgive me.'

Taking a deep breath I steeled myself for what I had to do next. This was against my nature, my breeding, everything I had been told – imagine what my mother would have thought had she been alive! – but I had no choice. 'No, it's you who should forgive me,' I said. 'I acted like a spoilt child.' I grasped for half-remembered images from romantic novels I had read long ago. I ran the tip of my tongue over my dry lips and tried to open my eyes a little wider. I thought of something that amused me: what was it my grandmother had said about that curious woman in her sewing circle? Was it something about her having only one passage like a bird? The memory made me laugh and I felt a sparkle return to my eyes.

'What's so funny?' asked Miller.

'I was just thinking about what a fool I'd been,' I said, placing a hand over his. 'I get so nervous at times. I'm not used to male attention, you see, especially from a man as handsome and as nice as you.'

Miller looked slightly taken aback. 'Well, I—'

'I'd hate it if I've given you the wrong impression,' I said, deliberately lowering my chin and raising my eyes so that I would appear a little more like a woman used to employing the dark arts of seduction.

'And what impression would that be?' he asked.

'That I was indifferent to your attentions.'

He took another step closer to me and inclined his head in such a way that I could feel his hot breath on my face. But just then – just as he reached out his hand to caress me – I thought about what Davison might have found in his room. Proof of Miller's original name? Or something even more sinister that would link the American with the death of Miss Bell?

I tried to banish these thoughts from my head, but they refused to be pushed away. I caught an uncertainty in

Miller's face that told me he had read something in my eyes. I thought again of what my grandmother had said about her sewing woman, but the spark of light amusement was impossible to rekindle. Miller's hand slowly retreated and, so as to not cause undue embarrassment, he pretended to brush a spot of dirt or sand from my shoulder before he stepped away from me and looked into the distance.

'But—'

'You're very sweet, but I can see that I was right,' he said, smiling kindly. 'Let's not hurry things. Now I really think we must be getting back. It looks as though there's a sandstorm coming.'

'Could we not just ...'

But Miller had already turned from me and had started to walk briskly back to the compound. As I rushed after him I thought about falling, claiming that I had twisted my ankle. But by the time the notion came to me Miller had reached the gates. Had he guessed that I had been trying to delay his return? I called out his name one last time, but I saw him disappear into the courtyard. I ran as fast as I could, but my skirts kept threatening to trip me up and send me crashing down into the sand. If I wasn't careful my idea of hurting my ankle might actually become a reality. I felt my heart race and my face redden. Beads of perspiration broke out across my forehead and then I felt something lodge in my left eye. I blinked and tried to rub the particle away with my finger, tears forming in my eyes as I did so. Miller had said that there was a sandstorm coming and, sure enough, the horizon had disappeared, to be replaced by an ominous dirty brown mass.

I had no way of letting Davison know of Miller's imminent return. I only hoped that he would have finished his search.

By the time I reached the compound I could not breathe. Feeling on the verge of collapse, I made my way to Miller's room. A stony-faced Davison stood in the doorway, holding out a handful of letters and other papers. Miller had pushed past him into his room, where he stood astounded and somewhat broken.

'So, Mr Miller, can I ask what you are doing at Ur?' asked Davison. 'Or should that be Mr Conway?'

'I don't know what you mean,' he said, in a rather half-hearted manner.

'Mr Harry Miller, as I'm sure you know, died in a car accident in 1925,' said Davison. 'And don't try and pretend that you are a different Harry Miller. It's a fairly common name, but you're not going to be able to get away with that.'

Miller – or Conway, as he was – looked sheepishly at Davison, before he spotted me standing outside the room. His eyes shifted from Davison back to me before he realised the truth of the scenario that had just taken place.

'I expected better of you, Agatha,' he said, picking up a stack of letters that Davison had left on his bed. 'I thought we had something special.' He sighed deeply. 'Was that – I mean – all that, out there,' he said, gesturing towards the direction of the ziggurat. 'Was that just a tactic to try and slow me down so my room could be searched? Didn't you mean any of those things you said?'

I could not answer him.

He looked at me with disappointment and despair in his eyes. He sat down on the bed, and let his head drop down, as if all the worries and problems of the world had suddenly been shifted onto his shoulders. A pallor now replaced his former healthy tan. 'Okay, you've got me,' he said. 'I suppose I may as well confess.'

Chapter Twenty-three

'Do you want to go and get Forster?' asked the man I now knew to be called Conway. His voice was heavy with resignation, as if he didn't care what happened to him.

'No, I think we can handle this for the time being,' said Davison. 'Agatha, why don't you come inside and close the door. Let's hear what the man has to say for himself.'

I stepped into Conway's room, took out my handkerchief and wiped my face of dust. I studied the details of the room. The white walls were festooned with hundreds of photographs: pictures of street scenes in Baghdad; an image that looked almost obscene but which I realised was a close-up of the inside of a pomegranate; the craggy, lined faces of Arabs; dozens of views of the ziggurat taken at different times of the day and seemingly hundreds of representations of the undulating sweep of the desert sands. Here too were snapshots of some of the team at Ur. A smiling Woolley, holding an ancient gold cup he had just removed from beneath the earth; Katharine Woolley with her cat; Father Burrows stooping over a cuneiform tablet, his brows knit in concentration as he tried to decipher its secrets; and Lawrence and Cecil McRae at work on their drawings. One wall was completely devoted to images of the artefacts unearthed at Ur: daggers, earrings, bowls, cups, necklaces, rings, headdresses, cylinder seals and cloak pins.

Davison walked over to a desk situated in the far corner of the room and picked up a maroon-coloured photograph album that was lying underneath a pile of papers.

'Now, this is interesting,' he said. I walked over to see what he had found.

As Davison turned the pages I saw image after image of Sarah Archer, photographs of her in every decent pose imaginable, together with close-ups of her neck, shoulder, wrists and mouth.

'I know what you're thinking, but it's not like that,' said Conway, standing up and reaching for the album.

'What is it like?' asked Davison, taking a step back.

'I liked the kid, of course I did,' replied Conway. 'She was beautiful. Everyone could see that. But I just wanted to try and see if I could capture that beauty.'

'Are you sure there was nothing more to it than that?' asked Davison.

'What do you mean?'

'Well, unfortunately we can't ask Miss Archer about the nature of your relationship and what really passed between you because she's dead,' said Davison. The words were harsh, if not a little cruel.

'You don't think that ... ' Conway could not complete the sentence.

'Think what, Mr Conway?' Davison took a step towards him. 'You see, we know now, thanks to these papers that I just happened to find, that your real name is not Harry Miller. That you stole that name from a dead man. My feeling is that if you could lie about that you could lie about almost anything.'

Conway turned to me, a pleading look in his eyes. 'Agatha – surely you don't think that . . . that I had anything to do with Sarah's death?'

'I'm afraid I don't know what to think any longer,' I said. 'But you said you wanted to confess.'

'Yes, but not – not to that,' he said. He looked as though he might be sick.

'If not that – then what?' I asked.

He took a couple of deep breaths and began. 'As – as you now know, my name is not Harry Miller. It's Alan Conway. Before I tell you any more I must say, once again, that I had nothing to do with Sarah's death. I could never have hurt that girl. You do believe me, don't you?'

'If you don't tell us what you are doing here, then I'm afraid we have no choice but to suspect you of the murder of Sarah Archer,' said Davison.

'But that boy, Cecil, he's already confessed to the crime,' said Conway.

Davison thought quickly on his feet. 'We have reason to believe that Cecil was not in his right mind when he made that statement.'

'So he didn't do it?' asked Conway.

'No, we believe he didn't,' said Davison. 'Which means that unless you convince us otherwise, then—'

Conway, panicking now, interrupted him. 'I know when

the game is up.' He ran a hand across his sweaty forehead. 'You may as well know that I got myself into trouble back in America. I owed some money, some big money, to a couple of shady characters, the Solomon brothers. You know the type, men you cannot afford to mess with. You see, my printing business went under back in New Jersey. The banks wouldn't lend me a cent, and so I borrowed some money from the Solomons. That was the worst decision I ever made. But I couldn't see a way out. I had bills to pay. And, if you must know, I had a wife and a child to support.' He looked at me. 'You've got every right to hate me, Agatha.'

'My feelings are neither here nor there,' I said sharply. 'What matters is the truth. How did you end up here, at Ur?'

'It was the Solomons' idea. They had seen the splashes in the press, all the write-ups about the treasures being pulled from the sands and how much they were worth. One newspaper said one object alone was valued at something like a hundred thousand pounds. So they came up with a plan. They gave me the passport of this Harry Miller, who had been a photographer, and told me that I was going to take a job half way round the world in the Near East.'

'You didn't object?' asked a sceptical Davison.

'Of course I did,' he replied. 'I tried to do everything to get out of it. I promised that I would pay the money back in instalments. That I would find a job in America. But they were very insistent. I thought about faking my own death. But I guess they'd come across my type before. They threatened all sorts of things. What they said they would do to Mary and Tabitha – that's my wife and daughter – well, let's just say that I had no choice.'

'So the story you told me about working for the *Middleton Bulletin* was a lie?' I asked.

'Yes, it was, but I had loved photography as a kid. I knew how to handle a camera, I'd done some photography during my work as a printer, so that part wasn't difficult. What was hard was what the Solomons asked me to do.'

'Which was?' I asked.

He looked up at the photographs of the artefacts on the wall. He took another deep breath. 'They wanted me to make copies of certain valuable pieces and ship the real treasures back to New Jersey. And I was to replace the originals with electrotype copies.'

'But how on earth did you do that here, without anyone realising what was going on?' I asked.

'I thought it would be difficult, but it wasn't really,' said Conway. 'I did it in the darkroom. I could work there for hours without being disturbed. I had a lock on the door, which people thought was reasonable because of course I couldn't have anyone just walk into the room in case the exposure to the light ruined the film. And when I was in Baghdad I would ship the artefacts back to America.'

'How many objects did you manage to copy? And which ones?' asked Davison.

'I couldn't get my hands on that scarab or anything similar,' said Conway. 'But I didn't do too badly. A couple of gold necklaces and bracelets, a fine gold bowl and a spectacular dagger.'

'Which would be worth – what?' I asked.

'I suppose altogether between ten or twenty thousand bucks,' he replied.

'And Mr Woolley has never suspected?' asked Davison.

'I don't think so,' said Conway.

'Well, he will have to be informed now,' said Davison. 'And the authorities will have to be made aware of it too.'

'Of course,' he said, looking defeated. 'In a way, it's come as a form of relief. I'd rather fall into the hands of the police than those brothers. The Solomons are incapable of mercy.'

Davison and I looked at one another; perhaps we were thinking along the same lines.

'And what about your camera?' I asked. 'Did someone else really smash it up or—'

'No, it was me,' he said, looking thoroughly ashamed of himself. 'I couldn't risk you seeing what was on the film, you see. I knew, because of your curious nature, that you'd want to see every single negative. And my game would have been up.'

'And that day in Baghdad?' I asked. 'When we first met?'

'I'm afraid that was all my doing too,' he said.

I turned away not so much in anger as in embarrassment and humiliation. To think I had fallen for his deceptive charms so easily. I felt my face flushing.

'But if you'll just let me explain,' he said. 'It wasn't because I was trying to hoodwink you.'

'No, because I certainly do feel—'

'No, it was – well, because I'd spotted you earlier that day at the hotel. I was there visiting a friend when I caught a glimpse of you. I asked at the desk and they told me who you were. I followed you along Rashid Street and – well, I couldn't think of a way of introducing myself without it sounding corny.'

'So you bribed a little Arab boy to snatch my handbag?' I remembered the look of hesitation and unease in that boy's eyes. 'Just so you could pretend to be a hero?'

'You make it sound so sordid,' he said, his voice rising in protestation.

I let his words speak for themselves.

'Look, the reason why I did that – and yes, it was unforgivable of me – was because I'd taken a shine to you. I thought – stupidly – that if I came to your rescue you'd . . . well, that I'd stand a better chance with you.'

The words were difficult to hear, but they must have been a thousand times more difficult to utter.

I was about to ask about Conway's wife, and whether she knew anything about the relationship he had enjoyed with Gertrude Bell. But Davison said, 'Would you excuse us for a moment? There's something I need to discuss with Mrs Christie.'

As we left Conway to contemplate his bleak future he looked like a mere shell of a man, broken and hollow.

'Do you believe his story?' whispered Davison.

'Yes, I do,' I said. 'It was humiliating to listen to.'

'Wasn't it just?' he said. 'And it had that dreadful ring of truth to it. But listen – I'm sure you had the same idea as me? About using Conway in some kind of way?'

'I wouldn't put it quite like that, but yes, I think there are a few things that he might be able to help us with. After all, I think he owes me a favour or two, don't you?'

'Indeed,' said Davison, a mischievous sparkle lighting up his eyes. 'Now what did you have in mind?'

Chapter Twenty-four

After having a long and detailed talk with Mr Conway all three of us walked back into the sitting room as if nothing untoward had happened. Fortunately, Conway was used to playing the part of Harry Miller and we instructed him to carry on doing so. Of course, there was a risk that he might make a run for it, but we doubted that he would. Davison had confiscated his passport and warned him that if he did try and escape then he would make sure that the authorities sought the heaviest sentence possible. The details of how to retrieve the stolen artefacts would have to be worked out later. There were more important things to deal with at the moment: the prevention of another murder.

'There's a sandstorm on its way, I'm afraid,' said Conway to the group comprising Mr Archer, Father Burrows and Miss Jones. 'Looks like a pretty nasty one too.'

'A sandstorm?' asked Mr Archer, standing up from the table. 'But we need to get out of here. We've got to get Sarah to Baghdad. And that boy needs to go into custody.'

'I don't think anyone is going anywhere for the next

day or so,' said Leonard Woolley, as he entered the room. I remembered how Woolley had told me that once, after returning from England to the compound, one part of the house had been covered with sand up to its roof and it had taken his men three days to clear it.

'How bad do you think it's going to be?' I asked.

'I've just been up to the roof and it's sweeping in from the south,' said Woolley. 'We've had them before at this time of year – luckily they're not as bad as the summer storms. It's nothing to worry about as long as we batten down the hatches. But obviously nobody can venture outside.'

'But that's just impossible,' said Archer, his face reddening. 'I told you, Woolley, we need to get my daughter's body to Baghdad. She needs to be buried, you idiot – don't you understand?'

'I'm afraid, Mr Archer, if you go out there, within a few hours you'll find yourselves buried too,' said Woolley, who had clearly lost all patience with his former patron. 'The sands are not discerning – they pay no attention to one's place in the world or how much money you have in the bank.'

'I don't need to stand here and be spoken to like this.' Archer looked at Woolley with contempt and walked away. As he left the room he turned and said, 'I intend to leave this godforsaken place just as soon as I can. There's nothing but evil here, I can see that now. And to think I was going to invest in you. Thank goodness I didn't – it would have been like giving money to the devil himself.'

Miss Jones jumped up from a chair. 'Shall I go and get him, Mr Woolley? Perhaps there's still time to try and salvage something from this.'

'No, let him go,' said Woolley, sitting down at the table and pouring himself a cup of tea.

'But what about our mission here?' she asked. 'Just think about what his funds would do.'

'Damn his money,' said Woolley, before realising what he had said. 'Sorry, ladies.' He took a sip of tea before he sprang up again. 'I can't sit around here, not with this storm coming in. Burrows, Miller, Davison – would you mind coming with me so we can secure everything outside. I've already got some of Hamoudi's men on the job down by the dig, but there are still a few pieces of equipment lying around here and there. Mrs Christie, Miss Jones, would you make sure all the windows in the house are closed and that the shutters have been secured?'

'Of course,' I said. 'But surely you're not going to let Mr Archer venture out there? Not if it's as dangerous as you say?'

Woolley walked over to the door and looked out at the sky, which was turning a sickly mix of sallow ochre, bright orange and dusty brown. 'No, by the time Archer's got his things together, he'll take one step outside and turn right around,' he said.

The men left Miss Jones and me alone in the room together.

'It's quite frightening,' I said. 'The sandstorm, I mean. Have you experienced one before?'

'Oh yes,' she said blithely, as we started to check on the windows and shutters. 'The house feels like it is going to come crumbling down around you. Sand gets everywhere. It makes a terrible racket. But then it passes and life, well, it gets back to normal.'

'But Woolley is right in saying that it would be dangerous to venture outside?'

'Indeed it would,' she said. 'I don't think you'd stand a chance, not against those sands. It's like an enormous tidal wave rolling in, only instead of water there's sand. I think the pressure would crush you – that or you'd die from taking too much sand into your lungs.'

'It sounds terrifying,' I said, drawing a deep breath. 'I can't believe you're not more scared.'

'I was the first time,' she said. 'I thought I was going to die of fright. I went to bed, but of course I couldn't sleep, what with everything rattling around me. It was as if an enormous giant had taken hold of the house and was trying to shake the life out of it. I had to keep singing nursery rhymes all night to try and comfort myself.'

We went around the house, duly securing each of the windows and shutters, until we came to Katharine's room. I knocked gently on the door and a voice told me to enter. Mrs Woolley was sat on a chair by the looking glass massaging some cream into her hands. She barely turned her head towards us as we stepped into the room. Despite the drama of the last few days, when she had been accused of the most terrible crimes, it was obvious that she still regarded herself as the Queen of Ur.

'Would you please pass me that bottle of perfume?' she asked, her hand gesticulating in the direction of the desk. Her haughty tone of voice gave the question the air of a command more than a request. 'It's just there, by my papers.'

By some sheets of what looked like Katharine's novel in

progress there was a beautiful black crystal perfume bottle with an enormous spray.

'Have you heard about the sandstorm?' I asked, as I passed the perfume to her.

'Yes, we do get them from time to time,' she said, spraying herself liberally with the musky aroma. 'Very inconvenient, of course, but nothing to worry about.'

'That's just what I was saying,' said Cynthia.

'Mr Archer seems intent on leaving,' I said, 'even though your husband has warned him against it.'

'I wish that beastly man and his awful wife would go,' said Katharine. 'After what they put me through. I wish they'd step out into the sands and never be seen again.'

'You don't mean that, surely,' said Cynthia.

'Why not?' asked Katharine, turning to us, her eyes blazing. She had that manic look about her again, an unnatural, detached expression that frightened me. She stared down at her hands and seemed to recoil, as if she had seen something that repulsed her.

'We've been told to make sure all the windows are closed and the shutters secure,' I said, hoping to distract her from thoughts of Mr and Mrs Archer. 'Would you mind if I check yours?'

Katharine did not respond, but continued to gaze at her fingers and palms with revulsion. As I walked over to one of the windows that looked out towards the courtyard Cynthia came to join me.

'Do you think she's all right?' she whispered. 'Why is she staring at her hands like that?'

'I have to admit I am concerned for her,' I said quietly.

'As you know, she has a delicate constitution and she may still be in shock.'

'What is that you are saying?' boomed Katharine.

'Nothing,' said Cynthia, turning back towards her. 'We were just talking about the approaching sandstorm and making plans for the clean-up afterwards.' She turned to me and said, 'You wouldn't believe how the sand gets into every nook and cranny in the house. It seems no matter how much you scrub you simply can't get clear of it.'

'The blood,' murmured Katharine. 'I can't wash it off.'

I went to her and took her hand. 'There's nothing there, my dear. Nothing at all. Your hands are clean.'

As I raised them up for her to see, she dashed them down with a fury. 'Get away from me,' she hissed. 'I can still smell it. It turns my stomach. I've got blood on my hands. I'll never be able to get rid of it.' She started to moan, a horrible, low moan that reminded me of the queer chant of the Arab workmen.

'I think we should fetch Mr Woolley, don't you?' said Cynthia.

'Yes, I think that would be a good idea,' I said. I tried to take one of Katharine's hands again, but she looked at me with poison in her eyes. It was obvious that I was causing her a great deal of distress.

'Perhaps it would be better if you went and found him?' suggested Cynthia. 'I can sit with her.'

'Of course,' I said.

Before I left the room I glanced back. What I saw nearly broke my heart. I had thought that Katharine had been making such good progress. But here she was, muttering to herself, looking like an inmate in some foreign asylum being tended to by a kind-hearted nurse.

I opened the door to a howling wind. This was no weather for a hat and so I left it inside. The air was thick with dust and sand and as I walked I had to shield my eyes and mouth with my handkerchief. It took me some time to find Woolley, who was down on a stretch of land, busy overseeing the collection of some odd pieces of equipment and tools. I didn't want to alarm either him or his men so I walked up to him and asked whether I could have a word. At first he could not hear me above the noise of the wind and so I had to shout. He noted the seriousness of my voice and the expression on my face and told the Arabs to carry on with their work while he stepped away from them for a moment.

'What's wrong?' he said.

'It's Katharine,' I said. 'I'm afraid she's having another one of her episodes.'

'A headache?'

'No, something altogether more worrying. She's got this fixation that she – that she can still smell the blood on her hands.'

'Oh no,' said Woolley, sighing. 'I thought, what with Cecil's confession, that she might have put that to the back of her mind.'

It was not yet safe to tell either him or anyone else in the compound the truth of Cecil's guilt or innocence. 'Yes, so did I. But she does seem very on edge. In fact, I fear for her sanity.'

Woolley returned to his men and shouted out instructions. On the way back to the house I told him something of what I had just witnessed, reassuring him that Katharine was being looked after by Miss Jones.

'That's all we need, just as this storm is coming,' he said.

He must have seen the minute changes in my face, a small lift of an eyebrow, a sudden blink of the eyes, because he then added, 'Sorry, I didn't mean for that to sound cruel.'

'It must have been difficult for you,' I said, remembering what Woolley had told me about the state of their marriage.

'It seems the events here have proved too much for her,' he said. 'I should never have placed her in such a vulnerable position. Of course, it's all my fault. I've been selfish, thinking about the contribution she made to the work here. I should finally face up to it. Perhaps it's time for Katharine to be taken somewhere safe, somewhere where she can be looked after properly.'

'You mean—'

But before I could finish my sentence Father Burrows came running out of the house. Panic haunted his eyes.

'Quick, Woolley!' he said breathlessly. 'You've got to come!'

'What's wrong, man?'

'It's Mrs Woolley – she's . . .'

Woolley pushed past him. I followed in his wake, running as quickly as I could across the courtyard towards Katharine's room. The scene that greeted me was even more disturbing than the one I had left. Cynthia Jones, her deathly white face streaked with tears, was already being comforted by Woolley. In the corner of the room, her head slumped forwards as if the life had been drained from her, was Katharine.

'What's wrong, Cynthia?' asked Woolley.

'It was awful, I can't tell you how – how awful it was,' she said, sobbing.

Woolley looked over towards his wife. 'Katharine – do you know anything about what happened?'

But she did not respond. I walked over to Katharine and knelt down besides her.

'Close the shutters,' she mumbled.

'Yes, that's right,' I said gently. 'We closed the shutters because of the storm, don't you remember?'

'Thirsty,' she said. 'Water.' As I poured her a glass I noticed that her pupils seemed dilated. After these few words she retreated back into unresponsiveness.

'I was only gone a few minutes,' I said, walking towards Cynthia. 'What happened?'

She took a few deep breaths and, encouraged by Woolley and me, began to tell us what had occurred.

'I thought that – that she would be a little calmer when you left the room,' she said. 'She was getting agitated by your presence, wasn't she? But, almost as soon as you'd gone she turned to me with that awful look in her eyes. Like – like she wanted to do me some harm, some real harm. I told myself everything would be all right. After all, Mrs Woolley had had these episodes before. But then, just as I was sitting there, trying to comfort her, she grabbed me. You can see how tight was her hold on me.'

Cynthia raised up her right arm to reveal a reddish-purple mark around her wrist and lower part of her arm.

'I tried to free myself, but her grip was too strong. I told her that she was hurting me, but she just carried on looking at me with those eyes – fierce, blazing eyes full of hatred. It was so painful that I started to cry. I heard some voices and called out for help. Thank goodness Father Burrows was nearby. But before he came into the room, Mrs Woolley said to me in a whisper, "Watch out – tonight . . ."'

Here Cynthia burst into tears once more. 'Sorry,' she said, as she tried to compose herself. 'But those words, you understand that they sent a chill straight through me. After everything that's gone on here.'

'What did she say, Cynthia?' I asked.

She hesitated for a moment, looked across at Katharine and took another deep breath. 'She said, "Watch out – tonight you're going to die. Tonight . . . I'm going to kill you."'

Chapter Twenty-five

Woolley led Cynthia, who was now so distressed that she was shaking with shock, out of the room. But then, just as I had given Katharine a glass of water and was guiding her towards her bed, I heard a loud cry. I made sure that she was sitting down and that she could do no harm to herself before I ran towards the source of the noise.

I arrived in the sitting room to see Mrs Archer in a state of hysteria, surrounded by her husband, Lawrence McRae and Father Burrows. After alerting us to what had occurred between Katharine and Cynthia, Burrows, who shrank from arguments, had retreated to the relative calm of the main room, only to be faced with yet more conflict.

'I told you it was her all along,' wailed Mrs Archer. 'If only you had all listened to me.'

'I don't think that's helpful, not in front of Miss Jones,' said Woolley, who stood in the corner of the room by Cynthia.

'Well, I'm afraid it's the truth,' said Ruth.

No doubt Mrs Archer had only recently learnt that she

could not leave the compound. This realisation that she would therefore be unable to take her daughter to Baghdad for burial was bad enough, but then she must have heard how Katharine Woolley had threatened to murder Cynthia Jones. The combination had proved too much for the grieving woman.

'Woolley, you can't make excuses for her any more,' said Mr Archer, as he tried to comfort his wife. 'You heard what Miss Jones just said, that your wife has got it into her head to kill her tonight.'

At this Cynthia started crying again and left the room, no doubt to return to her own quarters.

'Where's Captain Forster?' demanded Mr Archer. 'We need to get him here to sort out this madness.' He turned to find a servant, clapped his hands, but none came running. 'Burrows, would you go and fetch him. Tell him it's an emergency.'

Father Burrows stood there and, like a well-trained dog, looked to his master, Woolley, for guidance. Woolley no longer tried to defend his wife; after all, he had heard the words directly from Cynthia's mouth. 'I think Captain Forster is with Hamoudi,' he said, nodding at Burrows. 'He's trying to make sure everything is secure before the sandstorm arrives.'

'I can't believe we are going to have to stay here,' said Ruth Archer, looking at the sitting room as if it were a prison. 'Not after what happened. Not with her.'

'But you heard Cecil's confession,' said Woolley. 'All of us did.'

'But how do we know Katharine didn't do it all along?'

asked Lawrence McRae 'She could have killed Sarah and then made Cecil take the blame for it. You know how suggestible he is and how forceful she can be.'

'I doubt that very much indeed,' said Woolley, trying to remain calm.

'And where is she now?' asked Mr Archer. 'Your wife, I mean.'

'Mrs Woolley is in her room,' I said. 'I left her resting.'

'You left her by herself?' asked Ruth Archer, looking at me as if I were mad.

'She was lying on the bed, she seemed—'

But before I could finish my sentence, Ruth Archer pushed past me, quickly followed by Mr Archer and Lawrence McRae. There was a dangerous spirit in the air now, as if the threesome were unified by the unruly emotions of grief and anger. They hurried towards Katharine's room.

'Really, I don't think there's any reason for you to behave like this,' said Woolley, as he ran after them.

'I think there's every reason,' said McRae.

'Look here,' cried Woolley. 'You can't go forcing your way into my wife's bedroom. This is completely outrageous. Stop right there. Stop, I say!'

But his orders were ignored. Mrs Archer used her round form to push open the door and a moment later the three of them, shortly followed by Woolley, were standing in the room. They gazed at the bed. Katharine Woolley was exactly where I had left her, asleep. She had her hands crossed on her chest and there was something saint-like about her pose. Perhaps it was this – and the sight of her looking so peaceful – that made them turn away in embarrassment and shame and then walk out of the room. Woolley himself said nothing, but gave them a stern look as they filed past him.

We left Katharine sleeping and returned to join the group in the sitting room. Mr Archer was the first to speak.

'I'm sorry things got a little out of hand there,' he said. 'But, Woolley, you must understand why we feel as we do. Being unable to leave here, our poor daughter in that pantry, lying there like – like some slab of meat.'

'Hubert, no, please don't say that,' said Ruth Archer, her face creasing with grief.

'I'm sorry, dear, but some plain talking is what's needed here,' he said, reddening slightly in the face. He looked at us each in turn. 'Don't you all realise that we're living under the same roof as her killer?'

'Yes, but he's under lock and key,' said Woolley. 'Cecil is not going anywhere. I can assure you that you're safe.'

At this point the door opened and Father Burrows returned with Captain Forster, together with Davison and Conway, all of them looking distinctly weather-beaten. Their faces and heads were covered in a fine dust and Forster asked for a glass of water to clear his throat.

'It's getting quite lively out there,' he said, wiping some sand from the corner of his lips. 'Nearly everything has been moved to safety. The last of the men, including Hamoudi, have gone and we won't see them now until the storm's passed. Now what's this I hear about an emergency?'

Forster listened patiently as Mr Archer explained the situation. As he outlined what had happened – how Katharine had threatened to kill Cynthia Jones later that night – he did not gloss over the fact that he had stormed into Mrs

Woolley's room, accompanied by his wife and Lawrence McRae. None of them came out well from the story, but Forster made no judgement on their behaviour.

'Yes, I can see that you are worried,' said the Captain. 'Especially since we are all going to be cooped up here in the compound tonight. But what we cannot allow to happen, what must be prevented at all costs, is any kind of descent into mindless savagery.'

Various murmurs of encouragement and agreement came from Woolley, Burrows, Davison and Conway, the man everyone still knew as Miller.

'So what I suggest is this,' Forster continued. 'In order for everyone to feel safe I recommend that Cecil McRae and Mrs Woolley remain under lock and key tonight.'

At this, Mr and Mrs Archer and Lawrence McRae started to voice their objections, but Forster succeeded in shouting them down. 'And then, in addition, a man – who will be armed – is going to be stationed outside each of the rooms. Now, who among the men here knows how to handle a gun?'

Woolley, Davison, and McRae raised their hands. 'Thank you for volunteering, but I'm afraid we will have to discount Mr Woolley and Mr McRae as they have what are clearly personal interests in the case,' said Forster. 'So Davison, would you mind sitting outside Mrs Woolley's room? And I will station myself outside Cecil McRae's. Of course, although the courtyard is protected to a certain extent it's still open to the elements and we'll have to take certain precautions. Best to wrap ourselves up in blankets like Arabs, don't you think?'

Although it was clear Mr and Mrs Archer would much rather force both Katharine Woolley and Cecil McRae out of the house so that they could die in the desert storm, they had no choice but to accept Forster's decision.

'For those of you who haven't experienced one of these sandstorms before I would just say this,' continued Forster. 'You should be safe enough crossing the courtyard, but please on no account step outside. Yes, the house may shake and you may think it's about to fall down around you, but it will pass. I suggest that you try to take your mind off it – if you can't sleep, then read, play bridge or Patience, catch up on your correspondence.'

When Forster had finished his little speech the group dispersed into cliques: Mr and Mrs Archer continued to whisper about Mrs Woolley's instability of mind; Lawrence McRae still protested the innocence of his nephew to the man he addressed as Miller; Davison started talking to Forster about their forthcoming duties as sentry guards; Father Burrows questioned Woolley about the safety of the cuneiform tablets and the treasures of the *antika* room and I, unnoticed, slipped away to go and check on Katharine.

I found her still sleeping, her arms crossed. I walked over to her dressing table and selected two jars of cosmetics, as well as the pot of hand cream that she had been using earlier. I placed them carefully into my handbag and shut the door as quietly as I could. I returned to the sitting room where people were discussing how they were going to spend the night.

'You can count us out – we're in no mood for frivolities,' said Mr Archer. 'I trust all will be in order, Captain Forster.'

'Indeed it will,' he said. 'I can guarantee that you've got nothing to worry about.'

'Are you sure you don't want anything to eat before you go to bed?' asked Father Burrows, who, now that the servants had been dismissed, had been given the task of preparing a simple supper.

'No, we're not hungry,' said Mrs Archer. The couple said a plaintive goodnight and retired for the evening.

'Mrs Christie, would you care to join us at cards? Bridge, perhaps?' asked Woolley, who seemed to be in remarkably good spirits considering what he had witnessed at the camp.

'I would enjoy that, but perhaps a little later?' I replied. 'Mr Miller has been promising to show me the basics of film development in the darkroom, and I rather think now might be as good a time as any.'

'Oh yes, of course,' said Conway, acting on his cue. 'There's really nothing to it, once you've got the mix of chemicals right.'

A few eyebrows were raised, as no doubt the group had picked up on the friendly, even at times flirtatious, nature of our relationship. Perhaps they thought we were going into the darkroom to get to know each other a little better. Let them think that if they wanted. No, what we had planned was much more interesting. We were going to find out whether someone was being poisoned.

Chapter Twenty-six

'Is there a light we can use in here?' I asked as we stepped into the pocket of darkness.

'Sure, just give me a minute,' said Conway.

I felt him brush past me and, although I knew he was both a liar and a criminal, I still felt the leap of my heart and the quickening of my breath. Thank goodness the darkness hid my blushes.

'We can use an oil lamp,' he said, as he took out a flash-light to search for a box of matches. He drew out a match and lit one, and a moment later the lamp cast its gentle glow across the small room. 'Now, tell me what you need.' His tone was matter of fact and although he was not rude, there was none of that easy charm that I had come to know and, in some ways, enjoy. No doubt he was embarrassed by his recent exposure and humiliated by the way he had deceived me, but there was something in his manner which made me think I had behaved quite shabbily too. I would never forget that look of disappointment in his eyes when he realised that I had been trying to delay his return to his room.

'So this is where you did – all your work?' I asked.

'Yes, that's right, so we probably have everything you could want for your experiment,' he said, gesturing at a set of shelves on which stood various jars of salts and chemicals, a bunsen burner, a number of trays, a set of pipettes, glass jars, funnels, bottles of water and rubber pipes. Although Conway had kept the room locked, if a stranger had chanced to walk into it then they would assume everything here had something to do with the process of developing camera film rather than the expert copying of ancient treasures.

'When we've got more time I'd like you to show me how you went about it,' I said. 'But I suppose we'd better get down to the business in hand. Let's see. Yes, that bunsen burner, if you could set that up. Firstly, I need to make some fuming nitric oxide.' Fortunately, when I had been doing my nursing training, and in the dispensary in Torquay, I had always rather enjoyed the study of chemistry. All those lovely formulae, the way the elements could be combined to make something new; then there was the beautiful order of the periodic table. 'I need some potassium nitrate and a little sulphuric acid – you have those? And a couple of flasks.'

Conway scanned the shelves and placed everything on the table in front of me. I got to work, mixing the various chemicals until the ghostly form of the gas rose up from the beaker.

'Now, I need some potassium hydroxide, a three per cent solution in methanol,' I said.

Perhaps my knowledge of the wonders of the scientific

world astounded Conway because, as I busied myself about the makeshift laboratory, working with the various chemicals and flasks, he looked at me as if he were seeing me in a new light. I was no longer the helpless woman he had rescued from the backstreets of Baghdad, a lady grateful for a gentleman's attention, but someone with a surprising skill. Little did he know of the true extent of my mastery of poisons; and I decided that it would be best to keep this to myself for the present time.

'Yes, that's working very nicely,' I said. 'All this is in preparation for the real test, of course. Have you heard of the Vitali colour reaction?'

I was met by a blank stare. 'Oh, a very interesting experiment and so very pleasing when it works,' I said. I stood back and checked that everything I needed was laid out before me. It was, in a way, a little like cooking. When one was making a roast dinner there was little point in leaving everything – the roast potatoes, the joint, the bread sauce, the horseradish – to the last minute; the secret, as every good housewife knew, was in the preparation. So too with chemical experiments. Yes, everything was in order: the boiling water, various flasks containing the key ingredients, and all the equipment I would need. I took out Katharine's cosmetics from my handbag and spooned some of the pale cream into a flask. I then treated this with the fuming nitric acid, which was evaporated to dryness on the water bath. To this residue I added a small amount of the potassium hydroxide in methanol solution. As I did so the colour of the cream started to change from white to bright purple and then red, before fading to colourlessness.

'Yes, very interesting,' I said, as I watched the colours change.

'What's happening?' asked Conway.

'It's a positive result, just as I expected,' I said.

'Of what?'

'Of hyoscyamine.'

'Hyo . . . ?'

'It's a tropane alkaloid,' I said. 'Also known as daturine, found in the plants of the Solanaceae family.'

'Sorry, I don't understand.'

'Plants such as henbane, mandrake, jimsonweed and deadly nightshade. Of course, it has its uses, but it is also very dangerous. Hyoscyamine can cause a dry mouth, blurred vision and eye pain, dilated pupils, dizziness, restlessness, nausea, headaches, euphoria, short-term memory loss and, not least, hallucinations.'

It took a moment or two before Conway began to understand the implications of my little experiment. 'So you think that—'

'I don't think anything,' I said. 'Now, without a doubt, I have proof. Proof that Katharine Woolley is not mad. Her headaches were far from psychosomatic, as some people in the compound believe. Her visions were not those of someone who was deranged or mentally ill. Mrs Woolley has been suffering in the most awful way possible. She was being poisoned.'

'But how? Who by?'

'I cannot for certain answer your second question, but I do have an answer to your first one,' I said, as I spooned out a dollop of mixture from another of Katharine's jars so as to repeat the test. 'Katharine Woolley was being poisoned by means of her face and hand creams. Each day, each night, she would anoint herself in the belief that the creams would help moisturise her skin in this dry climate, the potions going some way to help keep her looking fresh and youthful. But, unknown to her, the creams were driving her to the point of madness. More importantly, the poison that slowly seeped into her system affected her behaviour in such a way as to give rise to the generally accepted notion that she was mad.'

'You mean someone deliberately put this substance into her night cream?' asked Conway. 'Into her hand cream?'

'Yes, and very effective it was too,' I said. 'Someone wanted her to appear as though she were insane.'

'So that's why she said that thing about threatening to kill Miss Jones,' said Conway. 'It was a hallucination brought on by that poison. But who would do that? Who would want to put poison in Katharine's face cream?'

I left the question unanswered and, instead, concentrated on the experiment. As I went through the steps of the procedure again – adding the fuming nitric acid to the cream, then the potassium hydroxide in methanol – I saw the mixture change its hue. For a brief moment, it turned bright red, a shade that reminded me of the colour of freshly spilled blood.

Chapter Twenty-seven

We returned to the sitting room to find Cynthia Jones setting the table for dinner. It was going to be rather a makeshift affair, she said; Father Burrows had told her that he was going to reheat some leftovers from lunch, together with a selection of cold dishes. She informed us that Davison and Forster had been given cups of tea before they had taken their respective positions outside the doors of Katharine Woolley and Cecil McRae.

'They were both adamant that they didn't want anything more than that,' she said. 'But I'll make sure to take them a little something else later, at least a little bread and cheese. It's going to be a long night for them.'

As Cynthia placed some knives and forks down on the table I noticed that her hands were shaking.

'You've no need to feel afraid,' I said.

'Afraid?'

'I can tell you're trying to put a brave face on things,' I replied. 'But I doubt Mrs Woolley knew what she was saying earlier.'

'I have been trying to put it out of my mind, but I must admit it hasn't been easy,' she said. 'It's good to know that there's someone stationed outside her door.'

Conway – who was pouring himself a generous measure of whisky topped with soda water – asked her if she would like a drink, but Cynthia declined.

'I'll stick to my water,' she said.

The table had been set for six and, over the course of the next few minutes, we were joined by Lawrence McRae, Leonard Woolley and Father Burrows, who came out of the kitchen bearing a tray of food.

'I can't make any great claims for the quality of the meal, I'm afraid,' said Burrows, as he placed the dishes on the table. 'It's all been flung together at the last minute.'

'We're very grateful for your efforts,' said Woolley. 'So what do we have?'

Father Burrows guided us through the plates of vine leaves, creamed aubergine, which he warned us had been made with a great deal of garlic, a hot pepper spread, a chick-pea salad, a minced meat stew, a plate of figs, and something that sat in a thick, congealed tomato sauce that I didn't quite catch. 'I suppose the Arabs have been eating like this for centuries,' he said. 'One only has to look at the cuneiform tablets to see that. I've studied one tablet which gives two dozen recipes for a type of stew cooked with vegetables and meat, and of course they've always liked their spices and garlic too.'

'It's certainly an acquired taste,' said Woolley, as he dipped a little of the bread into the garlicky aubergine dish. 'And it may be too much for some of the ladies here. Having said that, my wife has a fondness for strong-tasting Arabic cuisine, as did Miss Bell.'

At the mention of Gertrude Bell's name the table fell quiet and Conway looked down at his plate. Perhaps sensing that a spirit of melancholy was about to descend on the group, Woolley decided to lighten the mood by expanding the story.

'I remember on one occasion when Miss Bell was here, she polished off one particular dish that was so hot and spicy it brought tears to the men's eyes,' he said, smiling. 'We were all calling out for glasses of water after only a couple of mouthfuls, but she had a stomach like an Arab and she finished it off as if it were a dollop of mashed potatoes or custard.'

The statement presented me with the opportunity to ask a few questions. But I knew I would have to be very careful. 'I remember, many years ago, soon after I arrived in Cairo, falling very ill after sampling one of the local dishes,' I said. 'But I was very young and foolish. I could have done with someone like Miss Bell to show me the ropes.'

'Yes, she would have liked you,' said Woolley. 'Don't you agree, Miss Jones?'

'Oh yes, I'm sure she would,' said Cynthia, spooning a little of the minced meat stew on to her plate. 'Most women she regarded as – what was it? That's right – she called them "dull dogs".'

'I'm afraid I see myself as very dull indeed,' I said. 'Not like Miss Bell. So very clever.'

'She was extraordinary in so many ways,' said Woolley. 'Did you meet in Cairo?' I asked.

'Cairo?' Woolley looked askance.

I knew both he and Miss Bell had worked in intelligence in Egypt during the war – as had Katharine Woolley's first husband who had died in Cairo in 1919 – but I could not reveal this.

'I can't remember exactly how I first met her,' he said. 'In fact, I think I might have read her work, before I met her,' he said. He walked over to a bookcase in the corner of the room and ran his hands over the volumes before he found the one he was looking for. 'Yes, here it is,' he said, blowing some dust off its cover as he extracted it. *Amurath to Amurath*, published in, let's see, yes in 1911. Have you read it?'

'No, I'm ashamed to say I haven't,' I said.

'Listen to this from its dedicatory epistle,' he said. '"The banks of the Euphrates echo with ghostly alarums; the Mesopotamian deserts are full of the rumour of phantom armies; you will not blame me if I passed among them *trattando l'ombre come cosa salde*."'

'Treating the shadows as the solid thing,' said Father Burrows. '*Purgatorio*, if I'm not mistaken.'

'I see you know your Dante,' said Woolley.

There was something to be said for Miss Bell's words; indeed, there was something about the environment, this dry desert full of mirages, half-truths and buried secrets, that called out for such a method. However, it was just as important, I thought, to treat solid things – so-called facts and seeming certainties – as if they too were as insubstantial as shadows. Nothing was what it seemed here.

'Miss Bell once said to me something I've never

forgotten,' continued Woolley. 'We were at the end of a long day. We'd been digging in the heat of the sun. We walked up to the top of the ziggurat and looked out at the desert that stretched before us. She turned to me and said, "I wonder, are we the same people when our surroundings, friends and associations are changed?" I thought it a very interesting question, one I could not answer easily.'

Just then a blast of wind hit the house, shaking it to its very foundations. From somewhere outside, the courtyard perhaps, came the sound of a pot or a tile hitting the ground and smashing into fragments.

The noise made Cynthia Jones jump up from her chair, displacing a glass of water by her side as she did so. 'Oh, look – sorry,' she said. 'I'm so terribly clumsy.'

'Here, let me,' I said, taking up a napkin and dabbing the water from the table.

'Thank you,' she said. 'I don't know what's come over me. It's not as though I haven't experienced a sandstorm before.' She stared down at the food on her plate, most of which she had not touched. 'Would you mind awfully if I retired for the night? I'm not much company, I'm afraid.'

'Of course not,' said Woolley, standing up. 'It's been a long day for all of us. Now, will you need anything to help you sleep?' He looked at me for a response.

'Yes, I have a sleeping draught that you can take.' As I said this another gust of wind and sand buffeted the compound with such force that the glasses and cutlery on the table vibrated.

'Thank you, but I think I'll be all right,' she said. 'I'll just take some water with me.'

After Cynthia had said goodnight, talk returned to Woolley's question about identity.

Father Burrows believed that personality was something innate, something divine that one was born with, while McRae argued that surely it was formed by one's experiences. I'm afraid I sat on the fence rather and observed that it must be a combination of the two. Conway, who had remained quiet up to this moment, pointed out that the real issue at stake was not the source of one's personality, but the extent to which it fluctuated when one was wrenched out of one's normal environment.

'I'm sure all of us felt it when we first came here,' he said. 'Looking out at the stretch of desert sands. Feeling the dry air on our faces. Hearing that incessant chant of the Arab workers as they dig up the ground.' He talked with a passion which I knew came from personal insight. 'It's enough to drive even the sanest of men a little crazy.'

'Interesting you should say that, Mr Miller,' said Father Burrows, pushing his wire-rimmed spectacles back onto the bridge of his nose. 'I see it almost as a test, as though each of us were feeling a little of Jesus's experience when he spent forty days and forty nights in the wilderness. I always think it would have looked something like this, like the desert.'

Again the wind howled outside, shaking the house with an almighty force. 'It's like God is trying to talk to us tonight, don't you think?' Burrows asked in a voice that rang with a false brightness.

'If God does exist, which I seriously doubt, he's certainly not here,' said McRae, as he got up from the table and helped himself to a large measure of whisky.

'Now, now, McRae,' said Woolley, who went to join the architect by the drinks table. 'There's no point in letting gloom set in. Certainly, my father would never have allowed his spirits to sink in such a situation. He was a character, I can tell you.'

Woolley started to tell a series of amusing stories about his father, a vicar first in Clapton and then in Bethnal Green, a man who had a passion for collecting porcelain and pottery, Minton and Copeland and so on. He described to us his big black beard and dark eyes, and recalled how he had always insisted on taking a cold bath each morning. In the winter, Woolley said, his father would fill the bath the night before so that he could guarantee that it would have formed a crust of ice. 'Discomfort was almost holy to him,' he laughed.

Leonard was one of eleven children, he explained, and as a child he had been teased by his schoolmates that he was one of the flock of 'Woolly sheep'.

'But then, after a little reading, I discovered that the name Woolley actually came from "Wolf's Lea",' he said. 'And one day, in front of the group of boys, I proudly announced that my brothers and sisters were not sheep, but wolves in sheep's clothing!'

The comment made the group laugh and Miller, McRae and Burrows each revealed the names they had been called as children.

'And what of you, Mrs Christie?' asked Woolley. 'What were you called as a girl?'

I hesitated a moment – I never liked to reveal too much about myself – before I relented. 'Well, my brother, Monty, used to call me "Kid" when he was being nice and "Scrawny Chicken" when he wanted to upset me,' I said. 'Of course, now I would do almost anything to earn the same epithet.' The comment drew a laugh. 'But I suppose we all had funny nicknames. For example, my sister Madge, or Margaret, we called "Punkie". And my daughter, Rosalind, is sometimes known as "Teddy" or "the Tadpole".'

As I thought of the nicknames I had chosen for some of the characters in my books something stirred at the back of my mind. It had been such a trivial, silly exchange, but I was sure something of import had been said. Silently, I ran through the conversation. A nickname. A term of endearment. A sobriquet. The question that Miss Bell had posed to Woolley, and which he had related to us, about whether one was the same person when taken out of one's normal environment. *A sheep in wolf's clothing. A wolf in sheep's clothing.*

Another wall of wind hit the house at full force.

'Mrs Christie, are you all right?' asked Woolley. 'You're looking a little pale.'

I did my best to smile. 'Yes, just a little tired, I'm afraid.' I stood up from the table. 'Would you mind if I went to bed too? I think I'd like to try to sleep before the storm gets any worse.'

'Let me accompany you to your room,' said Woolley.

'No, I'm sure I can manage,' I said.

'But even negotiating the courtyard might be difficult in this weather,' he replied. 'After all, we wouldn't want any more accidents.'

'No, I'm quite confident, but – but thank you all the same,' I said.

Instead of going straight to my room I went to see Davison, who was stationed outside Katharine's quarters. As I stepped into the courtyard I felt myself being sucked into a whirlwind of sand. Particles of grit covered my face and I felt the dry, deathly taste of the desert on my lips. I closed my eyes and felt my way forwards, using the edge of the wall as my guide. Just as I thought I was nearing the room I felt something by my feet. I opened my eyes, but it was too late. As I fell I stretched out my hands and readied myself for the inevitable pain. I tried to imagine that I was surfing – oh, the fun I had had surfing during that glorious tour around the world I had made with Archie back in 1922! – and that I was just crashing into the sea. I think the thought helped as I landed quite lightly, with only a slight crush of palm against stone and a slightly grazed knee. It had been a tin bucket that had brought me down.

I sat there for a moment, my hair whipping about my face in a wild frenzy, the air thick with dust and sand, and realised that I was afraid. A deep sense of unease crept over me. I felt sick to my stomach. I began to understand some of the evil that pervaded Ur. Despite the fact that Davison was here, and that, if I so wished, I could spend the night making light conversation or playing cards with the people I had just left, I still felt desperately alone and exposed. A memory came to me then of a teacher I had had when I was young, at school in Torquay. During one of the lessons the teacher, whose name I could not recall, told us that, at some point in our lives, all of us would experience a crushing sense of despair, a feeling that there was no hope. She told us the story of Jesus in the Garden of Gethsemane and went on to talk about suffering and the importance of faith. I had to remember this. Without hope there was nothing. And yet, I felt frightened, so terribly afraid.

I pushed myself up and made my way towards Davison. Not only was I a stranger in a desert land, but there was a murderer very near. Someone who I was sure would want to kill again.

Chapter Twenty-eight

After a brief conversation with Davison, I made my way back to my room and despite my best efforts to stay alert, I felt myself drifting off. I couldn't be certain how long I had been asleep, but the sound of a scream woke me with a jolt. I pushed myself off the bed, grabbed a scarf, wrapped it around my head and ran into the courtyard. Visibility was next to nothing – all I could see was a whirling dervish of sand that danced in the air as if it were a malevolent spirit from the *Arabian Nights*.

I tried to shout, but the sand in the air reduced my voice to a rasp. 'Davison? Forster?'

There was no answer. Using strength I did not know I possessed I moved forwards, battling against the power of the wind. Above and around me echoed the terrible sound of the sandstorm, a noise that suggested that the vengeance of the heavens was being visited on the earth. I listened out for a scream. There was nothing. Had I been dreaming? Or had I mistaken the high-pitched whine of the wind for a cry ?

Then the scream came again, a sound that was unmistakably that of a woman. Once more using my hands to guide me, I edged my way around the courtyard until I came to its source.

'Help me! Oh, help!' cried the voice from inside.

Blinking through the sand that stung my eyes I saw the vague outline of a window. But the shutters were closed and I couldn't see inside the room. My fingers felt their way down and across against the mud-brick wall until I found the handle to the door. I grasped it and turned it, but it was locked.

'What's happened?' I cried. 'Are you all right?' But there was no response. 'Who's in there?'

I banged on the door, but this only prompted more unintelligible screams. I rattled the handle and pounded on the wood.

'Let me in!' I shouted.

'Please get help!' came the voice from the other side of the door.

I turned and edged my way around the courtyard until I came to what I thought was the main door into the house. I tried to open it, but it too was locked. Using my fist I knocked on the wood with all my strength and a few seconds later I heard the sound of footsteps and then the shifting of what sounded like a chair. The handle turned and a moment later I saw a sliver of light illuminate the inside of the frame. I pushed my way in.

'What the devil . . . ?' said McRae, as I fell into the room.

'There's something wrong,' I said, wiping the sand from my mouth. I felt like I wanted to gag, but it was important to get this out. 'There's a woman – in there, in one of the bedrooms. She says she needs help.'

'I told you to keep that door shut,' said Woolley, making his way through the living room towards us. 'What's going on?'

'It's Mrs Christie,' said McRae. 'She's saying something's the matter.'

'What?' asked Woolley.

'In her – her room,' I gasped. 'She can't get out.'

'Calm down,' said Woolley. 'Take a deep breath.'

'It's not Katharine, is it?'

'I lost all sense of whose room it was,' I said. 'She's in trouble. She says she needs—'

'McRae, Miller, come with me,' Woolley shouted. 'Bring those flashlights. Mrs Christie, you stay here with Burrows.'

I ignored them and rushed out into the courtyard followed by the plaintive cries of Father Burrows pleading with me to take shelter inside. Realising that I was not going to change my mind, the priest shut the door to the house.

'Hello!' Woolley shouted into the storm. 'Is someone hurt or in danger?'

He edged his way around the courtyard repeating the question. I followed behind the three men until we began to hear cries.

'It's coming from in there,' said McRae, who seized the handle of the door to find it was still locked. 'Can you get to the door?' he called to the person inside.

But his question was only met by another terrified cry.

'We're going to have to break it down,' he said.

Quickly, the men organised themselves. 'Stand back,' shouted Woolley. 'Are you ready?' he asked.

The three men seemed to come together to work as one and used all their strength to run at the door. I heard the splinter of wood and a cry of pain from one of the men.

'Nearly there, but not quite,' said the photographer.

'Let's give it another try,' said Woolley.

They stood back, regrouped and rushed towards the door once more. I heard something crack, the lock broke and, in an instant, they crashed into the room. A wave of sand followed them, making it hard to see inside. The wind unsettled a stack of paper and unused envelopes, sending them into a flurry of activity; the power of the storm even sent a pile of books crashing to the floor.

'Oh dear God,' said Woolley, as the beams from the flashlights began to illuminate pockets of the room.

'What? What is it?' asked McRae.

I stepped into the room and closed the door behind me. The men were gathered around the bed. I pushed past them to see Miss Jones, holding a glass, her face contorted in pain.

'What's happened?' I asked. 'Cynthia. Can you talk?'

She looked down at the bed with horror. A swirl of white fumes rose up from the bed sheet. As I bent down I noticed that there was a sharp smell that pinched my nose. On closer inspection I noticed that the patch by Cynthia's side seemed to bubble as if part of the mattress was beginning to dissolve.

'It looks – and smells – like hydrochloric acid,' I said. I remembered Woolley telling me that the archaeologists used the chemical to clean the cuneiform tablets.

'What?' said Woolley. 'Miss Jones – are you hurt at all?'

'The water – I went to take a drink,' she said, stumbling over her words. 'It was dark. I reached out for my glass by the bed, but . . .'

She started to shift her position, moving her legs dangerously close to the patch of bed that continued to fizz and dissolve.

'Don't move,' said Woolley. 'Cynthia – stay exactly where you are.'

'What?' she said, unable or unwilling to take in the horror of it all.

'It's dangerous,' said Woolley. 'There's acid on the bed. My God, look, it's eating away at the fabric.'

Cynthia looked down at the sizzling mass beside her and then up at her glass. 'I woke up with a dry mouth,' she said. 'I was about to take a drink and then I think the sound of the wind outside gave me a shock and I must have spilled it. I heard a fizzing noise. I felt something on my leg . . . and then pain. I managed to light a candle by my bed. That's when I saw this . . . that's when I started screaming.'

I looked down to see a circular patch of redness above her left knee. It looked as though the top layer of skin had been eaten away by a splatter of acid.

'We need to see to that straight away,' I said, finding myself slipping into nursing talk. 'Can you feel pain anywhere else?'

Cynthia shook her head.

'Miller, can you fetch me a pan of water and some clean towels?' Just as he was about to leave the room I called out, 'And I don't suppose you have any ice?'

'I'm afraid not,' said Woolley.

'Never mind,' I said. 'We can work with what we've got.'

As I was studying the burn I heard Cynthia's breathing become more erratic. Her face was white and she looked around the room with panic in her eyes.

'What would have happened if I had taken a drink?' she asked. She looked at the glass again.

'Try not to dwell on that,' said Woolley. 'Here, let me take that from you.' He took a pillow from the bed, stripped it and used the pillowcase to wrap around his hand and protect his fingers. 'There we are,' he said as he reached out and gently grasped the glass. 'Let's place it over on the bedside table.'

'To think that I could have swallowed it,' she said. 'How I would have suffered. With that – that acid burning its way through me.' Cynthia burst into tears and covered her face with her hands.

McRae leant down and studied the glass. 'The question is: how did the acid get into the glass and who put it there? Did you see anyone come into your bedroom?'

Cynthia looked confused, uncertain about what she had just been asked.

'What if . . . ?' said McRae to himself before he ran out of the room with no explanation.

'I nearly came head to head with McRae,' said the photographer, as he returned with a bowl of water and a pile of towels. 'What's got into him?'

'We can't worry about that now,' I said. I asked Cynthia to move to the other side of the bed, away from the nasty patch of acid on the mattress. 'Please, pass me the water and the towels.' I placed a hand on Cynthia's arm. 'Now, this is going to hurt so you will have to be brave,' I said. 'Can you do that for me?'

Cynthia nodded her head as she tried to wipe away her tears.

'I'm going to clean the affected skin,' I said, taking hold of a towel and dipping it in water.

I squeezed the towel over the acid burn so as to let some water drop onto the skin. Cynthia winced and automatically moved her leg away from me. As I began to dab and wash the site she clasped the bed sheets with her hands and gritted her teeth with a certain resolve.

'That's right,' I said. 'You're doing very well.'

'I heard the knocking and the banging at the door, but I couldn't get out of bed,' she said. 'I know I should have got up and let you in, but I felt incapable of moving. I felt so afraid.'

'Don't worry about that now,' I said, as I continued to clean the wound. 'You were probably in shock.'

I felt Cynthia's eyes on me as I dripped some more water onto her leg. I sensed that there was something she wanted to ask me. 'This won't take much longer, I promise,' I added.

Her eyes darted back and forth and she took a few deep breaths before she finally readied herself to speak. 'Do you think this was Mrs Woolley's doing?' she whispered. 'After all, that's what she threatened to do. But I don't understand. She was being watched, wasn't she?'

At that moment McRae entered the room shouting.

'Davison's out cold,' he cried. 'He was supposed to be on guard. And Mrs Woolley's door's unlocked.'

'What?' asked Woolley.

'He's still breathing, but I don't know whether he's been knocked out or drugged or what,' said McRae.

'Have you checked on Katharine to make sure she is all right?' asked Woolley.

'No, but I—'

Woolley cut him off. 'Stay here and look after the women,' he said, pushing past the architect. 'I'll go and see what's happening. What about Forster?'

'Yes, he's all right,' said McRae. 'At least I think he is. I called out to him and I heard him answer.'

'Why didn't he come earlier?' I asked. 'When I was calling?'

'Perhaps he fell asleep,' replied McRae. 'That or he didn't hear you because of the noise of the wind.'

With Woolley out of the room, McRae began to tell us his hypothesis. It was obvious who was behind all of this, he said. There was only one name under suspicion: that of Mrs Woolley. Had Katharine drugged Davison at some point that evening? Could she have persuaded him to open the door and then slipped something into his tea? Or had she simply hit him over the head?

'She would stop at nothing, of that I'm certain,' he said. 'With Davison unconscious all she needed to do was walk across the courtyard. Nobody would have seen or heard her what with the sandstorm raging. Then she stole into Miss Jones's room, who was sleeping. She simply replaced Cynthia's glass of water with one containing the acid and then slipped back to her own room.'

'I don't understand,' said Cynthia. 'What about the door? I locked it before I went to bed and it was still locked when you broke in.'

'Perhaps she has a spare key,' said McRae. 'Which she used to open the door and then lock it again from the outside.'

'But why – why would she want to do this to me?' asked the terrified woman.

'Maybe you saw something?' asked McRae. 'Or perhaps you overheard a snatch of conversation that incriminated Mrs Woolley in some way? Something that connected her to the murder of Miss Archer.'

'I don't know,' she said, as she started to cry once more. 'I'm so terribly confused.'

'Think!' insisted McRae. 'You must have seen something. Even if it seems inconsequential, trivial.'

'I don't know ... I , . .'

'Look, Woolley will be back soon and you may not get another chance,' said McRae. 'You know what he's like.'

'But surely not Mrs Woolley? We've always been friends, she would never do anything to hurt me,' said Cynthia.

'You may like to think that, but she's obviously lost her reason,' said McRae. 'What other explanation could there be? I'm certain that Cecil would never have hurt Sarah Archer. And if he didn't do it then who did? The answer is obvious, isn't it?'

'But what about the boy's confession?' asked the photographer.

'Just a smokescreen,' said McCrae in a dismissive manner. 'That, or as I've said before, it was forced out of him by that fool Forster. No, as far as I see it, the killer has to be—'

'You were saying, McRae?' said Woolley, stepping into the room. 'Who is the killer?'

McRae did not even pretend to cover up what he had been saying. Instead, he walked towards Woolley and faced him head on.

'Your wife!' shouted McRae. 'That's who.'

'We've been through all this before,' said Woolley. 'The accusation is ridiculous.'

'Can't you see what is right before your nose?' rallied McRae. 'She herself said as much earlier. She told Miss Jones that she would try to kill her and lo and behold, what happens?' He turned to each of us in turn, pausing for extra effect almost as if he were an actor on a stage. 'The poor woman wakes up with a glass of hydrochloric acid by her bed!'

'I'm sure there is some explanation for this,' said Woolley. 'Let's see . . . Perhaps, Miss Jones, you, you picked up the acid by mistake from one of the workrooms where the cleaning of the cuneiform tablets was taking place?'

Cynthia look bewildered. 'No, I don't think so,' she said.

'Or you . . . perhaps you experienced an episode of sleep-walking?' asked Woolley. 'I've heard it's more common than people think.'

'Really?' replied McRae. 'Is that the best you can do? It's pathetic.'

'I will not have you speak to me like that,' said Woolley, reddening in the face. 'Do you hear me?'

'What are you going to do about it?' McRae took another step closer to him. 'What?' he shouted, his spittle spraying the older man's face. 'Are you going to get your wife to try to kill me too?'

At this, Woolley clenched his fist and readied himself to

strike. I could bear it no longer. 'If fight you must, can you please take yourselves elsewhere,' I said.

The sharp remark brought both men to their senses. 'Of course,' mumbled Woolley. McRae did not respond, but turned away from the group in anger and no doubt embarrassment too.

'How was Katharine?' I asked.

'She was sleeping,' said Woolley. 'Looks like she's been in bed for hours. But it's true, her door was unlocked.'

My second question was about Davison.

'Yes, out like a light,' he said. 'Slumped in the chair outside her room. Mighty queer business. I can't make head or tail of it.' Bafflement was written across his face. 'I think it's best if I talk to Forster and tell him ... well, tell him about Davison and also what's gone on here.'

'And what about your wife?' asked McRae. 'You can't just leave the room unguarded.' Although his tone was more polite now, it was obvious that the architect still had his misgivings about Mrs Woolley. 'If nobody else is prepared to go and sit outside her room I will do it myself.' He walked to the door and, just before he left the room, he turned back to face us and said, 'I don't know about all of you, but I have no wish to be murdered in my bed.'

Chapter Twenty-nine

Just as the storm began to ease outside so the pressure inside the house intensified. As the windows stopped rattling and the force of the wind lessened, the atmosphere within the compound became even more tense.

Before he was due to take his place outside Katharine's door, McRae raised Mr and Mrs Archer from their beds and informed them that there had been an attempt on Cynthia Jones's life. Someone had swapped the secretary's customary night-time beaker of water for a glass of hydrochloric acid. Miss Jones was now in her room, where she was recovering from a minor burn to her leg and the impact of the shock. He also told them that Davison, the man in charge of guarding Mrs Woolley, had been rendered unconscious and that he had been carried back to his room. He pointed the finger of suspicion at the woman they had originally blamed for the death of their daughter: none other than the clearly deranged Katharine Woolley.

In the meantime, Woolley had given the facts of that

night's events to Forster. Although he tried to remain impartial and pronounced that it was important to wait for the appropriate evidence to come to light, it seemed that Forster too suspected Katharine Woolley. McRae appealed to the Captain to let him see his nephew. But the policeman was adamant. Nothing would be decided that night, he said. Instead, he thought it best if he took both Cecil McRae and Katharine Woolley back to Baghdad the next day for questioning. The storm should have cleared by first light and they would set out, together with the Archers and their daughter's body, in the morning. He guaranteed that he would get to the bottom of it all. He asked everyone to keep calm. It would be best, he said, if we all retired to our beds and tried to get a good night's sleep before we all met the next day. He suggested we could pull together to help clear up after the storm and then we could meet for breakfast. After that he would set out for Baghdad, where he was certain that justice would be done.

I had my doubts. Not only was there a very real possibility that the rule of law was in danger of being corrupted, but an innocent person would suffer in the most appalling way possible. If I did nothing I knew what would happen.

It was time for me to act.

I pretended to say goodnight, but instead of going to bed I slipped into Davison's room.

'Davison, it's me, Agatha,' I whispered. 'It's safe.'

I saw his body stir on the bed and a moment later he sat up.

'What's happening?' he asked.

'Just what we thought would happen,' I said. 'The Archers and McRae are baying for blood. Katharine Woolley is one step away from being hanged, or whatever barbaric form of punishment they choose to inflict upon her out here.'

I walked over to his bedside table and used a match to light a candle. 'Did you enjoy your beauty sleep?' I asked in a deliberately mocking tone.

'Pretending to be asleep is actually strangely exhausting,' he said, swinging his legs over the edge of the bed. 'And I must say having to be carried around like a sack of potatoes is most undignified.'

'Did you hear anything?'

'Nothing of any import,' he said. 'Just the usual grumblings about Woolley and his wife. Now, what's the next step?'

'We haven't got much time,' I said. 'What with the storm beginning to ease off, Forster has said that he plans to set off for Baghdad in the morning.'

'Do you think anyone suspects?'

'No, I don't think so,' I said. 'But maybe that's a bad thing.'

'What do you mean?'

'We know what's going on, but we still need proof,' I said.

Davison looked at me with a concerned expression. 'You're not proposing to . . . ?' He hesitated. He knew from past experience how far I would go in the search for truth. 'Please, Agatha. You're beginning to worry me.'

'It is a risk,' I said. 'But I cannot for the life of me think of another way.'

'What?' he hissed. 'Tell me what you've got in mind.'

Slowly, and as calmly as I could, I outlined my plan to catch the killer.

Chapter Thirty

During the night, as the scream of the storm reduced to a howl and then a mere whisper, we finalised and finessed the scheme. Just before dawn I crept out of Davison's room and retreated back to my own, where I waited for first light when everything would be exposed.

At dawn I made myself a cup of tea and stepped outside. The early morning sun caressed the desert with a delicate beauty that nearly brought tears to my eyes. I watched the ever-shifting sands turn from purple through violet and rose to ferric yellow, and finally towards something that approached a delicate shade of apricot. I breathed in the clean desert air and steeled myself for what was to come.

I listened as the house came to life and Woolley began to direct a superficial clear-up after the storm.

'I should think the servants, together with the rest of the workmen, will return from the village in a couple of hours,' he said. 'Hamoudi will tell them when he thinks it's safe to make the trek across the desert. After one of these storms one always has to be mindful of sand rivers and sinking sands. It can be quite perilous for those who don't know how to navigate the territory.'

Conway began to try to sweep up great piles of sand that had amassed in the corners of the courtyard. After making sure that Katharine could not escape, McRae left his position outside her room to climb onto the roof to fix some tiles that had come loose. Father Burrows had been given the job of preparing a substantial breakfast for all of us and from inside the compound came the delicious smell of baking bread and the cooking of fatty meat.

By seven o'clock everyone had started to drift into the main room for breakfast. When Davison appeared he was questioned about the events of the night before. What could he remember? Had someone bashed him over the head? Had he ingested something – perhaps a tincture or a tea laced with some drug – that had made him feel drowsy?

In response to the enquiries he said, 'I'm sorry, but I cannot for the life of me remember anything about it. I woke up this morning feeling a little groggy.'

'Perhaps you had too much sun,' said Father Burrows, as he placed a pot of coffee on the table. 'I did notice that you went out without your hat at one point yesterday.'

'Yes, that could well have been it,' said Davison.

It was then that Hamoudi knocked on the door. He said a string of sentences in Arabic which I think were related to the storm and then took out a letter and gave it to Davison. Apparently, it had been sent express from London, but had been delayed due to the weather. My friend did not open the message, but made his excuses and said he was still feeling a little off-colour. He retreated to his room, where I joined him a few minutes later. Both of us were a trifle nervous of reading the contents of the letter. We had been waiting for some information to come from London. Would the missive prove our theory or reveal that we were completely off the mark?

'Why don't you open it?' asked Davison.

'No, it's addressed to you, it could be anything – something personal.'

'I doubt it . . . here,' he said, passing the letter to me. 'It looks like it's from the department.'

I carefully opened the envelope and took out some documents, together with a photograph. My heart did not flutter so much as almost stop when I saw what lay before me. Here was the link to the past that had been eluding me.

On my way back to the sitting room I encountered Captain Forster in the courtyard and asked him whether it was possible to have a quiet word.

'I wonder if you'd mind bringing in Mr and Mrs Archer, Mrs Woolley and the young McRae,' I asked.

'I'm not sure about the Archers – they may have chosen not to come to breakfast after everything that's happened – but as far as the other two are concerned they are still held under lock and key,' he replied.

'Yes, I understand, but it is terribly important that everyone is here,' I said.

'As soon as Hamoudi tells me that the men have cleared a path to the station I plan to take Mrs Woolley straight from her room under guard,' he said. 'Of course, it's going to be difficult for the Archers, travelling with their daughter's body in the same vehicle as her suspected murderer, but I can't see any other way. Anyway, just time for a spot of breakfast before we set out.'

As he turned away from me to go back inside I placed a hand lightly on his arm. 'Captain Forster, I have to tell you that Katharine Woolley is not responsible for the murder of Miss Archer,' I said.

'But surely – what with the incident with Davison, who must have been drugged while guarding Mrs Woolley's room?' he said. 'I can't see any other explanation – the culprit has to be Mrs Woolley herself.'

I gave a little understated cough. 'If you'd be so kind as to go and gather everyone and bring them into the sitting room I will furnish you with all the details,' I said, almost as if I was about to run through a complicated travel itinerary or a particularly tricky recipe for a fruit cake.

'I'm really not sure that—'

Davison, who had silently stepped out into the courtyard, cut him off. 'Forster, please do as Mrs Christie says. She's the one who has been working on the case since she first arrived here at Ur. She is about to explain everything.'

Forster looked at me as if seeing an example of some exotic species for the first time before he flushed, no doubt embarrassed at the public wounding of his male pride. 'Of c-course, yes,' he said, stumbling over his words. 'W-whatever you say, Davison.'

The Captain left us alone. Davison's eyes were full of concern for me. I smiled gently at him, a smile that spoke of the depths of our friendship.

'You are quite certain of this,' said Davison softly. 'Of what you intend to do?'

'Yes, quite certain,' I said. 'In fact, it's the only course of action.'

'But what about the documents – the photograph?'

'It's something – but not enough. We need more.'

Davison looked at me in a new way, almost as if he were a visitor to a museum and I was one of the ancient marble sculptures on display. 'Why are you looking at me like that?' I said lightly.

'I'm just wondering how you became so brave, so fearless,' he said.

The comment made me laugh. 'I'm neither brave nor fearless, but good at pretending to be so,' I said. 'And you're forgetting my books.'

'Your books? What have they got to do with anything?'

They were not works of great literature, I knew that. But they had fast-moving plots that people could lose themselves in and characters who showed a great deal of pluck and courage. These were men and women who fought for truth and justice. Some of them were detective figures, of course, but others were ordinary enough individuals who found themselves in quite extraordinary circumstances. I saw myself in the same light. Although in the grand scheme of things I was nobody very special, I recognised that I did have a talent for rooting out evil.

'Everything,' I said. 'Now what do you say to a spot of breakfast?'

Chapter Thirty-one

'We're going to need some more chairs,' Davison said as we walked into the sitting room. 'And some extra places too.'

'What for?' asked a rather harassed-looking Father Burrows as he busied between kitchen and table.

'Mr and Mrs Archer are joining us for breakfast,' replied Davison. 'As well as Mrs Woolley and Cecil McRae.'

The mention of the two names sent a ripple of anticipation that stilled the low murmur of voices.

'But I thought that—'

Father Burrows was interrupted by Captain Forster, who appeared at the door like someone stepping onto a stage just on cue.

'Mr and Mrs Archer, please take your seats at the breakfast table,' said the policeman. 'I'll go and fetch, well, the rest of the party.'

The Americans looked wretched, pale and worn out. Mr Archer had tried to shave, but a series of nicks and cuts around his jawline suggested that the experience had been an unpleasant one. Mrs Archer had pouchy bags under her eyes and it appeared as though she had dressed in haste as her blouse was creased and a little dirty around the cuffs. Neither of them seemed to understand what was about to happen or that they were going to be joined at the table by the woman they blamed for the death of their daughter.

'Did you manage to get much sleep?' asked Woolley as the couple took their seats.

The archaeologist did not receive an answer and perhaps it was this that made him feel he should continue to talk over the awkward silence. 'At least the storm is over now,' he said. 'Terribly inconvenient, I suppose, but theirs not to reason why.'

I whispered the rest of the Tennyson couplet to myself, 'Theirs but to do and die.'

Understandably, everyone looked so miserable and downcast and Woolley realised at once he should have chosen a different line. 'Beautiful sky, though,' he said as he tried to ameliorate the situation. 'I don't know if anyone saw the dawn? Quite extraordinary colours cast over the desert, such a very pure quality of light.'

The comment drew a few murmurs of assent as people began to take their places at the table. Cynthia Jones chose a seat by Mr and Mrs Archer. I sat down opposite them, between Davison and Conway. Leonard Woolley waited for the arrival of his wife and Lawrence McRae stood by the door ready to welcome his nephew. Father Burrows continued to place plates and dishes on the table. People started to make small talk about storms, the sands, the possibility of other ancient sites lying buried under the surface of the desert. Yet apart from Mr and Mrs

Archer, who seemed to be still trapped in some kind of daze, everyone's real attention was directed towards the two people who were absent from the room. The collective air of expectation may have been invisible to the eye, but it was certainly a very real presence in that room. Those of us, such as Davison, Conway and myself, who had a view of the door, were waiting for the entrance of the two people who had been accused of the murder of Miss Archer.

As the minutes passed and the conversation ebbed and stalled I wondered what could be keeping Forster. All he had to do was unlock the doors and accompany Mrs Woolley and Cecil McRae from their makeshift prisons to the sitting room. I hoped the boy had not done anything stupid. We had tried to tell him that his imprisonment was just a temporary measure and that he was being kept in the room for his own safety, that he was in no way a suspect in the investigation into the death of Sarah Archer. But I knew he was in a terribly vulnerable state. His erratic behaviour when he had pointed a gun at Mrs Woolley and myself and contemplated killing himself, illustrated the unbalanced nature of his mind. I would never forgive myself if he felt he had no choice but to end it all. As I saw an image of him hanging from a length of sheeting I felt the familiar bird of panic flutter in my chest. *Please no, please not that.*

Just then I heard the sound of approaching voices. Captain Forster entered the room, followed by Katharine Woolley and Cecil McRae. The two made very different impressions. Katharine was immaculately turned out, dressed in her favourite shade of *vieux rose*. She greeted the room as if she were the star attraction and smiled demurely as Woolley walked forwards and pulled back a chair for her. The boy, meanwhile, was unshaven and unkempt and he appeared to have aged a good ten years. He could not meet anyone's eye and as he walked he stooped as if he feared the ceiling would collapse down on him.

The sight of Mrs Woolley and Cecil McRae wrenched Mr Archer out of his stupor. 'What is the meaning of this?' he demanded of Forster.

The Captain did not answer.

'Woolley, do you know what the hell is happening?' asked the American. But he received no answer from the archaeologist.

Mrs Archer reached out and grabbed the hand of a pale-looking Cynthia Jones. 'And to think that woman nearly succeeded in killing you,' she said. 'After what she did to poor Sarah I don't know how she's got the cheek to sit at the same table.'

'We must all remain calm,' said Forster. 'There's an important matter that needs to be discussed. Something that has come from quite unexpected quarters.'

'Don't be a fool,' said Archer. 'I appreciate the gesture of offering us breakfast before the journey, but if that means sharing a table with the likes of this – this creature then you can count me out.' He stood up to leave. 'It's a long way to Baghdad and, unless you've forgotten, we have to bury our daughter.'

'Please sit down, Mr Archer,' said Davison.

'And who are you to tell me what to do?'

Without a moment's hesitation Davison said calmly

and with a matter-of-fact tone, 'I'm with His Majesty's Government; more than that you need not know. Now, I'd like to invite Mrs Christie here to say a few words.'

'Mrs Christie?' asked a clearly appalled Mrs Archer. 'I don't think we're in any mood for frivolity. As you heard my husband say, we—'

But Davison cut her off. 'What Mrs Christie is about to tell us is far from frivolous,' he said. He paused for maximum effect. 'In fact, Mrs Christie was sent by the British government to accompany me to Iraq. And now she is about to reveal the real identity of your daughter's killer.'

'But we thought ...' said a bewildered-looking Mr Archer.

'Yes, you thought that Mrs Woolley here was the killer,' I said, standing up. 'And in many respects, it looked as though she must be the person responsible for the murder. After all, she was there at the scene of the crime. She was discovered kneeling by the body with blood on her hands. She was, people said, jealous of Sarah Archer's youth and her beauty. And the two women had had a disagreement just a matter of minutes before Sarah stormed off into the night, soon followed by Mrs Woolley. But more importantly, Katharine Woolley's behaviour was already – how shall I put it? – somewhat erratic. In effect, she needed little motive to kill Sarah – it was genuinely believed that her madness was enough of an explanation. Of course, she already had quite an unusual, dominant personality, but she started to complain of seeing things, hearing voices. Other people in the camp became understandably fearful of Mrs Woolley and her strange ways. Many of you also believed that she had killed her ginger cat too. But I can tell you that she did not kill that poor animal and neither did she murder Sarah.'

I had the attention of everyone now. I took a deep breath and continued. 'For a number of months Mrs Woolley has been subject to a most terrible form of poisoning.'

The revelation produced a series of gasps from around the room. I opened my handbag and took out one of the jars of night creams I had taken and tested from Katharine's dressing table.

'Mrs Woolley was in the habit of using a number of creams, some to moisturise her hands, others which she rubbed into her face last thing at night,' I said. 'It's a habit common to many of us women. And what with this very dry climate out in the desert it was almost essential. However, in the creams there was one ingredient that was responsible for Mrs Woolley's strange behaviour: hyoscyamine.'

'But how do you know?' asked Cynthia Jones.

'I tested them,' I said. 'You see, during the war I trained not only as a nurse but I also worked in the dispensary of my local hospital. I learnt a good deal about poisons and the knowledge proved very useful indeed. And not just for my books. You see, hyoscyamine produces a range of symptoms such as headaches, blurred vision, dizziness, disorientation, euphoria, short-term memory loss and, most importantly, hallucinations. Katharine Woolley here experienced many of these symptoms, symptoms which some of you believed were indicative of her mental instability.'

Katharine looked like she was too afraid to let herself

believe what I was saying – after all, for months she must have thought that she was going insane.

'She may have had this – this substance in her beauty creams, but why does it then follow that she is innocent of murder?' asked a clearly sceptical Mr Archer.

'A very good question, Mr Archer,' I said. 'One must learn not to take anything at face value. Let me explain.' I took another deep breath. 'As you now know, I did not come to Ur as a mere tourist. I came to investigate the death of Miss Gertrude Bell.'

'Miss Bell?' asked Woolley.

'Yes, I'm sorry for the deception, but I had to keep my real motives hidden. As you all know, Miss Bell died in Baghdad in July 1926. There was an air of mystery surrounding her death, but the official line was that she died due to a weakness of her system after an episode of pleurisy and then an attack of bronchitis. However, the doctor who examined her body discovered that she had died due to an overdose of Dial, a sedative similar to Veronal.'

'This is all very interesting, but what's it got to do with the death of my daughter?' asked Mr Archer.

'Absolutely nothing,' I said.

'Nothing?' he replied. 'Then why are you telling us all of this?'

'Quite recently two letters which seemed to have been written by Miss Bell came to light which stated that she had been in fear of her life,' I said. 'She added that if she were to die then it was likely that her killer would be found at Ur. But – and this is important – there is no connection whatsoever between the deaths of Miss Bell and your daughter.

The truth of the matter is simple: one was a case of suicide or an accidental overdose – I suppose we may never know – and the other was a cold and heartless murder.'

'Then why are you bringing it up here?' asked Mrs Archer, who had taken a handkerchief out of the sleeve of her blouse and had started to wring it between her fingers. She looked first at her husband and then at Cynthia Jones. 'I don't understand.'

'The killer thought they were being very clever, you see,' I continued. 'They laid the foundations of the case thoroughly, skewing one's perception of it from the very beginning. You see, someone took advantage of Miss Bell's death. They wanted it to be seen as a murder, when in fact it was nothing of the kind, just a rather wretched end after a life full of achievement.'

'But who would want to do that?' asked Father Burrows, who looked hot and exhausted after cooking a breakfast that no one was looking at, never mind eating.

'Indeed, who?' I asked, looking around the table. It always fascinated me how the innocent appeared the most guilty, while those who had committed the worst sins often seemed guileless.

'And what of Cecil?' asked Lawrence McRae. 'Surely it's time to clear the boy's name. He was foolish, wrongheaded, but he's suffered enough. Look at the state of him.'

Cecil could still not meet my gaze. It was time to tell the truth.

'What Cecil McRae did was unforgivable,' I said. The boy flushed red in the face and, as I spoke, he began to squirm in his seat. 'He threatened Mrs Woolley and me with a gun and we were in fear of our lives.' I paused. 'However, he acted in the mistaken belief that Katharine Woolley had killed Miss Archer. In a fit of adolescent passion he took it upon himself to try and enact some kind of revenge. Luckily for us – and for him – he had more sense than to let his hot-headed nature get the better of him.'

In fact, I believed it was Cecil's lack of courage that was behind this, but he was in too delicate a frame of mind to hear such an ungarnished interpretation of his behaviour. I could see the next question almost forming itself on the lips of my small audience and so I pre-empted it. 'And now we come to the matter of Cecil's confession.'

'Yes, why did the boy confess to a crime he didn't commit?' asked Mrs Archer.

'That was a necessary deception, I'm afraid,' I said. 'And I'm sorry, Cecil, that you've been locked up against your will for all this time. But really we had no other choice.'

'Locking up a vulnerable lad like a savage beast, it's criminal,' exclaimed Lawrence McRae. 'If we were so minded we could bring—'

'Yes, I can see your point, Mr McRae, but if you'd be so kind and let me explain,' I said in a calm and steady voice. 'After all, I'm sure Mrs Woolley here could easily pursue the matter herself if she so chose.'

The comment silenced McRae and I continued. 'The real killer of Miss Archer wanted to point the finger at Mrs Woolley. That had been the overarching motive from the very beginning. Everyone had to believe that it was Katharine who had killed Sarah Archer. After all, she had been found literally red-handed, next to the girl's body with blood on her hands. But what the killer had not expected – what no one could have predicted – was the way that Cecil McRae behaved after hearing the news of Sarah's death.' I addressed the boy directly. 'You truly loved that girl, didn't you?'

As Cecil nodded, a tear slipped down his cheek. He did not need to say any more.

'The killer panicked when Cecil threatened to shoot Mrs Woolley,' I explained. 'You see, the person behind all of this did not want Mrs Woolley to die. They wanted Katharine to suffer much more than that.'

'Suffer more than death itself?' said Woolley, looking at his wife with tears in his eyes.

For her part, Katharine listened to my account with a certain regal dignity, almost as if she occupied a different, higher plane. Although she had suffered, she would not let the group know quite how much she had endured. It was a trait that one had to admire.

'And so you see, after explaining the complex situation to Captain Forster, we fashioned a confession from Cecil in which he took responsibility for the crime,' I said. 'It seemed as though the motive was a clear one. He said he had been driven to kill Sarah Archer out of misdirected love. They had quarrelled that night on the ziggurat. He had reached out to touch her, but there was an unpleasant scuffle and Miss Archer fell back and hit her head. Seeing what he had done, Cecil decided he had to kill her and then went on to frame Mrs Woolley for the crime. This of course was absolute fiction. He played no part in Miss Archer's death. But Cecil's false confession was essential to drawing out the killer. I suspected that the person behind all of this would make a mistake. And Cecil's confession to the murder of Miss Archer forced the killer to make their first error.'

As I approached the final moment of revelation – a moment when anything could happen – I was conscious that my heart was beating faster than normal. My mouth felt dry and even though I took a sip of water – water I had brought with me from the safety of the kitchen – the liquid did nothing to ease my discomfort. All eyes were on me.

'What was the error?' asked Forster.

'The killer acted in such a way that they revealed their true identity,' I said, as I tried to clear my throat.

'So who is it?' demanded Mr Archer, looking around the table. 'If it's not Mrs Woolley and it wasn't the boy McRae, which one of us is the murderer? I demand to know who killed my daughter!'

Chapter Thirty-two

'I don't think I can stand this much longer,' said Cynthia Jones, standing up from the table. 'I think we could all do with a fresh pot of tea, don't you?'

The spinsterish woman shuffled off into the kitchen. 'Yes, a very good idea, Miss Jones,' I said, as I continued to look around the table at the increasingly uncomfortable faces before me. 'To begin with, it seemed we had our man. I discovered someone at Ur who was not who he said he was. Mr Miller – or should I say Mr Conway – could you please stand up?'

'Is this really necessary?' said Conway.

'Please do what Mrs Christie asks,' insisted Davison.

The photographer pushed himself out of his chair and nervously rubbed his moustache with the fingers of his right hand.

'Miller?' asked an astounded Woolley. 'What is this? What does Mrs Christie mean?'

Conway took a deep breath and on the exhalation began to make his confession. 'It's true,' he said. 'I'm not who

I said I was.' I nodded to signal that he should explain a little more. 'My name's not Harry Miller. I'm – I'm Alan Conway.'

'And tell everyone about your job,' said Davison. 'Your real purpose at Ur, I mean.'

Conway gave me a look that pleaded with me to intervene. I once had feelings for him, but I knew that it would be wrong to let my emotions stand in the way of justice.

'I think you owe Mr Woolley an explanation, don't you?' I said, perhaps more harshly than I intended.

'I'm – I'm here because I've been copying some of the treasures,' he said in a quiet voice.

'You've been doing what?' cried Woolley.

Conway described how he had been using his cover as a photographer to make electrotype copies of valuable objects so he could then ship the real artefacts back to America. As he listened to Conway's account Woolley's face drained of colour. Quite understandably, Woolley wanted to know the specifics, but while Conway was in the middle of detailing how he had made the copies and outlined who was behind the plan, Mr Archer lost patience and interrupted him.

'I don't care two hoots for you and your trinkets!' he shouted. 'I still want to know about my daughter. Who murdered my Sarah?'

'Yes, of course,' I said. 'I was just coming to that. As I said, I did have my suspicions about Mr Conway here, purely because of the evidence that showed he was operating under a false identity. But then something happened which proved beyond a doubt that—'

'Now who would like some tea?' asked Miss Jones as she returned to the table with a tray.

'Can't you be quiet, woman?' shouted Mr Archer.

Miss Jones looked shocked and upset. 'I'm sorry, but I just thought we could all do with—'

'Mrs Christie was about to tell us the name of the man who murdered my daughter,' he said.

I took another deep breath. The time had come for the final revelation. 'It was not a man,' I said.

There was a collective gasp and more than a few cries and exclamations, some of them quite earthy. Katharine Woolley placed a gloved hand over her mouth. Mrs Archer looked as if she was going to be sick. Father Burrows seemed so shocked he did not know what to do with himself. And Woolley was struck down by a strange paralysis, like one of those poor men at Pompeii whose bodies had been covered in volcanic ash all those hundreds of years ago.

'Yes, I'm sorry to say that the person behind all of this is a member of the so-called fairer sex,' I said. 'I may as well name her as she is standing before us. Have you anything to say for yourself . . . Miss Jones?'

Cynthia looked at me with a mix of innocence, amusement and astonishment. 'I really don't know what you're talking about.'

'Come now, you know very well that there's no point in pretending any longer,' I said. 'Yes, you've been very clever, particularly at the beginning of your plan, because you knew that Mrs Woolley already suffered from terrible headaches. You believed that people would think that

Mrs Woolley's other symptoms – which we'll come on to shortly – would simply be an extension of her illness. But you've also been so very wicked. How could you kill that poor cat? You knew that Mrs Woolley adored that creature. What did you do? Did you give it a special little something in its food? And I suppose you must have drugged Katharine at that time. She would have had to be sedated because you took something sharp to her arms to make it look as though the cat had scratched her. Is that what happened?'

'This is completely absurd,' she said. She looked around the table at her friends for signs of support. 'You must be out of your mind.'

'And, of course, you were the one who was responsible for spiking Katharine's creams with the hyoscyamine, weren't you?'

'The hyo . . . the what?'

'You know very well what I'm talking about. But I'll explain more in a moment.'

'I think you've been reading too many of those silly detective stories – that or writing one,' she said, pretending to laugh.

'I think this particular tale – and your involvement in it – is beyond anything even I could have dreamt up,' I said. 'Let's start at the very beginning, Miss Jones.' I could no longer bring myself to call her by her Christian name. 'With the letters that Miss Bell wrote to her father, unsent letters which described how she feared for her life, documents in which she said that if she were to be killed the authorities should look to Ur for her murderer.'

'But what's that got to do with the murder of Sarah?' asked Mrs Archer.

'Forgive me, but to understand Sarah's death we need to take a step back in time,' I said. 'It was perfectly natural that when those letters came to light the first line of enquiry would be to follow up Miss Bell's suspicions. And so that's why I was sent here, to investigate the matter.'

I paused for a moment as I cleared my throat. 'I always thought that there was something odd about those two letters and the accompanying drawing of the Great Death Pit at Ur, with Miss Bell's initials placed next to one of the stick figures. It was strange that they were never sent, that they were just waiting around for someone to find them, almost as if someone had placed them in that seed box, where the gardener would at some point come across them. And, despite the detailed nature of the letters, my first reaction was that they were forgeries. Yet apparently an expert had compared the handwriting of the letters to Miss Bell's own hand and had declared them to be one and the same. But then one night, during what seemed like a perfectly innocent conversation, I heard a comment that caused me to think. Do you recall what it was?'

Miss Jones did not answer.

'I thought not. Do you remember, Father Burrows?'

The priest looked bemused. 'No, I'm afraid I don't,' he replied.

'Let me refresh your memories,' I said. 'It was a conversation of all things about learning the cuneiform script. Father Burrows, you, very kindly, offered to teach me how to write it, and, in an aside, you said ... yes, this was it:

"When I first met Miss Jones she was a complete novice like you and she picked it up very quickly." You added, "Quite the natural, weren't you, Miss Jones?" That got me wondering. If you could master a script like cuneiform with such ease perhaps you were a natural at copying the handwriting of others. Not only that – you were a close companion of Miss Bell's and you would have been privy to certain pieces of information that she would only have shared with a very good friend. But I kept these thoughts in the back of my mind. You see, at that point it was still very early days. Everyone, I suppose, was a suspect.'

'Everyone knows I was a friend of Gertrude's and yes, I did pick up cuneiform easily – but so what?' asked Cynthia, looking at Mrs Archer for support. 'That means nothing.'

'And Mrs Christie, aren't you forgetting that poor Cynthia here was herself the victim of the killer?' asked Mrs Archer. 'She could have died by drinking that acid.'

'Thank you, Mrs Archer,' said Cynthia. 'At least somebody has some thought for my feelings. Now, why don't we all have a cup of tea?'

I watched as, with a steady hand, she lifted a large teapot and began to pour. There were, I noticed, eleven cups in total. But there were twelve of us in the room. She walked around the table, placing the cups before everyone apart from Katharine Woolley. I knew, when challenged and presented with the evidence, that Cynthia would plan something – and I had deliberately allowed her to go into the kitchen when she said she was going to make tea – but I could never have imagined that she would dream up such a wicked scenario.

As people lifted the cups to their lips I said, 'Don't touch the tea. Please, whatever you do, don't drink it. It's poisoned.' The group froze as if trapped in a grotesque tableaux.

It was in that moment that I saw Miss Jones show her true colours. She turned from the shy, retiring spinster – the kind of woman who would not say boo to a goose – into something else, something purely evil.

'Look at the cups,' I said. 'Look at where she's placed them. Each of you has one – including Miss Jones herself – everyone, that is, apart from Mrs Woolley.'

'So what?' said Cynthia, a darkness burning in her eyes.

'It's the death pit, isn't it?' I said, remembering the eerie place of human sacrifice that Woolley had shown me soon after I had first arrived at Ur. 'You were going to poison everyone – including yourself – and leave Mrs Woolley alive. When the authorities came they would find everyone dead apart from Katharine. No matter how hard Mrs Woolley tried to explain what had happened here her account would be taken as nothing more than the crazed rantings of a mad woman. Katharine Woolley would be surrounded by eleven bodies – all victims of poison – and she would be the only survivor. Of course, she would have to be the killer. No other explanation would make sense.'

'But why would Miss Jones want to kill herself, together with the rest of us, but leave Katharine alive?' asked Woolley, pushing his teacup away from him.

The hatred that had festered for so long in Miss Jones's breast began to show itself now. 'Just look at her, sitting there as if she were better than the rest of us, like some kind of queen.' Cynthia seemed to spit the words out of her mouth as she looked at Katharine. 'If only you knew the truth about her.'

At this Katharine started, as if woken from a dream.

'Yes, that got your attention, didn't it, Mrs Woolley?' continued Miss Jones. 'Or should I say Mrs Keeling?'

The mention of Katharine's first married name brought an expression of horror into her eyes.

'Why don't we talk a little of Mr Keeling and your love for him – or should I say your lack of love?' said Cynthia in a horrible, mocking voice.

'Oh, please no,' begged Katharine. 'Please not that.'

'Is there something you'd like to keep secret?' continued Cynthia in a nasty tone. 'People have always wondered why your husband killed himself after only six months of marriage. He shot himself, didn't he? At the foot of the Great Pyramid of Cheops. All very dramatic.'

'Cynthia, please, I'll do anything, anything you ask,' said Katharine, who had started to sob now.

'Stop this immediately!' shouted Woolley. 'I will not have you upsetting my wife.'

'Your wife?' replied Cynthia. 'Are you sure about that?'

It was then that Katharine reached out for her husband's cup and, so slowly as if to make the action almost indiscernible, began to lift it towards her. She brought it closer to her lips and just as she was poised to take a sip I said, 'The tea! Don't let her drink it.'

Woolley was quick to respond and dashed the cup from her hands, spilling the dark liquid across the table.

'I can't bear it,' cried Katharine. 'Oh, why didn't you let me end it all? I wish I were dead!'

'It's an interesting point,' I said, trying to sound as detached and unemotional as possible. Although I wanted to go and console Katharine, who was sobbing now, I had to steel myself to remain strong. 'Miss Jones could have killed Mrs Woolley at any time. After all, she had many opportunities. She was being poisoned over time with the small doses of hyoscyamine in her cosmetic creams and it would not have been hard to have given her a fatal dose. Indeed, she could have been the one to die that night out on the ziggurat instead of your daughter,' I said, addressing Mr and Mrs Archer.

'And by the sounds of it she could have died drinking this damned tea,' said Forster, who went to stand guard by Miss Jones. He cast a look of warning in her direction. 'Let's clear this lot up. Burrows, would you be so kind as to take the tea things away? Don't throw anything away though, we'll need to test it for poison.'

Slightly taken aback at being treated like a common servant, Burrows placed the cups, saucers and teapot back on the tray and then carried them back into the kitchen. I waited for Father Burrows to return before I continued.

'As I was saying, Miss Jones took advantage of what she thought were the perfect circumstances in which to commit the crime. It was night-time. The situation was chaotic. Her two victims – because there were two victims here, not just one – came together like two planets in an unholy alignment. There had been an argument between the two women and it was clear that Mrs Woolley did not care for Sarah Archer. Miss Jones saw Sarah rushing down the steps and then realised that, if she were to smash that rock over her head, the girl would collapse and be discovered by Mrs Woolley. Yet even she could not have envisaged the scene that followed with Katharine falling down by the girl's body, besmirching her hands in Sarah's freshly spilled blood.'

Mrs Archer looked at Cynthia Jones as if seeing her for the first time. No doubt she was forced to reassess the moments of friendliness and quiet intimacy that had passed between them. It had all been a sham, a performance. This quiet mouse of a woman was her daughter's killer. In addition, this unassuming spinster had prepared a pot of tea that had the capacity to kill everyone in the room, including Mrs Archer herself. As she came to this realisation hatred emanated from her eyes.

'Miss Jones wanted to exact a revenge that went beyond murder,' I continued. 'At the moment I merely want to set out the facts of the case – we'll come to the motive in a little while. When Katharine Woolley was discovered next to the body of Sarah Archer it seemed as though she must be the killer. After all, everyone in the camp believed Mrs Woolley to be unbalanced, if not insane. And so the stage was set. It looked as though Katharine Woolley would be hauled back to Baghdad where she would receive a terrible punishment for her crimes and, crucially, her reputation would be ruined.

'But then something happened which skewed Miss Jones's plan. When Cecil "confessed" to the crime, I knew the real murderer would have to do something desperate.

And indeed, Miss Jones was forced into a corner. First of all, there was the scene in Mrs Woolley's room the night of the sandstorm. Katharine had started to behave oddly. She was convinced that she could smell Sarah Archer's blood on her hands. Then she lashed out at me. I went to find Mr Woolley, but as I was speaking to him Father Burrows shouted for help. I returned to Mrs Woolley's room to discover Miss Jones in a highly distressed state. Her wrist and the lower part of her arm were all red, as if someone had twisted her skin, and she told me that Katharine had threatened her with the chilling words, "Watch out – tonight you're going to die. Tonight ... I'm going to kill you."

'The evidence seemed to suggest that Miss Jones was the victim. After all, Katharine Woolley was behaving very oddly. It looked as though she had attacked poor Miss Jones and had gone so far as to say that she was going to murder her. Then, later that night, we heard screaming coming from Miss Jones's room and, after breaking down her door, we discovered that someone had swapped her night-time water for a glass of hydrochloric acid. The conclusion was simple: Mrs Woolley had indeed tried to follow through on her threat to kill Miss Jones.

'And yet ... were there any other witnesses to the conversation between Miss Jones and Katharine Woolley? No, there were not. The account came entirely from Miss Jones herself. In addition, just before the episode when Mrs Woolley started to behave oddly, she had been massaging cream into her hands, one of the creams which I now know had been tampered with. I noticed that her pupils were dilated and she said she was desperately thirsty – both symptoms of hyoscyamine poisoning.'

Woolley could not take in the enormity of what I was saying. He had been betrayed by the man he knew as Miller already and that had hit him hard. I could see the desperate look in his eyes: please let this not be true, please let the murderer be anyone but dear old trusted Cynthia.

But it was Lawrence McRae who spoke up first. His suspicions of Katharine Woolley had not dissipated. 'What of the marks on Miss Jones's skin?' he asked. 'And the acid burns on her legs?'

'I'm afraid to say they were self-inflicted,' I said. 'Recognising that Mrs Woolley was suffering from hallucinations brought on by the hyoscyamine in her creams, she could do and say almost anything in front of her without impunity. My guess is that, while in Mrs Woolley's room, she repeatedly twisted and pinched the skin on her own wrist and arm until it became red and raw.'

'But what about the business with Davison here?' said McRae. 'He was sitting outside Katharine's room. She must have drugged him so she could escape from her room and place the glass of hydrochloric acid by Miss Jones's bed.'

Davison stood up from the table; it was time for him to speak. 'I must admit to a little deception on my part too,' he said. 'You see, before I went to take my position outside Mrs Woolley's room I was given a cup of tea by Miss Jones. I took the drink with me, but in the courtyard I poured the tea away as I knew it was likely to contain some substance that would make me go to sleep. And so you see, later that night, when Miss Jones came to check on me I was actually wide awake.'

He turned and looked directly at Miss Jones. He related what he had then witnessed. 'Although I appeared to be dead to the world, I did indeed see you turn the key in the lock and open the door into Mrs Woolley's room before you then made your way back to your own quarters. I stayed there, outside Mrs Woolley's room, all night and I can tell you that the lady never stepped outside.'

'So you see, Miss Jones, we have incontrovertible proof of your terrible crimes,' I said. 'There is no escape. Do you have anything to say for yourself?'

Just at that moment, as Captain Forster reached out to arrest her, Miss Jones started to cry like a little girl. Tears streamed down her face and she began to sniff. She searched the sleeves of her blouse for a handkerchief, and looked distraught at the prospect of not having one to hand to wipe her eyes and nose.

'Oh, dear,' she said. 'I appear not to have a . . .'

At that moment, Ruth Archer stood up and came towards her, proffering a handkerchief.

'Thank you for—' said Miss Jones, reaching out for the square of white cotton embroidered with forget-me-nots.

But before she could finish her sentence, Mrs Archer charged at her with the fury of a mother set on avenging a recently murdered daughter. All her pent-up aggression and anger found expression in that moment. With the handkerchief in her right hand she pushed the fabric hard into Miss Jones's open mouth. A muffled noise came from the secretary, as if she could not breathe. She clawed at her face, her eyes stretched wide in panic, but Mrs Archer pushed the cotton further in.

Captain Forster and Davison were quick to act. Forster separated the two women and Davison reached into Miss Jones's mouth and extracted the handkerchief. Miss Jones took in great gulps of breath and fell back into a chair.

Ruth Archer spat out her words as if they were poison. 'You deserve to die for what you did to my daughter!'

Nobody could argue with her. 'But w-why?' Ruth was hysterical now, shouting and crying at the same time. 'Why did you want to kill Sarah? What had she ever done to you?'

After she had recovered her breath, Miss Jones sat down in a chair and said in a quiet voice, 'Do you really want to know?'

'Of course I want to know,' replied Ruth Archer. 'I still don't understand . . . what was it all for?'

'It was a shame Sarah had to die, because I actually quite liked her – mainly because she stood up to that stuck-up bitch over there,' she said, directing the comment towards Mrs Woolley. 'But her death was convenient.'

Mrs Archer was appalled by the secretary's choice of word. 'Convenient?'

'Yes, convenient,' continued Miss Jones. 'She was in the right place at the right time, that's all.'

'So you're admitting to the murder?' asked Captain Forster. 'That you killed Sarah Archer?'

'Yes, I suppose I am,' said Miss Jones.

'But why?' pleaded Mrs Archer. 'Please tell me why.'

'*In a land of sand and ruin and gold,*' said Miss Jones. '*There shone one woman, and none but she.*'

That was obviously a reference to the regal Katharine Woolley. But was it not also a line from a poem. Who was it by?

'*I wish we were dead together to-day,*' she continued. '*Lost sight of, hidden away out of sight.*'

'What's she talking about?' asked Mr Archer. 'Has she lost her mind?'

Even in the grip of insanity there was a part of her that was able to tap into the poetry she had once learnt. I recognised the lines from Swinburne's 'The Triumph of Time'.

'... *Forgotten of all men altogether,*' she whispered. '*As the world's first dead, taken wholly away, made one with death, filled full of the night.*'

She gazed at some imaginary point in the distance and, in that moment, she seemed to glow with a kind of inner happiness. I knew then what she planned to do.

'Captain Forster,' I shouted. 'Davison – watch her ...'

But Miss Jones was too quick. She took out a small vial from her pocket and pressed it to her lips. As I ran towards her I smelt the slight aroma of bitter almonds. I knew my poisons. It was cyanide. The chemical started to do its sinister work quickly. She tried to quote another line from the poem, but as the life force ebbed away from her she could only whisper a few words. And so, as I stood over her body and watched her die, I completed another, more fitting couplet from the same poem which could serve as her epitaph.

'*At the door of life, by the gate of breath,*' I said. '*There are worse things waiting for men than death.*'

Chapter Thirty-three

Cynthia Jones's body had been taken away from the sitting room and both murderer and her victim, poor Sarah Archer, were hauled onto the same carriage that would soon take them back to Baghdad for burial. Alan Conway – the man who had pretended to be the photographer Harry Miller – had been handcuffed too and led away by Hamoudi. Davison and I would put in a good word for him – the American had helped us with part of our plan – and I hoped he would serve his sentence in his home country.

I had a quiet word with Cecil McRae and, during a tearful confession, he apologised again and told me how his parents had died. His father, in a fit of rage and jealousy, had shot his mother and then turned the gun on himself. The boy, who adored his father, had not wanted to reveal the sordid truth of the crime and had chosen to remain silent, leading to suspicion that he had been involved. I told him that I understood, and how noble I thought he had been.

Samples from the teapot would be taken and sent off to an expert in the capital for further testing, but Forster, Davison and I had all smelt the noxious brew that Miss Jones had prepared and we were certain that it too carried the distinctive whiff of cyanide.

Cynthia Jones's suicide had left many questions unanswered, of course, and I had to fend them off like flies around a decaying corpse. The main one was the issue of motivation: just why did Miss Jones want to frame Katharine Woolley? The line from the Swinburne poem echoed through my mind again. *'In a land of sand and ruin and gold . . .'* In this case the 'ruin' did not refer to the traces of ancient structures and temples found among the desert sands, but the destruction of a reputation. That was at the root of the case.

What was left of our little group remained in the sitting room and the time had come for me to finish the story.

'In order to see the whole picture, we have to take a step back in time once more,' I said. 'Before Katharine married Mr Woolley here, she was married to a Lieutenant Colonel Bertram Keeling.' I opened my handbag and took out a cutting from the *Times* newspaper. 'This is a brief report about his death, which doesn't give us very much detail about the manner in which he died, but we know that the unfortunate man did indeed commit suicide at the foot of the Great Pyramid in Giza, just outside Cairo, in September 1919.'

At this Mrs Woolley began to show signs of agitation. 'Do you really have to go into all of this?' she asked. 'Is it really necessary?'

'Yes, my dear, I'm afraid it is,' I said. 'Before his marriage to Katharine, Colonel Keeling had had a sweetheart, someone he had met while studying at Cambridge. That woman was a Miss Edgecombe, the daughter of one of his professors there.'

I could see Mr Archer looking at me with impatience; no doubt he thought what I was saying was nothing more than a needless digression. 'Please bear with me, because this is the key to the whole mystery,' I said, as I opened up my handbag once more. 'Please pass this photograph around. It's a picture of Colonel Keeling and Miss Edgecombe, taken in a punt on the River Cam.'

Woolley squinted at the picture. 'I'll be damned ... sorry,' he said. 'It's—'

'Yes, that's right, it's Cynthia Jones, in younger and no doubt happier days,' I said.

'I could barely recognise her,' said Katharine, taking hold of the snapshot. 'She looks – I don't know – so different. She was quite pretty, really. Nothing like – well, you know what I mean.'

'Grief and hatred and bitterness can eat up a person, masking whatever beauty they once possessed,' I said. 'But I also think Miss Jones went some way to disguise herself. I often wondered why she made herself so plain, taking little pride in her features. Then, Mr Woolley, do you remember that story you told about your name? How you were a wolf in sheep's clothing? The phrase started me thinking. What if someone was pretending to be meek and mild, but in reality they were vicious and cruel.'

'I knew Bertie had a girl before me, but he didn't talk

about her and I certainly never saw a photograph of her until today,' said Katharine. 'Where did you find this?'

'My colleague Mr Davison here did a bit of digging for me,' I said. 'It really was a matter of trial and elimination. I had an inkling that the past was significant in this case. But it was like a shadow always out of view. Each time I tried to get a fix on it then it seemed to slip away. At one stage, Mrs Woolley even thought that her first husband might be still alive and that he might have disguised himself. I quickly established this was highly unlikely. When I suspected that someone was trying to make it appear as though Katharine Woolley was mad – and then frame her for the murder of Sarah Archer – I thought about possible motivations. Of course, one's immediate suspicions always turn to the husband.'

'Me?' said an astonished Woolley. 'You really didn't think I would do anything to hurt my wife, did you?'

I thought it best that the conversation I had had with Woolley – about the lack of intimacy between the couple and his thoughts on divorce – remained private. I also decided not to reveal that I had made enquiries into the beneficiaries of the wills of Sarah Archer and Cecil McRae's parents. 'Please forgive me, it's just that husbands – or wives for that matter – are so often to blame for this sort of thing,' I said. 'But once I had ruled that out I had to think of other possibilities. I asked Mr Davison whether he could use his contacts back in London to do a thorough check of Colonel Keeling's background. It was the very helpful men back in England who discovered that the Colonel had had a sweetheart. Although by then I had a clear idea of the identity of the murderer, the photograph, which only arrived this morning, confirmed my suspicions.'

'So it was all because Miss Jones was – what? – jealous of Mrs Woolley?' Lawrence McRae sounded unconvinced. 'Because Katharine had stolen her boyfriend?'

'Not quite,' I said. 'You see, I think Miss Jones blamed Mrs Woolley for Colonel Keeling's death. Of course, she was not pleased by the fact that Katharine had taken away her lover. But that wasn't enough.'

Katharine blinked as if to warn me off the subject.

'She blamed Mrs Woolley for forcing Keeling to take his own life. In her own twisted way, she got it into her mind that Katharine was somehow responsible for his suicide. And so she wanted to make Mrs Woolley pay for the death. As I said earlier, she did not want to simply kill Mrs Woolley – that would be too easy, the suffering would be over too quickly. What she dreamt up was something much more wicked than plain murder. She wanted to bring about Mrs Woolley's downfall and ruin her reputation.'

'What a truly terrible thing to do,' said Mrs Archer. The sentiment was echoed by Father Burrows and Captain Forster, who was busy scribbling everything down. The American woman got up, walked over to Mrs Woolley and said, 'I think I owe you an apology.'

Katharine, no doubt remembering how Mrs Archer had treated her, simply nodded her head and held out her hand as if she were a queen forced to endure the pathetic attentions of a lowly subject.

'And I, in turn, need to beg for your forgiveness, dear

Agatha,' said Katharine, suddenly turning to me and flooding me with the light of her eyes and the brightness of her perfect smile. 'Would you mind coming with me? I have a small token I'd like to bestow on you.'

'Of course,' I replied.

As we entered her room and closed the door she turned to me and apologised. She did not have a present for me – this was a ruse to get me to herself – and she also said sorry for behaving so regally back in the sitting room. I understood that, for a great deal of her life, she too had been playing a part, a role that sometimes restricted and imprisoned her. But what she said next surprised me.

'I do have to thank you for so many things,' she said. 'One, for saving my life. Two, for sorting out all this awful business with Cynthia. And three – and most importantly – for not telling quite the whole truth in there.'

I did not respond.

'But you do know, don't you?'

'Not the full story by any means,' I said.

'I think I owe you an explanation then,' she said.

Katharine sat me on the bed and told me, as calmly as she could, the reason why her first husband had taken his own life. They had married in March 1919, but the wedding night had been a disaster. She did not quite know why she had been fooling herself, but she must have been blinkered, like one of those poor horses you see at the races who jump to their death. She'd had little expectation of that side of things, she said; her mother had said nothing to her about intercourse. After a few months of separate bedrooms, Bertie had insisted she go and see a doctor for a full mental and physical examination. She tried to avoid talking about the subject but finally, in Cairo, there had been a terrible bust-up. Either she had to see a doctor or he would call an end to the marriage on grounds of non-consummation.

That September, she swallowed her pride and submitted herself to an examination. She had felt a chill descend on the consulting room. The doctor, a horrid, rat-like kind of a man, looked at her as if she were some kind of freak. After she had got dressed again he told her the awful truth. And then he insisted on telling Bertie about it too. She pleaded with the doctor not to, but the medico said it was a husband's right to know.

The words were too terrible to hear.

'Your wife here will never have a child,' said the doctor.

'Is that all?' exclaimed Bertie, his face lighting up. 'There are ways around that, we can always—'

'It's not just that,' the doctor interrupted. 'Your wife is not – in the strictest sense – a woman.'

'What do you mean?' Bertie's face had drained of colour now.

'It's a very rare condition,' he replied, 'where a person shares certain aspects of the male with certain aspects of the female.'

'What the hell are you saying?' Bertie looked as though he might be sick. 'Y-you're saying that – that I've married a m . . . ?'

And with that Katharine had fainted. The next thing she had known she had woken up in her bedroom. She felt groggy, as if the doctor had given her something to make her sleep. She tried to get out of bed, but was unable to move. When she woke up again she was presented with the news that her husband had shot himself.

'That must have been awful for you to hear,' I said, placing my hand over hers.

She nodded. 'It was, but what I still don't understand is how Miss Jones knew about it,' she said.

'Perhaps your husband – the Colonel – wrote a letter to her before he died?' I suggested. 'No doubt he was feeling full of regrets and shame. It would have been natural for him to seek out the one person who said she had loved him all along – his first sweetheart. He probably poured his heart out to her, and told her what the doctor had related to him. It must have been awful for Miss Jones to receive that letter, no doubt days or weeks after first hearing the news of Bertie's death. In her view the Colonel killed himself because he considered that his reputation had been ruined; he could not bear the shame of continuing with the marriage and so he shot himself.'

'And her name – or names?' asked Katharine. 'Miss Edgecombe, Miss Jones?'

'Oh, that. Quite easy to establish, the information came from Somerset House, a birth certificate which arrived this morning. Although she was born Cynthia Edgecombe, Jones is her mother's maiden name. She simply started using that when she began work out here.'

Katharine swallowed nervously. 'And w-what are you going to do with all this information now?'

The final pieces of information began to fit into place.

I remembered the phrase that Betty Clemence, that overbearing woman on the Orient Express, had used when she had described Katharine Woolley to me: 'She's not all there,' she had said. At the time, I had assumed that she had meant that Mrs Woolley was not quite right in the head. Had she been referring to something else entirely?

'Of course, the police will have to be told about your connection to Miss Jones, but I can assure you there's no need for them to know about – well, about your own personal history and background.'

'And you don't think other people suspect?' she asked.

How had that old cat Betty Clemence guessed at the truth? Perhaps Cynthia Jones had initiated a vicious campaign of gossip, another weapon in her arsenal designed to destroy Katharine Woolley's reputation. 'No, of course not,' I said. 'How could anyone suspect such a thing? No, we'll keep this strictly between ourselves.'

'I don't know how to thank you,' said Katharine flatly. But then a thought occurred to her which brought life back into her face. 'I've got the most perfect idea!' she said, sounding like a schoolgirl. 'Although I've been enjoying working on my novel, I realise that I'm never going to make it as a proper author – I haven't got your talent. But one day, when I'm dead, perhaps you could write a book about what happened here and tell the whole story – like one of your mysteries.'

'I'm not sure whether I would be up to it,' I said, demurely.

'I can see it now,' she said. 'It's got everything ... murder, secrets, suspects, and all set against the backdrop of an archaeological dig. I've even got a title for it! *Death in a Desert Land*. What do you think?'

I thought about it for a moment before I replied yes, that might work very well indeed.

The Facts

- The body of the writer, traveller and archaeologist Gertrude Lowthian Bell – who helped found the modern state of Iraq – was discovered at her house in Baghdad on 12 July 1926. She appeared to have taken an overdose of 'Dial' (Diallylbarbituric acid), a sedative. While the two letters that open the novel are fictitious, some of the details contained in them are authentic and come from Bell's archive held at Newcastle University: http://www.gerty.ncl.ac.uk. I also drew on the excellent Gertrude Bell biography *Queen of the Desert* by Georgina Howell (Macmillan, 2006).

- In the autumn of 1928, after her divorce from Archie, Agatha Christie travelled by herself on the Orient Express to Istanbul and then another train to Damascus. From there, she journeyed across the desert to Baghdad in a bus organised by the Nairn Line, a trip that took two days. She then took an uncomfortable train south to Ur, where she met Leonard and Katharine Woolley and Father Burrows, a Jesuit priest and an expert on cuneiform tablets. 'I fell in love with Ur,' Agatha writes in her autobiography, 'with its beauty in the evenings, the ziggurat standing up, faintly shadowed, and that wide sea of sand with its lovely pale colours of apricot, rose, blue and mauve changing every minute.' Although Agatha acknowledged that Katharine could be a divisive figure, she remained friends with the Woolleys and she went on to dedicate her 1932 Miss Marple short story collection *The Thirteen Problems* (*The Tuesday Club Murders*, 1933, in the US) to the couple. It was in 1930 in Ur that the Woolleys introduced Agatha to the archaeologist Max Mallowan, who was indeed suffering from appendicitis during her first visit to the site in the autumn of 1928. The couple married in Edinburgh in September 1930. Christie's experience at Ur – and subsequent visits to the Near East – inspired her 1936 novel *Murder in Mesopotamia*, which features Louise Leidner, an archaeologist's wife who bears a striking similarity to Katharine Woolley.

- Leonard Woolley led a joint expedition at Ur organised by the British Museum and the University of Pennsylvania between 1922 and 1934. He worked in British Intelligence during the First World War, and was stationed in Egypt. It was there that he met Gertrude Bell, who also worked for the intelligence services. He met Katharine Keeling in the spring of 1924, when she came to Ur as a volunteer, and the couple married in April 1927. On their wedding night, Katharine locked him in the bathroom until he promised that he would not try to sleep with her. Leonard Woolley contemplated divorce, but thought that the scandal would ruin his career. For more information about the lives of Katharine and Leonard Woolley – including the detail of the boyhood taunt of Leonard being a 'Woolly sheep' and his subsequent retort that he was actually a wolf in sheep's clothing – see H. V. F. Winstone's superb biography *Woolley of Ur* (Secker & Warburg, 1990). A good online site is: www.penn.museum/blog/museum/adventure-calls-the-life-of-a-woman-adventurer/

• According to Max Mallowan, Gertrude Bell regarded Katharine Woolley as a 'dangerous woman'. (*Mallowan's Memoirs: Agatha and the Archaeologist* (William Collins, 1977, paperback reissue, 2010, p 36). She was born Katharine Menke in June 1888 in Worcestershire, but grew up in a German-speaking family. She dropped out of Somerville College, Oxford, after two years due to ill health. On 3 March 1919, she married Lieutenant Colonel Bertram Francis Eardley Keeling, at St Martin in the Fields, London. Six months later, on 20 September 1919, 39-year-old Colonel Keeling – who had also worked for British Intelligence during the First World War – shot himself at the foot of the Great Pyramid of Cheops, Cairo. There have been a number of theories surrounding the suicide of Colonel Keeling, but according to one report he killed himself after a meeting that he had with a doctor who had been called to examine his wife. According to Henrietta McCall – author of a biography of Max Mallowan (Agatha Christie's second husband) – Katharine may have had what is now known as Complete Androgen Insensitivity Syndrome, a condition which meant that although she was born genetically male, she was insensitive to male hormones and therefore appeared female. In 1929, John Murray published Katharine Woolley's novel *Adventure Calls* about a young woman, Colin, who disguises herself as her twin brother to travel through Iraq. The quote in Chapter Eleven is taken from page 99 of *Adventure Calls*, a copy of which I read in the British Library. Katharine Woolley died of multiple sclerosis at the Dorchester Hotel in London in November 1945.

- Many of the spectacular treasures of Ur unearthed by Leonard and Katharine Woolley can be seen on display at the British Museum, London. Other resources include: www.britishmuseum.org/research/research_ projects/all_current_projects/ur_project.aspx and www. ur-online.org.

- One large grave site at Ur was named the Great Death Pit because it contained the remains of sixty-eight women, and six guards or soldiers, together with four musical instruments, including a silver lyre. It was here that Woolley unearthed evidence of human sacrifice. As he writes: 'Each man and woman brought a little cup of clay or stone or metal, the equipment needed for the rite that was to follow. There would seem to have been some kind of service down there, at least it is certain that the musicians played up to the last, then each of them drank from their cups a potion that they had brought with them or found prepared for them on the spot – in one case found in the middle of the pit a great copper pot into which they could have dipped – and they lay down and composed themselves for death. Somebody came down and killed the animals (we found their bones on top of those of the grooms, so they must have died later) and perhaps saw to it that all was decently in order – thus, in the king's grave the lyres had been placed on the top of the bodies of the women players, leant against the tomb wall – and when that was done, earth was flung from above, over the unconscious victims, and the filling-in of the grave-shaft was begun.' (Woolley quoted in Winstone, p 153.)

Acknowledgements

The British Museum's Mesopotamian rooms are places of wonder and so I'd like to thank the curators and staff of this wonderful institution.

I would like to thank my fabulous agent and friend, Clare Alexander, as well as the whole team at Aitken Alexander Associates, in particular Lisa Baker, Lesley Thorne, Steph Adam, Geffen Semach, Anna Watkins, and Monica MacSwan.

At Simon & Schuster in the UK I would like to acknowledge Ian Chapman and my fantastic editor Suzanne Baboneau, both of whom have supported me throughout the writing of this series. In addition I would like to thank Bec Farrell, Jo Dickinson, Anne Perry, UK copy-editor Sally Partington, Justine Gold, Dawn Burnett, and the marketing department, Jess Barratt, Harriett Collins, Gemma Conley-Smith and everyone in publicity, Gill Richardson, Claire Bennett, Richard Hawton, Rhys Thomas, and the super-enthusiastic sales team. The cover was illustrated by the talented Mark Smith and was designed by Pip Watkins.

In the US I would like to thank the wonderful staff at Atria, particularly my editor Peter Borland, as well as Sean Delone and Daniella Wexler.

Thanks too to all the Agatha fans, scholars and academics who have embraced the series, particularly Dr John Curran, Mike Linane, Dr Jamie Bernthal, Scott Wallace Baker, Tina Hodgkinson, Emily and Audrey at The Year of Agatha blog, and many more.

Lastly, I would like to thank all my family and friends and Marcus Field.

I made my way down the grand central staircase, preparing myself to meet a murderer and his – or her – potential victim. The borrowed jewels and a beautiful emerald-green evening gown, a gift from my sister, gave me a false, hollow confidence. I forced a smile in the direction of my friend John Davison, a smile which masked not only a raft of uncertainties and doubts, but a deep sense of dissatisfaction, if not anger.

After all, I was supposed to be having a few relaxing weeks on the island of Skye before my wedding to my dear Max the following month. My plan was to travel to this romantic isle in the Scottish Highlands with my daughter, Rosalind, my secretary, Carlo, and her sister, Mary. The holiday would give me the chance to rest after a rather hectic year. I would go on some lovely, long rambles across the moors, perhaps even lose a few pounds. The fresh air would do me good and I would walk down the aisle of the chapel in Edinburgh with a clear head, a slightly better figure and a peach-like complexion. The last thing I wanted to think about was murder.

But that wasn't how it had turned out. The week before I was due to travel to Scotland, Davison had sent me a note, asking me to meet him at his London flat in Albany. After the preliminaries and polite small talk my friend, who worked for the Secret Intelligence Service, got down to business.

'I know the timing could be better,' he said, clearing his throat. 'But there's something that we would like you to do.'

'What is it?' I asked, taking a sip of soda water.

'I wouldn't ask you, but as I know you're going to Scotland, to Skye in particular, it seemed you were the ideal person for the job.'

I was so taken aback at his suggestion that I give up my precious holiday that I was lost for words.

'One of our former agents, Robin Kinmuir, who lives on Skye, believes that he is in danger. He's received a series of threatening letters informing him that his life will be taken from him.'

The situation intrigued me and, despite myself, I wanted to know more. 'I'm not saying I will agree to anything, but tell me a little more about Robin Kinmuir.'

'He's a friend of Hartford's, or was at one point,' said Davison. 'The head of the Service regarded Kinmuir as one of the best agents we ever had. He had a great brain, a photographic memory, and was incredibly brave and fearless. Then he experienced a run of bad luck. His only son, Timothy, was killed in the war, a loss which hit him hard. He took to drink. His wife Catherine, who was younger than him, left him – or rather, it seems she simply walked out of their house and disappeared.'

That word always made my heart miss a beat because of those 'missing' days in 1926 when the press reported that I had disappeared.

'And then ... well, in 1916, Kinmuir made a terrible

error of judgement out in Maastricht, which resulted in the deaths of eleven of our men,' Davison continued. 'Although he managed to make it back to Britain, he never forgave himself and his drinking became worse. In fact, I'm sure he would have died had it not been for Hartford, who sent him off to a place in the country.'

'And the letters? What do they say?'

Davison reached for his inside jacket pocket. 'I've got a copy of one of them here,' he said. 'Typed on what looks like an old machine, it warns Kinmuir that his time is up. Here, listen to this.' He unfolded a square of paper. '"I know what you did. Soon you will pay for your crimes with your life. Look over your shoulder – death will come when you least expect it."'

'Sounds a bit melodramatic,' I said. 'How do you know it's not from some crackpot? What makes you think there's any substance behind these threats?'

'One could easily dismiss it if we were dealing with an ordinary member of the public,' said Davison. 'It's Kinmuir's work for the agency that sets him apart. Of course, there may be nothing to it, but I said to Hartford that we – or I, at least – would investigate. Although Kinmuir lives in this country, it's not strictly within the remit of the Secret Intelligence Service – as you know, the SIS is a primarily a foreign intelligence service. But what happened abroad may well have a bearing on the case.'

'I see. And are there any suspects?'

'One line of enquiry is that it's something to do with the failed mission in Maastricht. Either that or ... well, let's just say that Kinmuir has hardly led a blameless life.'

'In what respect?'

Davison paused as he considered how best to express himself. 'He had a certain reputation with the ladies. Several affairs and indiscretions, particularly during the time he was drinking heavily. There are also a number of suspect or failed business deals that we're looking into.'

'And who would benefit from his death?'

'Ah, the perennial question! Kinmuir has a substantial estate. Although it goes by the name of Dallach Lodge it's actually a huge old pile and sits in hundreds of acres of land. On his death that would all pass to his next of kin, his nephew, James, his late brother's son.'

'Well, he's the obvious suspect,' I said, standing up. 'There, you've heard what I've got to say on the matter. Now, I really must go and finish packing for my trip.'

'You mean, that's it?' said Davison, barely able to disguise the disappointment in his voice. 'You're not prepared to help?'

'Davison, I don't know if you heard, but I'm getting married,' I said, rather more impatiently than I meant to. 'I really haven't got time for this at the moment.'

'So you're prepared to let a man die?'

'Oh, stop,' I said, reaching for my coat. 'Don't try to blackmail me into this.'

'Listen – if it's any comfort, I would come with you. Kinmuir now runs his country house, Dallach Lodge, as a hotel. Of course, he would never call it that – his guests are supposed to feel like old friends who have just come to stay for a few days. Although the estate is a quite valuable one, recently Kinmuir has suffered from a lack of funds. It seems he can't afford the upkeep of the house without taking in paying guests. A number of people are about to arrive at the lodge and we suspect that one of them may want to kill Kinmuir.'

'Why doesn't he simply cancel their stay? That would be the easiest thing, surely?'

'That's what he wanted to do, but we pointed out to him that the best way to catch his murderer – if indeed he or she does exist – would be to act normally and open his hotel as usual.'

'Yes, I can see that,' I said. 'But why can't you just go on your own – or get someone else to accompany you?'

'We've been through all the different possibilities. You're the best person suited for the job. What with your sharp mind and your ability to get to the—'

I cut him off. 'I don't need to hear your flattery, Davison. The truth of the matter is you've got no one else. I'm right, aren't I? Inform Hartford from me that he really needs to sort out his recruitment, especially of those of the fairer sex.' I paused. 'Tell me this – and I want you to answer honestly, or with as much honesty as you can muster: what would happen if I didn't choose to help you?'

A grave expression crossed Davison's face. 'I'll be frank,' he said. 'Yes, you are the only woman we have free at the moment. If I were to go by myself, or take another man with me, the murderer might suspect that we were from the police, if not the agency.' He hesitated for a moment. 'But if we were to register *together*, you and me, well, it would give us quite a good cover, don't you think?'

'Are you really suggesting that we turn up there as ...' I could hardly get the words out. 'As a *couple* ... when I'm preparing to get married next month?'

Davison must have seen the horrified expression on my face because the lines of seriousness around his mouth melted away and he burst out laughing.

'Sorry – but your face was a picture!'

'How could you be so—'

'I know – it was beastly of me. I couldn't help myself.'

'Look – if you want me to assist you then the least you can do is show me a little respect.' The words made me sound like a prig, and even though I was beginning to see the funny side of Davison's teasing, I wasn't going to let him off the hook that easily. 'Really, it's quite uncalled for.'

'I apologise,' he said, trying to compose himself. 'No, of course I realise that would be quite an unsuitable arrangement. What I propose is for the two of us to turn up at Dallach Lodge as cousins. You register as yourself – and tell the truth about your forthcoming marriage to Mr Max Mallowan. You'll have to do this anyway because you want to register the banns in Scotland. And it will be easy for me to be plain old John Davison, a middle-ranking civil servant. I am here to be your chaperone in the weeks leading up to your nuptials. So what do you say?'

He looked at me with a mischievous glint in his eyes. 'You never know, you might help save an innocent man's life,' he said. 'Surely that's worth sacrificing a little of your holiday?'

'Oh, very well,' I sighed. 'But on condition that it's for

a short amount of time only. I won't do anything that puts the date of my wedding at risk. That wouldn't be fair on Max.'

Of course, the man I was due to marry could never know. Max – who himself was seeing various friends in the south of England – thought that I was simply taking a holiday in Skye with my daughter, my secretary and her sister. But what could I tell my travel companions? It was too late to cancel the trip and they were so looking forward to journeying north with me.

It was no surprise to learn that Davison had already thought of this. The four of us would check into the hotel in Broadford, on Skye, as arranged, and after a day or so, I would receive a telegram from my literary agent Edmund Cork requesting my presence back in London. I would say that there were serious problems with the script of my forthcoming play *Black Coffee*, which was due to open in December. The director was on the point of quitting the production. My return to London was a matter of urgency: there was too much work to do from such a distance. Although I was sure that Carlo and Mary would offer to accompany me, I would persuade them to stay on in Skye so that they could give Rosalind the holiday she deserved. I would travel back to Scotland as soon as I could.

What Davison had not counted on in his plan was the sense of guilt I felt at abandoning Rosalind. No doubt this was made worse by the fact it was not the first time. I had left her in 1922 during my ten-month-long trip around the world with Archie, who was then my husband. There was the unfortunate episode of my disappearance in 1926 and then the separation when I travelled to Ur, in southern Iraq, in 1928. During that time in the desert there were times when I doubted I would ever see my daughter again.

The few days we had spent in Broadford with Carlo and her sister, Mary, were a delight. We had enjoyed a picnic in a field overlooking the bay and played endless games of cards and cat's cradle. As I watched my daughter explore the beach I had to keep telling myself that she was a resilient child. After all, she was now eleven years old, away at boarding school, and her headmistress told me she was a happy and self-sufficient girl. Yet I knew that my separation from her would leave me feeling wretched. I could not allow Rosalind to see my distress and so, when the day of our parting arrived, I had to steel myself as best I could. As the taxi pulled away from the hotel in Broadford, I pretended to myself that I was just popping to the shops and I would see her in a couple of hours. However, instead of taking me to the ferry and then to the railway station, the car drove me across the island to a deserted spot to pick up Davison.

Once we had collected Davison, we travelled towards the south-eastern side of Skye, past a number of abandoned cottages, moors covered in purple heather and long stretches of bogland. The drama of the island was on an epic scale: here was a true example of the sublime that I had only previously seen in paintings or photographs.

Since arriving on the island I had been transfixed by the sky. The atmosphere was constantly changing: one moment the clouds were black and ominous, the next the sun would cast its light down and transform the harsh landscape into a sight of savage beauty. Before I journeyed to Scotland, friends had told me to expect four seasons in one day and they had not been wrong. During the course of twenty-four hours one could experience rain, wind and mist but also the most beautiful pure sunlight I think I had ever seen.

Eventually, we turned off the road and down a lane until we reached a set of gates leading to a drive lined with rowan and birch trees to Dallach Lodge. It was a handsome, gabled, three-storey baronial house, built from red sandstone, standing on the banks of a sea loch, set within a lush garden with monkey puzzle and red cedar trees. Above it, on a higher ridge that dominated the bleak landscape, was a ruined castle. In the distance, across the Sound of Sleat, one could see the rocky peaks of the Knoydart Peninsula on the mainland that stretched up into the clouds.

'It's all very *Castle of Otranto*,' whispered Davison as we stepped out of the car. 'In fact, it looks the perfect place for a murder.'

Make sure you've read the first mystery featuring Agatha Christie, in the acclaimed series by Andrew Wilson ...

A TALENT
FOR
MURDER

'You, Mrs Christie, are going to commit a murder. But before then, you are going to disappear ...'

Agatha Christie, in London to visit her literary agent, boards a train, preoccupied and flustered in the knowledge that her husband Archie is having an affair. She feels a light touch on her back, causing her to lose her balance, then a sense of someone pulling her to safety from the rush of the incoming train.

So begins a terrifying sequence of events. Her rescuer is no guardian angel; rather, he is a blackmailer of the most insidious, manipulative kind. Agatha must use every ounce of her cleverness and resourcefulness to thwart an adversary determined to exploit her genius for murder to kill on his behalf ...

AVAILABLE NOW IN PAPERBACK AND EBOOK

SIMON &
SCHUSTER

Read the second mystery featuring Agatha Christie, in the acclaimed series by Andrew Wilson . . .

A DIFFERENT KIND OF EVIL

'Do you have a secret side, Mrs Christie?'

In January 1927, Agatha Christie is on an ocean liner bound for the Canary Islands. She has been sent there by the British Secret Intelligence Service to investigate the death of an agent, whose partly mummified body has been found in a cave.

Early one morning, Agatha witnesses a woman fall from the ship into the sea. At first, nobody connects the murder of the young man on Tenerife with the suicide of a mentally unstable heiress. Yet, soon after she checks into the glamorous Taoro Hotel, Agatha uncovers a series of dark secrets . . .

AVAILABLE NOW IN PAPERBACK AND EBOOK

SIMON &
SCHUSTER